C. Vincent · B. Panneton · F. Fleurat-Lessard (Eds.)

Physical Control Methods in Plant Protection

Springer
Berlin
Heidelberg
New York
Barcelona
Hong Kong
London
Milan
Paris
Tokyo

C. Vincent · B. Panneton · F. Fleurat-Lessard (Eds.)

Physical Control Methods in Plant Protection

With 102 Figures and 34 Tables

Springer

INRA EDITIONS

Dr. Charles Vincent
Dr. B. Panneton
Horticultural Research
and Development Centre
Agriculture and Agri-Food Canada
430 Gouin Blvd.
Saint-Jean sur Richelieu
Quebec, Canada J3B 3E6

Dr. F. Fleurat-Lessard
INRA-Laboratoire sur les
Insectes des Denrées Stockées
Domaine de la Grande Ferrade
BP81, 33883 Villenave d'Ornon Cedex
France

Cover photo: Post-emergence mechanical (between rows) and thermal (on the row) weeding in a cornfield (photogr. C. Laguë).

ISBN 3-540-64562-4 Springer-Verlag Berlin Heidelberg New York

Library of Congress Cataloging-in-Publication Data
Physical control methods in plant protection/C. Vincent, B. Panneton, F. Fleurat-Lessard (eds.)
p. cm.
Includes bibliographical references.
ISBN 3540645624
1. Insect pests – Control. 2. Plants, Protection of. I. Vincent, Charles, 1953 – II. Panneton, B. (Bernhard), 1959 – III. Fleurat-Lessard, F. (Francis), 1950 –
SB931.P49 2001
632.7–dc21 2001032275

Springer-Verlag Berlin Heidelberg New York
a member of BertelsmannSpringer Science+Business Media GmbH

http://www.springer.de

Printed in Germany

Cover design: design & production GmbH, D-69121 Heidelberg
Camera ready by INRA, Versailles

SPIN 10668533 31/3130xz 5 4 3 2 1 0 – Printed on acid-free paper

Charles Vincent dedicates this book to M. Serge Ruchonnet (Rolle, Switzerland) who, while residing at Hemmingford, Quebec, Canada, gave him his first research contract regarding the control of insects with physical methods.

Acknowledgements

The editors gratefully acknowledge the following persons: Jean Bélanger (Premier Tech, Rivière-du-Loup, Québec, Canada) for critical review of the manuscript from an industrial point of view; Benoit Rancourt (Agriculture and Agri-Food Canada, Saint-Jean-sur-Richelieu, Qc), for work in computer-assited edition; Jeannine Hommel (INRA, Versailles) and her team for their excellent work related to the edition of the book.
The following managers financially supported the project: Alain Coleno (INRA, Paris), Yvon Martel (Agriculture and Agri-Food Canada, Ottawa), Denis Demars (Agriculture and Agri-Food Canada, Saint-Jean-sur-Richelieu, Qc). Finally, we thank the authors who had confidence in us.

Contents

Mechanical Control

Pneumatical Control

Plant Protection and Physical Control Methods The Need to Protect Crop Plants

Bernard PANNETON, Charles VINCENT and Francis FLEURAT-LESSARD

1 Introduction

Plants, like all living organisms, are preyed upon by various parasites of plant or animal origin, which either directly attack their tissues (fungi, insects, etc.) or compete with them for resources (air, water, nutrients in the soil, etc.). For practical as well as economic reasons, crop production has evolved toward monocultures of species grown on relatively extensive surface areas. This has simplified ecosystems, often eroding the inherent complexity of the natural environment. Since pest populations tend to expand under such circumstances, this situation has magnified the risk of serious repercussions on crop plant species (Metcalf and Luckmann 1994 ; Chap. 2).

The direct cause of the Irish potato famine in the mid-19th century was an uncontrolled infestation of potato fields by the pathogenic fungus *Phytophthora infestans* (Mont.) de Bary. Occasionally, insect damage is so severe that the harvest is completely spoiled (Matthews 1992) and competition between cultivated plants and weeds results in major yield losses ranging from 24 to 99% (Lacey 1985). On a worldwide basis, preharvest losses amount to roughly 20 to 40%, whereas post-harvest losses (stored food products) account for 10 to 20% of the total (Riba and Silvy 1989).

In view of the need to make a profit by producing high yields of quality plant products, combined with the trend toward regional specialisation in crop cultivation, crop protection is a crucial activity in agriculture and forestry. Tremendous progress has been made in crop pest management during the 20th century, thanks to scientific and technological breakthroughs, notably in chemistry (analytical and synthetic) and in biology (population dynamics, ecosystem analysis, biological control theory and practice, and biotechnology).

The rapid demographic growth which has characterised the final part of the second millennium has raised concern about potential food security problems. It has been predicted that the world population, estimated at 5.7 billion in 1995, will stabilise between 7.9 and 11.9 billion around 2050. Rapid population growth is the main factor determining world demand for food (FAO 1996) and will necessitate increasingly more effective methods of crop protection. Urbanisation and soil erosion have exacerbated the problem by reducing the amount of arable land available for cultivation (Fig. 1). The factors governing the balance between agri-food supply and food demand are both numerous and complex. Therefore, there are many possible avenues of intervention for establishing and maintaining a balance between supply and demand. However, a consensus exists at present that "*increasing the efficiency of agricultural and food production is the key to ensuring an adequate food supply... Attainment of food security requires sustained action... Such action should be aimed at ... the development of environmentally sound new technologies*" (UN 1996).

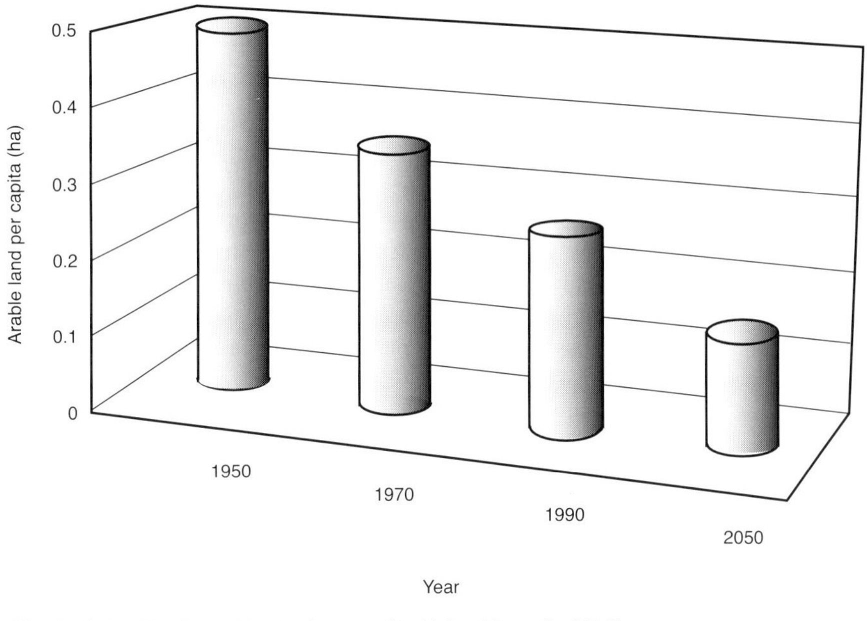

Fig. 1. Reduction in arable land per capita (After Novartis 1997).

2 Crop Protection

Crop protection can be classified according to five approaches (Fig. 2). A majority of agricultural commodities are protected using systems relying primarily on chemical control. Human factors come into play when efforts are directed at either curbing or promoting the use of chemicals. For example, owing to the stringent requirements associated with marketing standards, particularly those imposed on the cosmetic appearance of agricultural products (without considering the objective nutritional qualities of those products), producers are compelled to employ efficient control measures against harmful organisms. In the present context, this essentially means using chemical pesticides. On the other hand, exacting, and sometimes subjective, food safety standards curb the use of chemical pesticides. In such a context, biological control and biopesticides represent tools that can facilitate the implementation of pest management programs, providing a more acceptable balance between the critical need to protect crops and the need to address ecotoxicological concerns. For the same reason, crop protection based on the application of physical control methods during production or storage is gaining in popularity. Indeed, most physical control techniques have no deleterious environmental effects and they are generally limited to the site of treatment and the period during which it is applied. Furthermore, physical control methods bring no chemical or biological substances into play and therefore do not leave undesirable residues on food commodities intended for human or animal consumption.

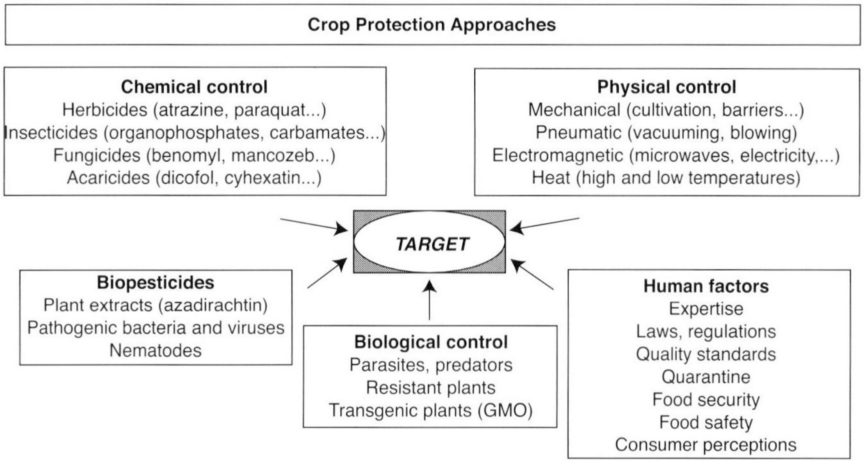

Fig. 2. Five approaches to crop protection.

3 Use of Pesticides

When synthetic pesticides first appeared on the market about 50 years ago, many people imagined that crop enemies would be completely eliminated (Metcalf 1980). That is clearly not the case. However, the increase in the quantity and quality of agricultural food products has certainly been aided by pesticide use, and usually, producers who have easy access to synthetic pesticides do not have to cope with devastating infestations (Hislop 1993).

According to Metcalf (1994), the advantages of chemical pesticides can be summed up by the following qualities: convenience, ease of use, effectiveness, flexibility and reasonable economic cost. In the case of insecticides, for example, the benefits can be summarised as follows:

1. Insecticides represent the only effective control measure when pest populations in crops approach the economic threshold. In such a case, action must be taken without delay, and the control tactic employed must have an immediate effect to ensure that the threshold is not exceeded.
2. Insecticides have a rapid, curative action and are likely to stop the dynamics of a pest population within a few days, even a few hours.
3. Synthetic organic insecticides belong to different categories of chemicals and feature various modes of action and methods of application; hence, collectively they cover a remarkable diversity of situations in which plants are attacked by pest species.
4. The cost-benefit ratios for insecticide treatments are generally favourable if the product involved is suited to field crops. Indeed, insecticides for field crops are manufactured in large quantities for use around the world, and so can be obtained at a low cost. Nonetheless, the costs are rising because of the need to continually synthesize new, more complex chemical compounds and to conduct in-depth research on their toxicology for registration purposes. This has resulted in an exponential increase in research and development expenses.
The market trend toward developing formulations that can be applied in liquid form has permitted uniformity in methods of application. Today, in most agricultural operations, sprayers are the only equipment needed for most crop protection treatments. This standardisation has brought about a reduction in machinery costs and simplified the task of workers who carry out treatments.

3.1 The Pesticide Market

In 1996, the world pesticide market expanded for a third year in a row. The British Agrochemical Association (BAA) estimated 1996 growth at 3.6%, which would make the world market for pesticides worth about FF 185 billion (US$31.25 billion) (Anonymous 1997). According to the BAA, the breakdown of total sales in the different regions of the world was as follows: 30.6% in North America, 26% in Western Europe, 22.5% in East Asia and 11.9% in South America. The growth in pesticide use can be explained largely by the increased acreage of crops to which large quantities of pesticides are applied. In 1996, herbicides accounted for 48% of the world market, compared to 28% for insecticides and 19% for fungicides (Anonymous 1997). Despite the problems that have arisen in relation to pesticide use, the market has grown steadily since 1960 (Fig. 3). The market for herbicides has grown the most. As we will see, physical methods hold considerable promise for controlling weeds and thus for reducing pesticide use, in furtherance of the objectives established by many countries and agencies, such as the European Commission, which are seeking to decrease the inputs used in field crop production.

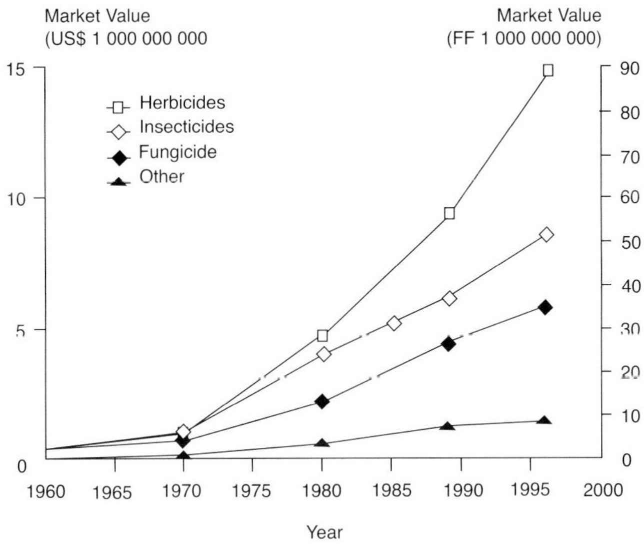

Fig. 3. Trends in the world pesticide market (Data from Matthews 1992 and Anonymous 1997).

Some factors make it more difficult to interpret the statistics on pesticide use. In the United States, for example, cropping techniques based on minimum tillage are being applied increasingly. As a rule, these methods boost herbicide requirements (Anonymous 1997). The BAA believes that pesticide sales will rise in the United States as more and more herbicide-resistant plant varieties are derived through genetic engineering. However, some genetically modified varieties may lead to a reduction in pesticide use (e.g. transgenic tomatoes that synthesise the *Bacillus thuringiensis* Berl. toxin). In the United States, the Freedom to Farm Act should boost the total acreage devoted to cereals, which could result in higher pesticide use (Anonymous 1997). In the European Economic Community (EEC), the rate of pesticide application dropped from 4.2 to 3.3 $kg.ha^{-1}$ between 1991 and 1995. In Sweden, critics have attributed this decline to improved application techniques and the introduction of new pesticides applied at lower doses (Anonymous-b 1997).

From a quantitative standpoint, the pesticide market appears to be buoyant and this is reflected in the substantial sales volume along with recent developments like those mentioned above, which should promote further expansion of their use. However, from an operational standpoint, resistance problems and serious secondary effects, including pollution, reduced food safety and poisonings (Table 1), together with consumers' unfavourable perceptions, are constraining pesticide use.

Table 1. Disadvantages of pesticide use.

1	Pesticide resistance
	Decline in the useful life of a pesticide
	Increased cost of pesticides associated with managing resistance
	Ongoing monitoring of the level of resistance
2	Resurgence and outbreaks of secondary pests
3	Harmful effects on non-target organisms (lack of "ecological selectivity")
	Natural enemies
	Bees and other pollinators
	Wildlife
4	Hazards of pesticide residues
	Chronic intoxication of beneficial species
	Alteration in the reproductive potential of pests
	Alteration in the biotopes of natural enemies of pests
	Human exposure to residues
	Food safety problems
5	Hazards directly associated with pesticide use
	Occupational illnesses in workers
	Pesticide drift during application

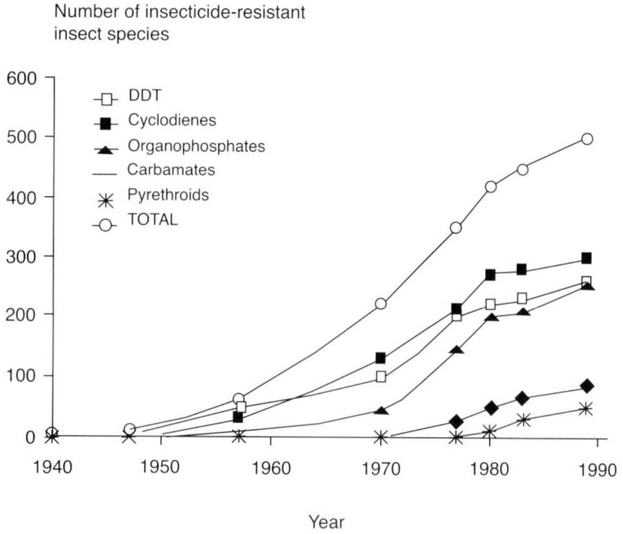

Fig. 4. Time history of the number of insecticide-resistant insect species in the world (After Metcalf and Luckmann 1994).

3.2 Resistance

Pesticide resistance involves insecticides, herbicides and fungicides (See Box 1). This phenomenon is not recent; insecticide resistance was first reported in 1914 (Metcalf and Luckmann 1994). Fungicide resistance has been of concern since modern fungicides came onto the market (Urech 1994). In the case of herbicides, a few cases of resistance had already been reported before the discovery in 1968 that *Senecio vulgaris* L. was resistant to triazines (Heap 1997).
The sharp rates of increase in the number of insect species (Fig. 4) or weed species that have become resistant to pesticides paint a telling picture. Since the 1970s, these rates have been fairly steady, despite awareness of the phenomenon of resis

Box 1. Pesticide Resistance

It is important to distinguish between tolerance and resistance in individuals. Whereas tolerance occurs naturally, resistance is induced. Pesticide resistance results from the selection of individuals that are genetically predisposed to survive a pesticide. Cross-resistance is the phenomenon whereby a resistance mechanism for one pesticide provides resistance to a different pesticide having a different mode of action. In multiple resistance, several mechanisms simultaneously operate to produce resistance to a specific pesticide. The mechanisms involved may include:

- Behavioural resistance, which consists in avoidance of certain chemicals.
- Physiological changes, such as sequestration (accumulation of toxic substances in certain tissues), reduced cuticular permeability or accelerated excretion.
- Biochemical detoxication (called metabolic resistance), often mediated by enzymes such as esterases and mixed-function oxidases.
- Decreased sensitivity to the pesticide at the site that it normally targets (called target site resistance).

Resistance may be mono-, oligo- or polygenic. Although resistance is reversible, the rate of regression to wild levels of susceptibility depends on the mechanism involved and the degree of selection pressure. Resistance management programs are based on determining the mechanism(s) of resistance, monitoring resistance levels, implementing control methods whose mode of action differs from that of pesticides and creating refuges that will attract wild (and therefore susceptible) populations which will dilute the genome of the resistant populations.

More than 550 species of insect pests have acquired some form of resistance to chemical insecticides. Natural enemies (parasites or predators) are often more susceptible to insecticides than are insect pests, and Croft (1990) reports that only 30 species have developed resistance of some type.

tance and efforts to implement strategies to check the development of resistant populations. Between 1980 and 1990, the average rate of increase in resistance was seven insect species per year (Fig. 4) and ten weed biotypes per year (Heap 1997). Today, it is clear that resistance can arise in response to any type of pest control product. The occurrence of cases of cross-resistance to active ingredients that act on different physiological target sites, fortunately a rare occurrence, is nonetheless a factor which aggravates the risk of failure. This limits the possibilities for prolonged use of new active ingredients designed for crop protection purposes (Urech 1997). Consequently, prolonged use can no longer be envisaged outside integrated pest management programs designed to prevent the development of resistance. Physical control methods need to be developed and integrated to enhance such programs, which cannot at any rate be based solely on pesticide use.

3.3 Secondary Effects

The secondary effects of pesticide use are numerous: impacts on the health of humans, flora and fauna and contamination of water, soil and air. Pimentel et al. (1997a) report that, based on official statistics from the World Health Organization (WHO), 3 million poisonings and 220 000 deaths occur around the world annually from the use of pesticides. Back in 1969, Simmons (as cited in Bouguerra 1986) estimated that there were 150 000 poisonings and 150 fatalities annually in the United States due to insecticides. Although the corresponding figures are slightly lower nowadays, the most reliable estimates cover only industrialized countries. Elsewhere in the world, fatal accidents are probably a more common occurrence. In France, more than 8500 cases of poisoning by pest control products were recorded in 1991 (Harry, 1993).

Table 2. Social and environmental costs of pesticide use in the United States (Adapted from Pimentel et al. 1992).

Impact	Relative cost (%) (Total cost: US$8 123 000 000 year^{-1}) (Total cost: FF 48.7 billion)
Wild birds	25.8
Groundwater contamination	22.2
Costs associated with pesticide resistance	17.2
Crop yield losses	11.6
Public health	9.7
Decrease in natural enemies	6.4
Honeybees and effect on pollination	3.9
Government oversight (e.g. regulatory framework)	2.5
Other	<1

Economic indicators, compiled by Pimentel et al. (1992), are useful for comparing the magnitude of the different secondary effects (Table 2). Environmental costs are predominant, while costs with a direct impact on agricultural production account for 39.1% of the total (e.g. resistance, crop losses, decrease in natural enemies and pollination). The public health costs make up nearly 10% of the total.

Although economic indicators are inherently flawed (what cost can be assigned to the acute poisoning of an agricultural worker?), they make it possible to reevaluate the cost-effectiveness of employing pesticides. In the United States, $4 billion (24 billion FF) worth of pesticides is used in producing $16 billion (FF 96 billion) worth of agricultural products. While this outcome may appear very cost-effective, when social and environmental costs are factored into the picture, the profit margin decreases considerably (Pimentel et al. 1992).

In recent decades, the attention devoted to the secondary effects of pesticides has profoundly modified perceptions of these chemicals. Once hailed as miracle products, pesticides have come to be viewed by some people as hazards that should be banned outright or, at best, as a necessary evil. The pesticides industry has sought to modify this perception by disseminating information that puts the secondary effects into perspective. For example, it has been claimed that a glass of apple juice made from fruit treated with Alar poses a risk 58 times lower than the carcinogenic substances found in a single edible mushroom (http://www.cropro.org/history.html ; December 1997). The industry has also responded by developing new products that meet the increasingly stringent standards designed to protect public health and the environment. These more exacting ecotoxicological criteria have driven up the development costs of new pest control products. For new chemical compounds undergoing registration, the cost of environmental impact studies tripled between 1982 and 1992 (Silvy 1992). As a result of the growing costs, few new pesticides will be registered for crops grown on relatively small surface areas because the world market is too small, nor will registration be sought where the higher profit associated with pesticide use does not offset the cost of the product concerned. For the crops involved, physical control techniques might represent a promising alternative in the near future.

3.4 Rational Pesticide Use

In all situations necessitating the use of pesticides, efforts must be made to minimise secondary effects through the wise application of chemical control, a goal which entails augmenting selectivity. In fact, lack of selectivity is the most serious drawback of broad spectrum pesticides. Selectivity can, however, be achieved in several ways (Metcalf 1994).

1. *Physiological selectivity*. Certain molecular structures augment pesticide specificity and reduce the toxicity of residues. For example, insect growth regulators, a class of natural chemicals which disrupt insect developmental patterns or cuticle synthesis, are highly effective against arthropods and non-toxic to mammals.

2. *Ecological selectivity.* Knowledge of a given pest's behaviour and dispersal and its areas of aggregation can be used to more effectively time applications so as to protect beneficial species and reduce the quantities of pesticide applied. Localized application of herbicides is based on the same rationale. This type of approach is viewed as highly promising, considering the possibilities offered by the global positioning system (GPS).
3. *Selectivity through improved application.* Reduced-dosage programs have been put in place over the past 10 years. This approach centres on using chemicals that degrade quickly (non-persistent) and applying them at the most appropriate time of day. Selectivity can also be achieved by adapting the formulation (granular applications at planting and micro-encapsulation) and choosing pesticides with the most appropriate mode of action (systemic or contact type). Thanks to technological improvements in spraying equipment (e.g. air-assistance, specialized nozzles, electrostatics), operators now have a range of options to choose from in carrying out various treatments.
4. *Behavioural selectivity.* This strategy involves using traps in order to make applications at the most effective time.

4 From Pesticides to Integrated Pest Management

In light of the pesticide resistance developed by many plants and animals and the deleterious effects of these chemicals on the environment, human health and agriculture, it is now widely acknowledged that pesticides are not a panacea for protecting plant life. Control strategies for crop pests incorporating a variety of complementary techniques appear to offer much better chances of success than does exclusive reliance on synthetic pesticides (Metcalf and Luckmann 1994). Integrated pest management (IPM) is an approach that involves selecting, integrating and applying control methods for crop pests based on predictions of the economic, ecological and social effects (Anonymous 1986). This integrated approach should to take into account the concerns associated with the use of toxic chemicals in the natural environment.
The concept of integrated pest management centres on the ecosystem, that is, the relationships between living organisms and their environment or habitat. In the beginning, this strategy was aimed at reducing pesticide applications and minimizing secondary effects. In Quebec apple orchards, for example, the number of insecticide applications per season has fallen from over six to at most three without increasing the average damage level at harvest. Subsequently, the progress achieved in harnessing natural enemies for biological control has resulted in a shift from simple pest management to integrated pest management. Introductions of natural enemies of pests, for example phytoseiid mites which prey on harmful spider mites (Tetranychidae), played a decisive role in the implementation of integrated pest management strategies (IPM programs). Typhlodromous strains that are resistant to organophospho-

Box 2. Integrated Management of Weeds

Integrated management programs should be developed for weed control in order to reduce chemical inputs and curb the amount of herbicide residues in soil, water and agricultural products. It is also imperative to avoid inducing resistance through the prolonged use of chemical agents. Application of the principles of sustainable agriculture in weed control has been slow because of the difficulty of making generalized predictions like those derived for insect pests primarily on the basis of climatic data. General advices cannot be produced for weed species because predicting the infestation dynamics of a given field depends above all on its past history (Debaeke 1988). In addition, the diversity of weed species is enormous and cannot be compared with the small number of truly harmful insect pests. This flora is the site of diverse crop plant-weed species interactions and interactions between different weed species, which complicate the task of setting action thresholds.

Achieving a better understanding of the biology, life cycles and dynamics of weed populations is the foundation of integrated weed management (Debaeke 1997). To implement this approach, it is necessary to provide training for IPM managers and to make decision-support tools available to decision-makers (e.g. software programs for identifying weed species) along with tools that provide information on species ecology (Marnotte 1995). Expert systems are being developed that cover all the stages involved in making treatment decisions: inventory of species, evaluation of yield loss, choice of chemical, cost of treatment (Stigliani et al. 1993). Criteria have been devised for assessing the environmental risks associated with each type of active ingredient (Gillet and Dabene 1994); these could soon be incorporated into decision-support expert systems (Debaeke 1997). The main stumbling block in terms of moving from today's chemically based weed control practices toward an integrated management approach is the difficulty of incorporating various control measures into a technically coherent program, guided by an expert system. Work is still under way to attain this goal.

rous insecticides have made it possible to use chemical and biological control together. This combination of techniques has proven successful in previously uncontrollable situations where the prevailing conditions were too unfavourable for introduced beneficial species. More recently, integrated pest management has given way to sustainable agriculture (Cross, 1997), particularly in orchards, by incorporating social aspects (e.g. consumer demands, health protection issues related to workers who apply pesticides and environmental issues) and ecotoxicological concerns, such as preventing adverse effects on pollinators. These considerations has helped to further reduce reliance on pesticides.

Theoretically, integrated pest management can accommodate any type of crop protection technique (Fig. 1) based on its particular merit in a known situation. The move to integrated pest management began with classical biological control. The development of biopesticides is a distinct component of biological control. As we will show in this book, a number of physical control techniques likewise offer sufficiently attractive advantages and benefits to enrich the IPM arsenal.

4.1 Biological Control

In a sustainable agriculture context, biological control (taken in its broadest sense) provides many alternatives to the use of synthetic insecticides (Vincent and Coderre 1992; Hokkanen and Lynch 1995; Van Driesche and Bellows 1996; Jervis and Kidd 1996). Biological control is underpinned by several fundamental theories, including population dynamics theory and optimal foraging theory. In practice, the application of biological control often hinges on a multitude of actions and complex and detailed information.
Biological control gathered a great deal of enthusiasm in the early 20th century because of the success obtained with *Rodolia cardinalis* (Mulsant) in California (Caltagirone and Doutt 1989). This type of approach is attractive from a scientific standpoint and it is viewed favourably by the general public. In spite of this, biological control has generated only a few commercial successes during the 20th century because of its own limitations (Greathead 1995).

Table 3. Biological control strategies (After Cloutier and Cloutier 1992).

1. Use of inert biocides (biotoxins of microbial origin)
2. Use of autonomous biocides (microbial or animal control agents)
   A. Direct manipulation through the release of natural enemies into the environment
      1. Introduction of exotic control agents (classical biological control)
      2. Mass release of control agents with demonstrated efficacy
         a) Inoculative release (preventive control)
         b) Inundative release (curative control)
   B. Indirect manipulation by changing the environment
      1. Protection of natural enemies through specific measures
      2. Provision of supplementary hosts/prey insects at a low density
      3. Provision of food resources or favourable niches (shelter)
      4. Crop rotation (weeds, nematodes, disease)
      5. Chemical stimulation of the activity of control agents

There are several different biological control strategies (Table 3). The first, which involves the use of inert biocides, is the biopesticide approach (next section). The second strategy consists in employing autonomous biocides, which can be released using a classical control approach (i.e. inoculative or inundative releases). Biological control can also be enhanced through environmental manipulation.

4.2 Classical Biological Control

Variability is an inherent and fundamental characteristic of biological organisms and constitutes one of the major limitations of biological control. For example, Lewis et al. (1990) have stated that the performance of parasites as biocontrol agents is often erratic. Among other things, this situation is linked to the parasites' foraging behaviour, which is a function of their genotypical and phenotypical diversity, their physiological condition and the environment in which they are living. These variables have a limiting effect in a commercial context where standardization and quality control are acquiring ever-greater importance.

The augmentation of natural enemies as part of a biological control program can lead to undesirable side effects if insects shift to a non-target species. In biological control of weeds, for example, if the biocontrol agent damages the cultivated plant species as well, the net result may be negative. A change in the hierarchy of risks is another potential consequence that must be carefully weighed. Once a crop pest has been brought under control, another species which is normally a secondary pest may benefit from the absence of competition. In such a case, this pest may cause serious damage that can no longer be ignored.

A major problem with biological control is its incompatibility with chemical control, since natural enemies of insects are often more susceptible to the pesticides applied than are the insect pests themselves (Croft 1990). Releases of natural enemies can, however, be alternated with insecticide treatments.

Biological control is not suitable in all crop protection contexts. There are many situations in which this type of approach does not work. For example, it is not effective against the tarnished plant bug, *Lygus lineolaris* P. de B., as discussed in Chapter 19 of this book. The cost-benefit ratio is the main drawback, since it is much higher for biological control than for chemical pesticide.

4.3. Biopesticides

Among the different tools of biocontrol, biopesticides (Table 4) hold a leading position because they often can be mass produced, which is a prerequisite for industrial use, and they can be applied with a conventional sprayer, which facilitates their adoption by agricultural producers. They are generally compatible with classical biological control methods (e.g. releases of predators or parasites), although they may have adverse effects on beneficial organisms (Giroux et al.

1994, Roger et al. 1995). Biopesticides may be composed primarily of bacteria, fungi, viruses, nematodes or plant extracts.

Until the early 1980s, most studies of biopesticides were based on the classical principles of pathology, as described in Tanada and Kaya (1993). The advent of molecular biology techniques gave a tremendous boost to biopesticide development. The creation of transgenic plants, for example those that may manufacture their own protective toxins from *Bt* crystal proteins, represents an important achievement. It was announced in a fall 1984 press release by the Belgian firm Plant Genetic Systems (PGS), and subsequently detailed in a scientific publication (Vaeck et al. 1986). Transgenic corn (a genetically modified organism, or GMO), in which the Cry1A(B) gene that codes for a protoxin protein has been inserted, is now commercially available in Europe and North America (Novartis 1997).

Table 4. Examples of biopesticides developed for use in controlling pests and weeds.

Class of biopesticide	Organism
1. Entomopathogenic bacteria	*Bacillus thuringiensis* Berl.
2. Mycopesticides	*Metarhyzium anisopliae* and *M. flavoviride*
	Beauveria bassiana, B. brognardtii, etc.
3. Entomopathogenic viruses	Baculovirus (e.g. Carpovirusine)
4. Bioherbicides	*Colletotrichum* spp. (Waage 1995).
5. Micro-organisms antagonistic to pathogens	*Trichoderma viridae; Gliocladium* spp.
	Pseudomonas spp.
6. Entomopathogenic nematodes	Genera *Steinernema* and *Heterorabditis*

There are several ways to enhance the efficacy of biopesticides. One method consists in finding more virulent strains of the candidate species. Alternatively, efforts can be devoted to developing formulations with extended persistence in the field or incorporating synergistic chemicals which are non-toxic at the doses applied but greatly increase the toxic effect of the biopesticide (Bernard and Philogène 1993). Finally, transgenic plants may offer continuous or modulated protection (depending on the genetic expression systems employed) against insects.

The biopesticide that has had the greatest commercial success, *Bacillus thuringiensis*, now accounts for about 1.5% of the world insecticide market (Peferoen 1991). Bt is not a panacea, given the small number of species against which it is effective and its lack of stability in the field. Since it has to be ingested, it is only useful once pests have begun feeding. Furthermore, the development of *Bt*-resistant populations has been observed in several insect species (Tabashnik 1994).

The scientific criteria used to determine the safety of synthetic pesticides are also applied to natural pesticides (Coats 1994) within the same regulatory framework for registration. The legislation in industrialized countries has become so constraining and restrictive that private-sector companies are now reluctant to work on registering new synthetic pesticides. Riba and Silvy (1993) and Powell and Justum (1993) suggested that pathogen-based biopesticides should be subject to less stringent standards than synthetic insecticides in terms of the information required for registration. According to Powell and Justum (1993), the four elements governing the commercial success of a biopesticide are:

1. occupation of a commercial niche in which synthetic chemicals are ineffective or unacceptable owing to specific production principles (for example, in organic farming),
2. an environmental niche where there is protection from UV radiation, desiccation and extreme temperatures,
3. an environmental niche where the biological agent provides an advantage in terms of colonization,
4. a commercial niche where a small amount of damage can be tolerated.

Box 3. Biological Control of Fungal Diseases

Soil-dwelling bacteria of the genus *Pseudomonas* are beneficial to plant health and provide effective protection against certain root diseases (e.g. *Thielavopsis basicola* in tobacco). Genetically modified (GMO) strains of *Pseudomonas* have been introduced that have a greater root colonizing ability than antagonistic flora which have not received a boost from genetic engineering. Initial applications of this biocontrol agent to wheat roots have shown that the risk of indigenous (natural) flora being displaced is low (at season's end, less than 2% of roots were colonized by the genetically modified *Pseudomonas*; Weller et al 1983). However, the results may vary depending on the soil type, crop type and possibly the variety.

Biological control of leaf diseases may be feasible using natural antagonists of the causal pathogen, following an approach similar to that used for insect pests. The phyllosphere is composed of yeasts (mainly the genera *Sporobolomyces* and *Cryptococcus*), bacteria and a few fungal hyphae (e.g. the genus *Cladosporium*; Blackeman 1985) that are potential candidates for that purpose.

Trichoderma, a soil fungus, has proven effective in checking the development of *Botrytis cinerea* on grapes, strawberries, apples and tomatoes (Dubos 1987). *Gliocladium roseum* can also be used for preventive control of *Botrytis* in strawberry (Peng and Sutton 1991). In biological control of fungal pathogens of grapevines, fungicide use can be reduced by about 50% even though it is impossible to reduce fungicide use below five to eight applications during the vegetative phase.

Some fungi that are hyperparasites of pathogens are also the focus of intensive research activities: *Ampelomyces quisqualis*, *Tilletiopsis* spp., *Verticillium lecanii* (A. Zimmerm.) Viégas and *Stephano ascus* are the most common ones. Drought-resistant mutants are also being sought to stimulate the action of these hyperparasites on pathogenic fungi.

5 Physical Control Methods in Agricultural Production

Physical control in crop protection comprises all pest management techniques whose primary mode of action excludes biological and biochemical processes. In contrast, all other methods of control are only effective where an interaction is established between the target pest (through physiology, behaviour, ecology) and the control agent. Sometimes, in physical control, the primary action has a direct impact, such as when insects are killed immediately by mechanical shocks. In other cases, the desired effect is attained through the stress responses that are induced.

The use of physical control methods needs to be incorporated into an integrated pest management strategy. Like any pest management approach, physical methods have certain strengths and weaknesses, and some of them are likely to have secondary effects on fauna and flora. In an IPM context, the decision to employ a physical control tactic must therefore be made on a case-by-case basis according to the same criteria as in decision making regarding the appropriateness of pesticide applications: efficacy, cost-effectiveness and undesirable effects. In addition, no physical control technique is sufficient on its own for all pest control treatments in a given crop. While the self-sufficiency of pesticide applications for crop protection systems is the primary strength of chemical control, it is probably also the Achilles' heel of the chemical approach, given the fact that chemical applications tend to accelerate the development of resistance and slow down the introduction of alternative control tactics. Integrated pest management offers the only hope of avoiding the pitfalls of relying on a single approach. Moreover, it simultaneously opens up possibilities for the commercial application of physical control methods.

It is important to distinguish between the two basic types of physical control methods: active and passive methods. Active methods consist in using some form of energy to destroy, injure or induce stress in crop pests or to remove them from the environment. This type of approach has an effect at the time of application, with virtually no residual action. Passive methods, on the other hand, cause changes in the environment and have a more lasting effect. Physical methods of control can also be classified according to the mode of energy use. That is the classification which has been used in this book. Four categories have been established: mechanical control, thermal control, electromagnetic control and pneumatic control. It would be quite conceivable to add other categories, as appropriate, such as acoustic control and thermodynamic control.

There is no clearcut relationship between categories of physical control methods and the broad groups of crop pests: weeds, insects, mites and microscopic pathogens. Obviously, there are some natural associations, such as the use of mechanical control against weeds (tillage); however, in almost every case at least one method in the different control categories applies to a specific group of crop pests. Pneumatic control is an exception to the rule, however, because it is limited to insect control.

5.1 Physical Control of Weeds

Physical control of weeds is based on the use of several techniques. Active methods include manual weeding, hand pulling, mowing, thermal methods (electricity, microwaves, and heating or cooling) and tillage. Mulching and flooding are classified as passive techniques.

Manual weeding and hand pulling are commonly used around the world (see Chap. 13). An estimated 50 to 70% of agricultural producers use this type of weed control (Wicks et al. 1995). Pulling is generally done by hand, although various tools are available for mechanically removing weeds that are higher than the crop plants (Wicks et al. 1995).

Grazing and mowing can also be used for weed control. This approach is often applied in controlling orchard vegetation and in promoting the establishment of forage crops. Mowing allows forage species to grow taller than the weeds and become better established. In orchards and vineyards, this technique is useful for controlling weed height and minimising competition with the crop plants (Kempen and Greil 1985).

Mulches (see Chap. 15), which are frequently used to suppress weed growth, are applied before or after crop establishment. Mulches can be divided into two groups: mulches made of natural materials and synthetic mulches. These types of products prevent weed emergence by forming a physical barrier and blocking sunlight. In hot, sunny regions, plastic mulches can also be employed to destroy weeds through solarization (Braun et al. 1988; Silveira et al. 1993).

Thermal methods consist in applying high or low temperatures in order to destroy weeds. Electricity (see Chap. 12), infrared radiation, microwaves, hot water, steam and flame weeding (see Chap. 2 and 3) are all techniques that employ high temperatures to kill weeds (Ascard 1995). Flame weeding can be done in a non-selective manner throughout the field, or selectively by directing the burner at weeds while taking care not to damage crop plants.

Water can be used for weed control in rice crops (*Oryza sativa* L.) and cranberry crops (*Vaccinium macrocarpon* Ait.). Total submersion of weed plants kills them through asphyxiation. This technique is very effective in limited agricultural contexts and under suitable conditions (Schlesselman et al. 1985).

Tillage is the second most widely used physical control technique for suppressing weeds. Tillage is divided into primary, secondary and tertiary methods. Whereas primary methods of tillage are designed to prepare the field for crop establishment, secondary methods prepare the seedbed. Tertiary techniques consist of hoeing and cultivation operations performed throughout the season. There are many types of cultivators, which have proven effective for weed control in the majority of crops; in some situations, they are the only tools available for this purpose or the primary alternative to herbicide use (Chap. 13 and 14).

5.2 Physical Control of Insects and Mites

Physical control methods for insect pests comprise a number of technologies, some of which are based on active methods: thermal shock (heat), electromagnetic radiation (microwaves, infrared and radio frequencies), mechanical shock and pneumatic control (blowing or vacuuming tools). In the field, physical barriers represent the only passive technique available.

Various applications that employ thermal shock for crop protection in the field have been developed and research is continuing in this regard (see Chap. 4). This type of approach is nonetheless based on the premise that the commodity or crop to be protected will be less sensitive than the target pest to an abrupt change in temperature. Research on thermal sensitivity thresholds (Chap. 2) and physiological reactions to short-duration thermal stresses is central to the development of control methods based on thermal shock.

Several avenues have been explored for applying the different forms of electromagnetic radiation as a tool for controlling insects. Non-ionizing electromagnetic radiation (see Chap. 7, 8 and 11) kills insects by raising their internal temperature. The utilization of radio, microwave and infrared frequencies is based on a principle similar to that of thermal shock methods except that, with applications involving electromagnetic radiation, the transfer of energy occurs without using a heat transfer fluid. Technologies that harness electromagnetic radiation are often too expensive for use in the field. Furthermore, existing regulations restrict the available frequency bands, either for reasons of user and environmental safety, or because certain frequency bands have been set aside for specific applications that do not tolerate interference (e.g., microwave-based landing guidance systems for aircrafts).

There is a wide variety of physical barriers used as physical control techniques (see Chap. 15 and 16). The technologies associated with physical barriers can be applied in the field or in greenhouses. Barriers used in the field can take several forms (trenches, vertical nets, etc.) and can be deployed on a range of scales to protect either a complete field, a crop row or a group of plants.

Pneumatic control consists in using an airstream to dislodge insect pests (see Chap. 18, 19, 20 and 21). Insects that are removed by vacuum are killed when they pass through the moving parts or the blower (mechanical shock). After being dislodged by a blowing device the individuals are injured and die because they are unable to climb back onto the host plant. Other machines are equipped with a device for collecting the dislodged insects, which are subsequently killed. Sound knowledge of the target insect's behaviour is necessary in order to enhance the effectiveness of this type of approach.

5.3 Physical Control of Microscopic Pathogens

There are fewer scientific studies dealing with the application of physical control in the area of plant pathology. The research described in Chapter 9 demonstrates how a physical control approach can be used against *Botrytis* in the greenhouse. The treatment employs polyethylene films that have the ability to filter out specific parts of the solar radiation spectrum. It is therefore a passive technique based on the use of a physical barrier. Researchers have tested a microwave treatment for eradicating *Fusarium gramineaum* Schw. in wheat seed (Reddy et al. 1996), and similar trials have been conducted by the same team with the aim of suppressing *Ustilago nuda* (Jens.) Rostr. on barley seeds. In potato growing, a fungicide is often used in combination with a chemical defoliant as a sanitary measure to prevent the spread of *Phytophthora infestans* (Mont.) de Bary to the next crop grown in the infected field. Thermal top killing, used instead of chemical defoliation, significantly reduces the viability of *P. infestans* present in the foliage at the time of treatment (Désilets et al. 1996).

6 Post-Harvest Physical Control

In long-term storage of non-perishable agricultural commodities (seeds and grain, dried fruits, by-products, dried and dehydrated plants, spices, herbs, coffee, cocoa, etc.), the most serious losses are due to the action of insects and mites or the spoilage by certain microorganisms (fungi). All other organisms have a negligible effect (Fig. 5). Chemical control using highly persistent insecticides is the most commonly used approach at present for preventing damage by insect pests to grain and seed. Some of the benefits of this strategy are its low cost, ease of implementation and protection that lasts for several months, until the quantity of active residues falls below the lethal threshold for the target species. However, regular use of insecticides entails serious problems such as the possibility of creating resistant strains and the risk of exceeding maximum residue limits (MRL) owing to multiple applications of insecticides by different players in the commercial grain system. This situation is problematic, because pesticide residues resulting from improperly applied procedures create a bad impression among final consumers. When residues are found, pressure is exerted by both the processing industry for which the treated grain is destined and by informed consumers to ensure that chemical treatment is restricted to situations of absolute necessity and to guarantee residue-free finished products. Strategies for using insecticides with an extended residual action (these products are authorized solely for bulk grain) involve risk assessment prior to each application and an evaluation of the results taking into account the economic cost and the potential loss of markets.

This situation has spawned many studies worldwide over the past 20 years on physical procedures for preventing infestations and protecting inventories of stored agricultural products from insect pests, and on the use of such methods in food processing plants. Physical control applications developed for post-harvest treatments have focussed on procedures for controlling physical conditions in stocks of stored grain (temperature and water content; see Chap. 5, thermal (Chap. 5 and 11) or mechanical shock (Chap. 17), the establishment of extreme conditions for insect

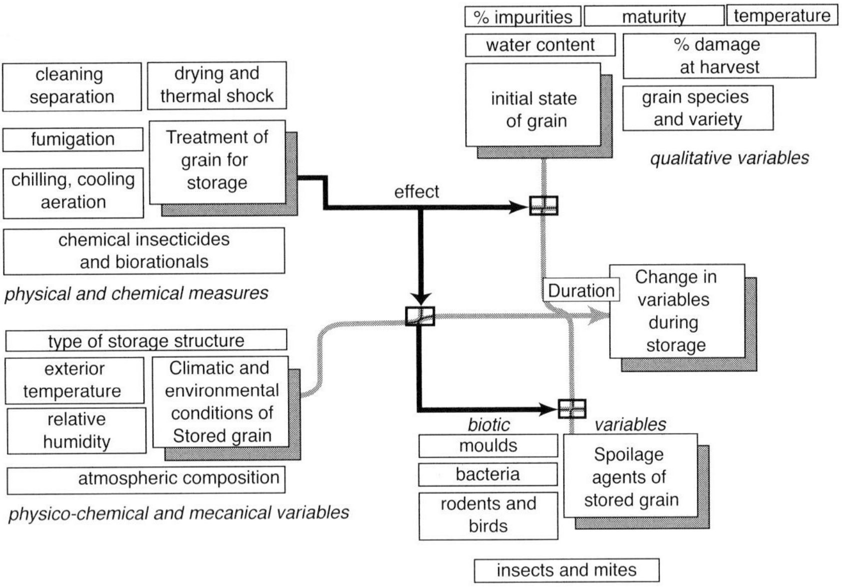

Fig. 5. Key components of the stored grain ecosystem and associated interactions, along with effects on spoilage agents and final quality (for industrial processing).

pests (anaerobiosis, pressure and modified atmospheres; Chap. 5), the use of abrasive or dehydrating minerals (Chap. 17) and the erection of physical barriers to keep insects out (Chap. 17). Post harvest control approaches are essentially based on passive methods, although thermal and mechanical shock treatments are exceptions. In post-harvest situations, most physical methods are suitable solely for prevention against pest and hence cannot be compared with classical chemical control. A thorough knowledge of integrated pest management strategies is required, since these techniques afford no protection following application, unlike persistent insecticides. Practical use of physical control necessitates multiple verifications and supporting data, particularly in relation to secondary effects on the quality of treated products (for example, the germinating power of malting barley or the baking quality of bread-making wheat). Nonetheless, the prospect of registration reviews for several insecticides (e.g. dichlorvos, methyl bromide) that are currently used in post-harvest protection of raw foodstuffs or in processing and storage facilities for intermediate or finished products has revived interest in post-harvest physical control. Furthermore, since the use of persistent contact insecticides on processed food products (e.g. wheat flour, semolina, dried fruit) is prohibited worldwide, the industry has to rely exclusively on fumigants or physical procedures to eradicate insects in such products. Fumigants are used only marginally in many developed countries like France, the United Kingdom and Canada, and their use is likely to decline further when methyl bromide production and use are phased out, sometime between 2002 and 2005. In view of this situation, physical control is the only strategy likely to ensure successful outcomes in the highly specific context of pest and mite suppression for post-harvest food stocks (Jayas et al. 1995). Physical methods hold promise as a complement to chemical pesticides, and a means of moving away from the excessive reliance on chemical control, which is being called into question increasingly and is no longer justified today. Furthermore, the most accessible physical methods for this pest management sector, such as dry heat treatments or airtight storage using inert gases, should help to diminish the secondary risk of spoilage of raw foodstuffs by storage moulds, which is something that cannot be achieved with chemical control measures.

7 Conclusion

Physical control deserves to be recognised as a well-defined area of expertise as is the case for biological control. This recognition is bound to come as the quest for alternatives to chemical pesticides intensifies. Although physical control went out of use with the advent of chemically based pest control methods in the middle of the 20th century, the limitations of pesticide use, coupled with the difficulties of implementing biological control, have created a crossroads for the renewed development of physical control.

Practical application of physical control methods requires collaboration among professionals trained in fields that are often quite distinct and separate. For example, a physicist specializing in electromagnetic energy might be expected to work with an expert in plant physiology or an entomologist. These individuals have acquired expertise in different branches of science and they use different languages and work tools. Furthermore, research activities are organized in such a way that these persons undoubtedly work in different laboratories, pursuing goals that are often incompatible. In spite of this state of affairs, most of the studies described in this book are the fruit of collaborative efforts among persons working in very different areas of research.

It is our hope that this book, the first of its kind on the use of physical control methods in crop protection, will help to raise the status of physical control to that of a true field of research, development and application; promote fruitful initiatives and collaborative work; and provide guidance to individuals interested in making it their field of endeavour.

References

Anonymous, (1986). Toward an IPM Strategy for Canada, Research Branch, Agriculture Canada. Ottawa. 233 p.

Anonymous, (1997). World pesticide market expands. URL: gopher://gopher.igc.apc. org:70/00/orgs /panna/panups/panups_text/275, 2p.

Anonymous-b, (1997). European Pesticide Update. URL: gopher://gopher.igc.apc.org: 70/00/orgs/ panna/panups/panups_text/261. 3 p.
Ascard J., (1995). Thermal weed control by flaming: biological and technical aspects. Swedish Univ. of Agric. Sci., Dept. Agric. Engin., Report 200.
Bernard, C.-B., Philogène B. J. R. (1993). Insecticide synergists: role, importance and perspectives. J. Toxicol. Env. Health 38:199-223.
Blakeman J.P., (1985). Ecological succession of leaf surface microorganisms in relation to biological control. *In* C.E. Windels et S.E. Lindow (eds) *Biological control on the Phyllophane*. American Phytophatological Society, St Paul, MN, pp. 6-30.
Bouguerra M.L., (1986). Les pesticides et le tiers monde. La Recherche, avril 86, n°176.
Braun, M., W. Koch, H.H. Mussa, Stiefvater M., (1998). Solarization for weed and pest control - Possibilities and limitations, pp. 169-178, In R. Cavalloro and A. El Titi (eds). Weed control in vegetable production, Proc. EC Experts Group, Stuttgart, 28-31 Oct. 1986, A.A. Balkema, Rotterdam, 303 p.
Caltagirone, L.E. , Doutt R.L., (1989). The history of the Vedalia beetle importation to California and its impact on the development of biological control. Annu. Rev. Entomol. 34: 1-16.
Cloutier, C., Cloutier C., (1992). Les solutions biologiques de lutte pour la répression des insectes et acariens ravageurs des cultures, pp. 19-88 *In* C. Vincent et D. Coderre (eds.) La lutte biologique, Gaëtan Morin (Boucherville, Qc) et Lavoisier Tech Doc (Paris), 671 p.
Coats, J. R., (1994). Risks from natural versus synthetic insecticides. Annu. Rev. Entomol. 39: 489-515.
Croft, B. A., (1990). Arthropod biological control agents and pesticides. Wiley, New York, 723 p.
Cross J.V., (1997). L'état actuel de la protection fruitière intégrée de pomacée en Europe de l'Ouest et ses réalisations. Adalia 34 :12-21.
Debaeke P., (1988). Effets de l'histoire culturale d'une parcelle sur la composition qualitative et quantitative de la flore adventice. C.R. Acad. Agric. Fr., 74 :21-30.
Debaeke P., (1997). Le désherbage intégré en grande culture : bases de raisonnement et perspectives d'application. Cahiers Agriculture, 6 :185-194.
Désilets H., Coulombe J., Gill J., (1996). La répression du *Phytophthora infestans* dans les tissus végétaux lors du défannage thermique de la pomme de terre. Symposium : Lutte physique en phytoprotection, 88e Assemblée annuelle de la Société de protection des plantes du Québec. 6-7 juin 1996. Québec.
Dubos B., (1987). Fungal antagonism in aerial agrobiocenoses. *In* I. Chet (ed.). *Innovative approaches to plant disease control*, Wiley & Sons, New-York, pp. 107-135.
FAO, (1996). Food Requirements and Population Growth. Technical Background Document 4, World Food Summit. Rome. 62 p.
Gilet H., Dabène E., (1994). Contamination des eaux superficielles par les produits phytosanitaires. In La protection des végétaux et de l'environnement. Bull. Tech. Info. Minist. Agric. Fr., 17-19 :34-50.
Giroux S., Côté J.-C., Vincent C., Martel P., Coderre D., (1994). Bacteriological insecticide M-One effects on the mortality and the predation efficiency of adult spotted lady beetle *Coleomegilla maculata* (Coleoptera: Coccinellidae). J. Econ. Entomol. 87:39-43.
Greathead D.J., (1995). Benefits and risks of classical biologial control. *In* H.M.T. Hokkanen et J.M. Lynch (eds.). *Biological control: benefits and risks.* Cambridge University Press et OCDE, Paris, pp. 53-63.
Harry, (1993). Intoxications humaines par produits phytosanitaires. Conférence séminaire européen COHETT, Angers, 11-14 oct. 1993 " Pesticides agricoles et environnement ".
Heap I.M., (1997). The occurrence of herbicide-resistant weeds worldwide. Pest. Sci. 51:235-243.
Hislop E.C., (1993). Application Technology for Crop Protection: An Introduction In G.A. Matthews and E.C. Hislop eds. Application Technology for Crop Protection. CAB Int., UK. pp: 3-11.
Hokkannen, H.M.T., Lynch J.M. (eds.), (1995). Biological Control, Benefits and Risks. Cambridge University Press, 304 p.
Jayas D., White N.D.G., Muir W.E., (1995). Stored-grain ecosystems. M. Dekker Inc., New York, 757 p.
Jervis M., Kidd N. (eds.), (1996). Insect Natural Enemies, practical approaches to their study and evaluation, Chapman & Hall, New York, 491 p.
Kempen H.M., Greil J., (1985). Chapter 4: Mechanical control methods, pp. 51-62, In E.A. Kurtz and F.O. Colbert (eds.). Principles of Weed Control in California, Thomson Publications, Fresno, 474 p.

Lacey A.J., (1985). Weed control pp: 456-485 In P.T. Haskell (ed.) Pesticide Application Principles and Practice. Oxford Science Publications, Oxford, U.K..
Lewis W.J, Vet L.E.M., Tumlinson J.H., van Lenteren J.C., Papaj D.R., (1990). Variations in Parasitoid Behavior: Essential Element of a Sound Biological Control Theory. Environ. Entomol. 19: 1183-1193.
Longchamp J.P., Barralis G., Gasquez J., Kerguelen P., Le Clerch M., Maillet J., (1991). MALHERB, logiciel de reconnaissance des mauvaises herbes des cultures : approche botanique. Weed Res. 31:237-245.
Marnotte P., (1995). Utilisation des herbicides : contraintes et perspectives. Agriculture et Développement, 7 :12-21.
Matthews G.A., (1992). Pesticide Application Methods, 2nd edition. Longman Scientific and Technical. UK. 405 p.
Metcalf R.L., (1980). Changing role of insecticides in crop protection. Annu. Rev. Entomol. Vol. 25:219-256.
Metcalf R.L., Luckmann W.H., (1994). Introduction to Insect Pest Management, 3rd ed., Wiley Interscience, New York, 650 p.
Metcalf R. L., (1994). Insecticides in pest management pp. 245-314 In R. L. Metcalf and W. H. Luckmann 1994. Introduction to Insect Pest Management, 3rd ed., Wiley, New York, 650 p.
Novartis, (1997). Le livre vert du maïs Cb. Novartis, Saint-Sauveur, France, 109 p.
Peferoen, M., (1991). *Bacillus thuringiensis* in crop protection. Agro-Industry 2(6):5-9.
Peng G., Sutton J.C., (1991). Evaluaton of microorganisms for biocontrol of *Botrytis cinerea* in strawberry. *Canadian Journal of Plant Pathology*, 13, 247-257.
Pimentel D., Acquay H.A., Biltonen M., Rice P., Silva M., Nelson J., Lipner V., Giordano S., Horowitz A., D'Amore M., (1992). Environmental and economic costs of pesticide use. Bioscience 42:750-760.
Pimentel D., Culliney T.W., Bashore T., (1997a). Public health risks associated with pesticides and natural toxins in foods. http://www.ent.agri.umn.edu/academics/classes/ipm/chapters/pimentel.htm.
Powell K.A., Justum A.R., (1993). Technical and commercial aspects of biocontrol products. Pest. Sci. 37:315-321.
Reddy M.V.B., Raghavan G.S.V., Kushalappa A.C., Paulitz T.C. , (1996). Effect of microwave treatment on survival of *Fusarium gramineaum* in wheat seed and seed quality. Symposium : Lutte physique en phytoprotection, 88e Assemblée annuelle de la Société de protection des plantes du Québec. 6-7 juin 1996, Québec.
Riba G., Silvy C., (1989). Combattre les ravageurs des cultures, enjeux et perspectives. INRA, Paris, 239 p.
Riba G., Silvy C., (1993). La lutte biologique et les biopesticides, pp. 49-64 in La lutte biologique. Dossier de la cellule environnement no. 5, 238 p.
Roger C., Vincent C., Coderre D., (1995). Mortality and predation efficiency of *Coleomegilla maculata lengi* Timberlake (Coccinellidae) following application of Neem extracts (*Azadirachta indica* A. Juss., Meliaceae). J. Appl. Entomol. 119:439-443.
Schlesselman J.T., Ritenour G.L., Hile M.M.S., (1985). Chapter 3: Cultural and physical control methods, pp. 35-49, In E.A. Kurtz and F.O. Colbert (eds.). Principles of Weed Control in California, Thomson Publications, Fresno, 474 p.
Silveira H.L., Caixinhas M.L., Gomes R., (1993). Solarisation du sol, mauvaises herbes et production, pp. 6-13, in J.-M. Thomas (ed.). Maîtrise des adventices par voie non chimique; errata et complément, Thème 3, Comm. Quatrième Conf. Internationale de l'International Federation of Organic Agriculture Movement, ÉNITA, Quétigny, 38 p.
Silvy C., (1992). Quantifications… Info Zoo, Bulletin d'information des zoologistes de l'INRA. No 6: 90-103.
Stigliani L., Resina C., Cardinale N., (1993). An expert system for crop weed management. *In Proceedings 8th EWRS Symposium on quantitative Approaches in Weed and Herbicide Research and their Practical Application.* Braunschweig, pp. 855-862.
Tabashnik B.E., (1994). Evolution of resistance to *Bacillus thuringiensis*. Annu. Rev. Entomol. 39: 47-79.
Tanada Y., Kaya H.K., (1993). Insect pathology. Academic Press Inc., New York, 666 p.
UN, (1996). Declaration adopted by consensus at the Parliamentarians' Day. Report of the World Food Summit. Rome, November 1996.
Urech P.A., (1994). Fungicide resistance management: Needs and success factors In S. Heaney et al. (ed.) Fungicide resistance, BCPC Monograph No 60. pp: 349-356.

Vaeck M., Hofte H., Reynaerts A., Leemans J., Montagu M., Zabeau M., (1986). Engineering of insect resistant plants using a *B. thuringiensis* gene in C.J. Antzen et C. A. Ryan. Molecular strategies for crop protection, A DuPont-UCLA Symposium, Steamboat Springs, Co. (March 30 – April 6, 1986).
van Driesche R.G., Bellows T. S., (1996). Biological Control, Chapman & Hall, New York, 539 p.
Vincent C., Coderre D.(eds.), (1992). La lutte biologique, Gaëtan Morin Editeur (Montréal) et Lavoisier Tech Doc (Paris), 671 p.
Waage J., (1995). The use of exotic organisms as biopesticides: some issues, pp.93-100 In H.M.T. Hokkanen and J.M. Lynch (eds) Biological Control: Benefits and Risks. Cambridge Univ. Press. OECD, Paris, 304 p.
Weller D.M., (1983). Colonization of wheat roots by a fluorescent pseudomonad suppressive to take-all. Phytopathology, 73, 1548-1553.
Wicks G.A., Burnside O.C., Warwick L.F., (1995). Mechanical weed management, pp. 51-99 In A.E. Smith (ed.). Handbook of Weed Management Systems, Marcel Dekker Inc. New York, 741 p.
Winston M.L., (1997). Nature Wars, Harvard University Press, Cambridge, Mass. 209 p.

Thermal Control Methods

Thermal Control in Plant Protection

Claude LAGUË, Jacques GILL, Guy PÉLOQUIN

1 Introduction

This chapter presents the status of the use of thermal control in plant protection. After a brief presentation of the operating principles of thermal control, a review of the most important milestones in the evolution of this technology is presented. The different uses of thermal control in plant protection are presented along with the criteria that are used when evaluating the performances of the different types of equipment used. A description of the different modes and strategies of application of thermal control follows. A discussion about the environmental impacts of thermal control, as compared to those of pesticides, closes this chapter.

2 Principles of Thermal Control

2.1 Effects of Temperature on Living Organisms

For plant protection purposes, thermal control of insects is generally aimed at inducing internal injuries that will lead to death over a short period of time. Heat treatments leading to internal temperature increases of 50 to 100 °C for at least 0.1 s will result in either heat-induced rupture of the cell walls or in the coagulation of cell proteins (Pelletier et al. 1995; Morelle 1993). Similarly, reducing the temperature of these living organisms below freezing will lead to similar injuries caused by the crystallization of the cells content (Fergedal 1993) (See Box 1).

2.2 Thermal Sensitivity

The efficiency of a thermal control treatment aimed at a specific target can be evaluated using two parameters: (1) the amount of heat transferred between the ther-

Box 1. Heat and Temperature in Thermal Control

Thermal control in plant protection relies on heat transfer between two bodies (the thermal control apparatus and the agricultural pest to be controlled) that have different temperatures. Temperature is an indicator of the quantity of thermal energy contained within a physical body. One will refer to heat treatments when the temperature of the thermal control apparatus is greater than that of the pest to be controlled and to cold treatments in the opposite situation. In the first case, the purpose of thermal treatments is to increase the thermal energy of the pests while cold treatments are intended at reducing their thermal energy. In both cases, the heat transferred between the thermal treatment apparatus and the pest is equal to the quantity of energy effectively exchanged during the heat transfer process.

mal control equipment and the targeted organisms and (2) the duration of exposure of the targeted organism to the thermal control treatment. For a given equipment, the heat transferred is proportional to the combustion rate of the fuel used (for heat treatments) or to the refrigerating power of the cooling apparatus (for cold treatments). The exposure time depends on the field travel speed or on the operating time of the apparatus if it is stationary (e.g. during crop storage). The increase or decrease in temperature of the targeted organisms is a result of both parameters. The exposure temperature can be used as an indicator of the thermal sensitivity of different living organisms (Laguë et al. 1997).

Different species of weeds, insects or crops respond differently to a given thermal treatment due to physiological differences (Ellwanger et al. 1973a,b). Morelle (1993) reported that broadleaf weed species are more sensitive to thermal treatments than grasses and that perennial weeds are the most resistant. The thermal sensitivity of most living organisms varies according to their stage of growth (Leroux et al. 1995; Ascard 1994). Since young weed plants are generally more heat-sensitive, this can translate into substantial energy savings when thermal weeding is done early in the season.

An appropriate knowledge of the thermal sensitivity thresholds is thus a prerequisite to the development of efficient thermal control systems for plant protection (Leroux et al. 1995). For some applications, only the thermal sensitivity of pest organisms (e.g. pre-planting or pre-emergence thermal weeding) or of the crops (e.g. thermal top killing of mature potato plants, thermal renovation of lowbush blueberry plants) need to be determined. For selective applications (e.g. post-emergence thermal weeding, thermal control of pest insects present on the crop), the crops must tolerate the temperatures required for control of the pests. Many selective applications of thermal control have been succesfully implemented: control of alfalfa weevil (Blickenstaff et al. 1967); post-emergence thermal weeding in corn (Lien et al. 1967), potato (Hansen et al. 1968) and in some horticultural crops (Chappell 1968). An appropriate knowledge of the thermal sensitivity thresholds of the crops and of the pest organisms is required for development of an optimal thermal control strategy (Leroux et al. 1995). When these thermal sensitivity thresholds are very close, it may be necessary to refine the heat transfer between the thermal control equipment and the target or to physically protect the crops from the heat (Parish 1995).

3 Background

Thermal control in crop protection was introduced almost a century ago (Edward 1964). This author reports that many patent applications were filed from 1900 to 1940 for thermal weeding equipment and systems for thermal control of pest insects. Large scale use of thermal control was initiated in 1940 in US cotton production (Kepner et al. 1978). It was also at that time that liquified petroleum gases, mainly propane and butane, started to replace the kerosene and oils that were then used as fuels for the thermal control systems (Edward 1964).

Significant research on practical applications of thermal control in plant protection was conducted between 1940 and the mid-1960s, mainly in the USA. Thermal weeding was investigated for peanuts (Parks 1964), cotton (Matthews and Tupper 1964; Parks 1964), beans (Parks 1964), alfalfa (Thompson et al. 1967), corn (Albrecht et al. 1963; Parks 1964; Reese et al. 1964), sorghum (Reese et al. 1964), and soybeans crops (Matthews and White 1967; Parks 1964). Parks (1964), Chappell (1968) and Hansen et al. (1968) reported that thermal weeding of vegetable crops such as lettuce, broccoli, cauliflower, onions and potatoes was also studied. Work was also done for some small fruit production, mainly blueberry, strawberry and grapes (Hansen and Gleason 1965), and in ornamental horticulture (Wolfe and Horton 1958). Thermal weeding of irrigation and drainage ditches has also been considered (Lowry 1965). Thermal defoliation of cotton plants prior to harvest, either by direct exposure of the plants to the flames (Batchelder et al. 1973; Kent and Porterfield 1967) or by infrared radiation (Reifschneider and Nunn 1965), pre-harvest field drying of sorghum (Parks 1964; Reese et al. 1965), and thermal control of pest insects in alfalfa fields (Harris et al. 1970; Thompson et al. 1967) are among the other applications investigated during that period.

At the same time, many efforts were devoted at evaluating and optimizing the design parameters of thermal control equipment. Research concentrated on the general design of thermal weeders (Batchelder et al. 1970), the study of heat distribution in the vicinity of different types of burners (Harris et al. 1969; Page et al. 1973), the economic evaluation of thermal weeding (Parks 1964), and the use of liquid or gaseous shields to concentrate the action of the burners and to protect the crops from the heat (Kepner et al. 1978). Parks (1964) estimated that there were 15 000 thermal control units in use for different row crops in the USA alone in 1964.

The introduction of efficient and economical pesticides started in the 1950s and has been directly linked to the loss of interest for thermal control (Daar 1987; Kepner et al. 1978). These authors reported that only a limited number of cotton growers in the southwestern part of the USA were still relying on thermal weeders to control herbicide-resistant weeds. Lambert (1990) also reported that thermal renovation of lowbush blueberry plants has continued to a certain extent until now.

4 Modes of Intervention in Thermal Control

Three different techniques may be used to expose pests to high temperatures: direct exposure to flames, use of infrared radiation or steam projection. For cold temperature control, powerful refrigerating systems are needed. The remainder of this chapter deals only with the use of high temperatures.

4.1 Direct Exposure to Flames

The equipment used for directly exposing pests to flames are similar to agricultural sprayers. They incorporate a pressurized liquid fuel reservoir, a network of pipes and hoses that carries the fuel to the burners, pressure regulators and flow controllers, and a number of individual burners where the chemical energy of the fuel is converted into heat (Kepner et al. 1978). The pests to be controlled are directly exposed to the flames generated by these burners.

Two types of burners may be used: liquid burners that incorporate their own fuel atomiser and vapor burners requiring an external atomiser located upstream from the burners. Well-designed burners generate thin flames of constant cross-section having a relatively uniform temperature profile. This allows for constant and uniform temperature rises corresponding to the thermal sensitivity of the organisms to be controlled.

Different types of thermal weeders may be used, depending on the characteristics of the crops and weeds to be treated. For field crops or row crops that are more resistant to heat than the weeds, it is possible to use thermal weeders that generate a uniform temperature rise at the soil level across the whole width of the implement. For more sensitive row crops, in-row thermal weeding can be achieved by orienting the flames of the burners toward the base of the rows, thus reducing the exposure of the crop canopy to heat. Precise guidance systems are often required to maintain proper alignment of the implement relative to the crop rows during field operation (Kepner et al. 1978). Ascard (1992) and Parsish (1989a) reported that in-depth evaluations of different types of thermal weeders were conducted in Europe.

4.2 Exposure to Infrared Radiation

In this technique, the burner flames are directed toward a metallic or ceramic surface that reflects infrared radiation toward the organisms to be controlled (Parish

1989b). This prevents direct exposure of the crops to the flames. For maximum efficiency, the generated infrared radiation must be concentrated within a narrow spectral band corresponding to maximal absorption by the pest insects or weeds targeted by the treatments. According to Parish (1989b) and Lewandowski (Chap. 7), infrared radiation in the 1.44 - 1.93-µm band is very efficient against organisms with a high body water content since it corresponds to infrared absorption band of water. However, the heating rate is not high and a rather long period of exposure is needed to control the pests. This limits the possible operating speed of such equipment in the field, which, in turn, limits the interest for this technique.

4.3 Use of Steam

Thermal control equipments using steam as the heat transfer medium have also been developed by the Aquaheat corporation of Minneapolis, Minnesota. These machines use burners that evaporate water in order to generate steam that is then projected towards the pests to be controlled. These machines are mainly used for weeding applications. In addition to fuel, they also require a large water supply, which increases their complexity and operating costs.

Kepner et al. (1978) reported that the fuel consumption of most burners used in the 1960's ranged from 7.5 to 15 $l.h^{-1}$ of liquid propane, either for burners used on flaming implements or on those used on infrared machines.

5 Performance of the Field Equipment

The various field equipment used in thermal control must be designed and evaluated using three main criteria: (1) ability to transfer heat uniformly and in a controlled manner to the targets, (2) energy requirements, and (3) emission of pollutants.

5.1 Heat Transfer

The ability of thermal control equipments to maximize the transfer of heat toward the targeted pests while minimizing the negative impacts of the heat treatments on the crops is their first performance criterion. Data relative to the temperature profile within the flames generated by different burners used on flamers allow for the evaluation of the uniformity of heat distribution, the geometry of the flames and the maximal temperatures of exposure in relation to the operating fuel pressure (Laguë et al. 1997). Such data can be used to identify the most appropriate type of burner and the proper burner arrangment on the machine for either localized or full coverage treatments.

The determination of the thermal sensitivity thresholds of the different living organisms of interest (crops, weeds, insects) requires data on the temperature of exposure recorded during field or laboratory experiments. For example, full coverage thermal weeding requires data on the temperature of exposure of both weed and crop plants recorded at the soil level. For post-emergence treatments, it is also required to record the temperatures at which the different parts of the plants are exposed. It is thus important to determine the temperature rises at the different locations corresponding to the targeted organisms as a function of the operating parameters of the thermal control equipment (type, location and orientation of the burners; operating speed of the implement; fuel pressure). This allows for the optimization of the design of the equipments. For each type of equipment, charts that relate the temperature rises at the target level to the operating parameters of the machine can be developed to assist the operators in the proper adjustment of their equipment (Laguë et al. 1997).

5.2 Energy Use

The operating costs of most thermal control equipments depend largely on their fuel consumption. Fuel consumption varies with the type of burner used and their operating pressure. Laguë et al. (1997) have demonstrated that fuel consumption is directly proportional to pressure according to:

$$Q = ap, \tag{1}$$

where Q : unit fuel consumption of one burner (kg h^{-1})
p : operating pressure of the burner (kPa)
a : constant (kg h^{-1} kPa^{-1}),

The unit fuel requirements for a given thermal control treatment can then be easily determined:

$$q = 10\frac{Q}{ve} = 10\frac{ap}{ve}, \tag{2}$$

where q : unit fuel requirements of the thermal control treatment ($kgha^{-1}$)
v : field operating speed of the machine (kmh^{-1})
e : spacing between the burners (m)
10 : conversion factor.

For banded thermal control treatments, the spacing between the burners, e, is obtained by dividing the row spacing by the number of burners used on each row. Laguë et al. (1997) have determined that the value of the constant a, found in eq. (1) and (2), ranges from 0.019 to 0.025 for three different types of open-flame burners.

5.3 Emission of Pollutants

Propane, which is currently the fuel used by most thermal control equipments, is generally considered to be a clean fuel since its complete combustion generates carbon dioxide and water vapour. Under field conditions, propane combustion is often incomplete, resulting in the emission of significant amounts of carbon monoxide, sulphur dioxide, and nitrous oxides, especially if the supply of oxygen to the burners is insufficient. It is thus important to identify the optimal operating zones for each type of burner in order to limit the emission of these pollutants.

6 Application Modes and Strategies

Among the different uses of thermal control in plant protection, one may include: integrated weeding (mechanical cultivation between the rows and thermal weeding on the rows) for row crops, as performed on corn in post-emergence (Fig. 1); thermal control of Colorado potato beetle early after emergence of the potato plants (Fig. 2); thermal top killing of mature potato plants that may also have a favourable impact on mildew (Fig.3), and thermal renovation of lowbush blueberry plantations (Fig. 4). In most cases, timeliness of the thermal treatments is critical to maximise their efficiency against the pests to control while minimizing their negative impacts on the crops to be protected.

7 Environmental Impacts

Since thermal control in agriculture represents an alternative to chemical pesticides, it is important that the global environmental impacts of this technology be

Fig. 1. Integrated mechanical and thermal weeding in row crops.

Fig. 2. Thermal control of Colorado potato beetle on young potato plants.

Fig. 3. Thermal top killing of mature potato plants.

Fig. 4. Thermal renovation of lowbush blueberry plants.

lower than that of the chemicals. The evaluation of the environmental impacts of thermal control has been the object of research, especially in Canada and in Switzerland. A multidisciplinary research team from Université Laval has conducted an extensive comparative study of the respective impacts of thermal control and of chemical control of pests in agriculture on soil, water, air and energy resources. This study has dealt with many different applications of thermal control: thermal weeding, potato top killing, thermal control of pest insects.

The main potential impacts of thermal control on the soil are: (1) soil compaction resulting from increased traffic of tractors and field implements in thermal control as compared to chemical control and, (2) the momentary increase in surface temperature of the soil during the treatments. It has however been demonstrated that traffic-induced soil compaction caused by the thermal treatments was not important and that it was limited to the surface layer of the soil. The soil temperature is only increased for a very short time and over a depth of only a few millimeters.

Thermal control has more negative impacts on the air than does chemical control. These impacts are directly related to the combustion by-products (CO, CO_2, nitrous and sulphur oxides) that are important atmospheric pollutants directly related, in many cases, to the global warming of the Earth's atmosphere. These impacts are deemed more important than those associated with the use of pesticides (volatilization, drift of the sprayed products).

As opposed to most chemical pesticides, thermal control has no negative impacts on surface or underground water. On the other hand, the energy balance of most applications of thermal control in agriculture is negative when compared to that of pesticides, since thermal control requires large quantities of non-renewable fossil fuels.

This study showed that even though thermal control presents environmental benefits in terms of impacts on soil and water, it shows a negative balance for air and energy. The environmental benefits of thermal control are thus mitigated.
Jolliet (1994) conducted a similar study for the specific case of potato plant top killing. He determined the impacts on soil, air and water of five different top killing techniques: use of chemical defoliants, mechanical, thermal, mechanical-chemical, and mechanical-thermal. His conclusions were that mechanical top-killing presented the least negative environmental impacts while chemical top-killing had the most. Top-killing techniques making use of heat presented intermediate environmental impacts between these two extremes.
One may thus conclude that thermal control may, in some cases, provide environmental benefits compared to the use of chemical pesticides. These benefits will vary depending upon the type of application, the performances of the equipments used to apply the thermal treatments, the timeliness of these treatments and also the toxicity of the pesticides that can be substituted by the thermal treatments.

8 Conclusion

Recent R&D work on thermal control in plant protection have contributed to a much better understanding of this technology. Work targeted at the determination of the thermosensitivity levels of crops, weeds and pest insects coupled to determination of the operating characteristics of thermal control equipments have yielded much needed data for optimizing the different applications of thermal control in agriculture. As a result, many applications of thermal control can now be considered. In order for this technology to become more attractive to the end user, it is now necessary to better and more thoroughly integrate the agronomical, economical, environmental and technical aspects of thermal control in plant protection.

References

Albrecht K.J. et al, (1963). Flame cultivation of corn and related crops. ASAE Paper 63-606. American Society of Agricultural Engineers. St. Joseph, MI, USA.

Ascard J., (1992). Weed control effect of different flaming systems, pp. 580 - 581, in Proceedings of the International Conference on Agriculture Engineering AgEng 92, Swedish Institute of Agricultural Engineering.

Ascard J., (1994). Dose-response models for flame weeding in relation to plant size and density. Weed Research 34 : 377 - 385.

Batchelder D.G., Porterfield J.G., Taylor W.E., Moore G.F., (1970). Thermal defoliator developments. Transactions of the ASAE 13 (6) : 782 - 784.

Batchelder D.G., McLaughlin G., Porterfield J.G., (1973). Cotton thermal defoliation as affected by relative humidity and time of application, pp. 78 - 84, in Proceedings of the Ninth Annual Symposium on the Thermal Agriculture, Natural Gas Producers Association et National LP-Gas Association.

Blickenstaff C.C., Steinhauer A.L., Harris W.L., Clark N.A., (1967). Flaming for control of the alfalfa weevil in Maryland in 1966, pp. 41 - 44, in Proceedings of the Fourth Annual Symposium on the Thermal Agriculture, Natural Gas Producers Association et National LP-Gas Association.

Chappell W.E., (1968). Flaming in corn, soybeans and vegetable crops, p. 55, in Proceedings of the Fifth Annual Symposium on Thermal Agriculture, Natural Gas Producers Association et National LP-Gas Association.

Daar S., (1987). Update: Flame weeding on European farms. The IPM Practitioner IX (3) : 1 - 4.

Edward F.E., (1964). History and progress of flame cultivation, pp. 3 - 6, in Proceedings of the First Annual Flame Symposium, Natural Gas Processors Association.

Ellwanger T.C. Jr., Bingham S.W., Chappell W.E., (1973a). Physiological effects of ultra-high temperatures on corn. Weed Science 21 : 266 - 269.

Ellwanger T.C. Jr., Bingham S.W., Chappell W.E., Tolin S.A., (1973b). Cytological effects of ultra-high temperatures on corn. Weed Science 21 : 299 - 303.

Fergedal S., (1993). Weed control by freezing with liquid nitrogen and carbon dioxide snow: A comparison between flaming and freezing, pp. 153 - 156, in J.M. Thomas (éd) Proceedings of the Fourth IFOAM International Conference.

Hansen C.M., Gleason W., (1965). Flame weeding of grapes, blueberries and strawberries, pp. 11 - 12, in Proceedings of the Second Annual Symposium on the Use of Flame in Agriculture, Natural Gas Producers Association and National LP-Gas Association.

Hansen C.M., Chase R.W., Tabiszewski A., Bowditch G., (1968). Flame weed control in potatoes, pp. 63-64, in Proceedings of the Fifth Annual Symposium on Thermal Agriculture, Natural Gas Producers Association et National LP-Gas Association.

Harris W.L., Buttiglieri D.A., Marchello J.M., (1969). Techniques to evaluate combustion characteristics of flaming equipment. Transactions of the ASAE 12 : 212 - 215.

Harris W.L., Marchello J.M., Coan R.M., (1970). Effects of thermal energy treatments on the alfalfa weevil. Transactions of the ASAE 13 : 799 - 802, 805.

Kent J.D. , Porterfield J.G., (1967). Thermal defoliation of cotton. Transactions of the ASAE 10 (1) : 24 - 27.

Kepner R.A., Bainer R., Barger E.L., (1978). Principles of farm machinery. AVI Publishing Company, Inc., Wesport, CT, USA.

Jolliet O., (1994). Bilan écologique de procédés thermique, mécanique et chimique pour le défanage des pommes de terre. Revue suisse d'agriculture 26 (2) : 83 - 90.

Laguë C., Gill J., Lehoux N., Péloquin G., (1997). Engineering performances of propane flamers used for weed, pest insect and plant disease control. Applied Engineering in Agriculture 13 (1) : 7 - 16.

Lambert D.H., (1990). Effects of pruning methods on the incidence of mummy berry and other lowbush blueberry diseases. Plant Diseases 74 : 199 - 201.

Leroux G.D., Douhéret J., Lanouette M., Martel M., (1995). Selectivity of propane flamer as a mean of weed control. Hortscience 30 (4) : 820.

Lien R.M., Liljedahl J.B, Robbins P.R., (1967). Five year research in flame weeding, pp. 6 - 20, in Proceedings of the Fourth Annual Symposium on Thermal Agriculture, Natural Gas Producers Association et National LP-Gas Association.

Lowry O.J., (1965). Weed and woody plant control on irrigation and drainage ditches, pp. 16 - 18, in Proceedings of the Second Annual Symposium on the Use of Flame in Agriculture, Natural Gas Producers Association et National LP-Gas Association.

Matthews E.J., Tupper G., (1964). Flame cultivation in mechanized cotton weed control, pp. 36 - 38, in Proceedings of the First Annual Flame Symposium, Natural Gas Processors Association.

Matthews E.J., White J.H., (1967). Technical and economic considerations in high-speed flame weeding of cotton and soybeans. ASAE Paper 67-643. American Society of Agricultural Engineers. St. Joseph, MI, USA.

Morelle B., (1993). Le désherbage thermique et ses applications en agriculture et en horticulture, pp. 109- 115, in J.M. Thomas (éd) Proceedings of the Fourth IFOAM International Conference.

Page F.J., Harris W.L., John J.E., (1973). Energy characteristics of an impinged flame. Transactions of the ASAE 16 (2) : 195 - 199.

Parks J.H., (1964). Progress of flame cultivation in the Texas high plains and Rio Grande valley, pp. 8-17, in Proceedings of the First Annual Flame Symposium, Natural Gas Processors Association.

Parish R., (1995). Flame weed control for southern peas and lima beans. Louisiana Agriculture 38 (4): 23 - 24.

Parish S., (1989a). Investigations into thermal techniques for weed control, pp. 2151 - 2156, in V.A. Dodd et P.M. Grace (éds) Land and Water Use, Proceedings of the 11th International Congress on Agricultural Engineering.

Parish S., (1989b). Weed control - testing the effects of infrared radiation. Agricultural Engineer, Summer 1989 : 53 - 55.

Pelletier Y., McLeod C.D., Bernard G., (1995). Description of sublethal injuries caused to the Colorado Potato Beetle by propane flamer treatment. J. Econ. Entomol. 88 (5) : 1203 - 1205.

Reese F.N., Fitzgerald L.W., Anderson L.E., Larson G.H., (1964). Flame cultivation of corn and grain sorghum, pp. 22 - 25, in Proceedings of the First Annual Flame Symposium, Natural Gas Processors Association.

Reese F.N., Stickler F.C. , Anderson L.E., Larson G.H., (1965). Production of grain sorghum using flame for weed control and desiccation, pp. 7-10, in Proceedings of the Second Annual Symposium on the Use of Flame in Agriculture, Natural Gas Producers Association et National LP-Gas Association.

Reifschneider D., Nunn R.R., (1965). Infrared cotton defoliation or desiccation, pp. 25-29, in Proceedings of the Second Annual Symposium on the Use of Flame in Agriculture, Natural Gas Producers Association et National LP-Gas Association.

Thompson V.J., Scheibner R.A., Thompson W.C., (1967). Flaming of alfalfa in Kentucky for weevil and weed control - Results of 1967 winter. ASAE Paper 67-645. American Society of Agricultural Engineers. St. Joseph, MI, USA.

Wolfe J.S., Horton D.E., (1958). Investigations on the clearing of weeds from bulb beds by flaming. Journal of Agricultural Engineering Research 58, 324-335.

Flame Weeding in Corn

Gilles D. Leroux, Jocelyn Douhéret and Martin Lanouette

1 Introduction

Grain and silage corn (*Zea mays* L.) are grown in several countries throughout the world. Herbicides are currently the main method of weed control used worldwide. In Quebec, for example, one out of every three agricultural pesticides sold is a herbicide used for corn (Leroux et al. 1990). However, the intensive use of herbicides in corn results in serious environmental problems. According to several studies, herbicides used in corn have made their way into streams, rivers and drinking water in a number of countries (Ayotte and Larue 1990; Giroux and Morin 1992; Economic Commission for Europe 1993). Furthermore, owing to the appearance of a growing number of weed biotypes resistant to traditional chemical herbicides, farmers have been forced to resort to new, more expensive treatments. The increase in consumer demand for food free of chemical residues, and the desire of agricultural organizations to promote sustainable agriculture highlight the need to develop alternatives to chemical weed control. Thus, we propose the use of thermal weed control (or flame weeding) alone or in combination with traditional mechanical methods (rotary hoe, sweep cultivator).
Flame weeding was very popular in the first part of the century, until 1940. The first known prototype flamer, used for sugar cane (*Saccharum officinarum*), dates back to 1852 in the United States (Vester 1986). The use of flame weeding ceased in the late 1960s with the advent of cheap and effective selective chemical herbicides. The renewed interest in flame weeding has encouraged us to work on improving this method through research in the laboratory and in the field.

2 Principles of Flame Weeding

Theoretically, thermal weed control is a method that can comprise various forms of electromagnetic radiations such as microwaves, thermal radiations, laser beams and gamma radiations to kill weeds (see Chap. 7). Often, this radiative transfer is enhanced by heat transfer by conduction when hot gases come into contact with

plant tissues. In practice, however, only thermal radiation from flame, steam, hot air or infrared sources and heat transfer by conduction are of interest.
Open flame weeding kills weeds by heating plant tissues rather than burning them. Temperatures must be at least 95–100 °C and should be applied for at least 0.1 s (Vester 1986; Parish 1989). The resulting thermal shock destroys the above-ground parts of weeds. A study on the effects of ultra-high temperatures on corn by Ellwanger et al. (1973) showed alterations to cellular membranes leading to dehydration. Dehydration results from the expansion of the cell contents (made up of 95% water) and subsequent bursting of the cell membranes, and from the coagulation of proteins at temperatures of 50-70 °C (Morez 1985). Both effects damage plant cells to the point that normal growth is impossible. Hot, dry weather accelerates this phenomenon, which takes a few hours (Vester 1986). The resulting increase in evapotranspiration causes plant desiccation in 2 or 3 days.

3 Laboratory Studies

Thermal weed control is not a selective method. Therefore the tolerance of weeds to heat must be balanced against that of corn. Two experiments were set up. The first aimed at determining the minimum temperature required to kill weeds and the second at identifying the maximum temperature supported by corn without negative effects on yield.

3.1 Burner Configuration

Two round burners were used in the lab experiments. They were placed 8 cm above the ground surface, 20 cm apart, shifted 18 cm one from the other along the rows and tilted 30° relative to the vertical (Fig. 1). These values were dictated by the configuration of the cultivator used in the field (Fig. 2). The configuration is important since it defines the extent of temperature rise (Parish 1989; Ascard 1995).

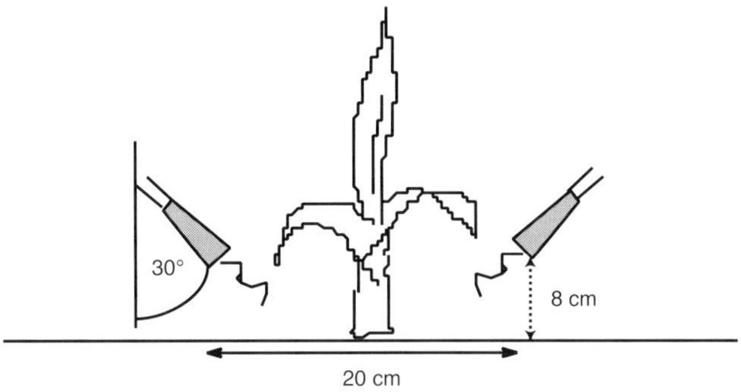

Fig. 1. Placement of burners in relation to soil and corn plants.

Fig. 2. Rear view of flame cultivator used in tests.

Since, in this case, the type of cultivator used limited the burner angle, height and spacing, the temperature rise was determined solely by the gas operating pressure in the burners and the exposure time (i.e. the travel speed of the burners in relation to the plants).

3.2 Effects of Flaming on Corn

A corn hybrid (Pioneer 3979) was sown in metal flats (30 cm wide x 75 cm long x 20 cm tall), six plants to a row. Four growth stages were studied: (1) coleoptile stage; (2) 5–8 cm tall or 2–3-leaf stage; (3) 20–25 cm tall or 4–5-leaf stage; and (4) 45–50 cm tall or 6–7-leaf stage. The corn plants were exposed to ten temperatures between 110 and 390 °C; produced by combining various travel speeds and gas operating pressures (Gill et al. 1995). The temperature probe was located 20–30 cm from the soil surface (Laguë et al. 1997).

In general, temperature rises resulted in decreased dry biomass 2 weeks after treatment, regardless of the growth stage at time of treatment (Fig. 3). However, corn growth was least affected when treatment was at the coleoptile stage, with a decrease in growth of 50% at 390 °C, compared to an average decrease of 80% at later growth stages. The 20–25-cm stage (which corresponds to the 4-5-leaf stage of corn) was the most sensitive: above 160 °C, dry biomass was reduced by at least 50% (data not shown). For the two other stages (5–8 and 45–50 cm), the threshold was around 175 °C. Below this temperature, there was little damage to corn, while above this temperature, leaf necrosis and reduced growth were observed. These results, similar to those reported at the California Weed Conference (1985) and by Daar (1987), indicate that corn is very sensitive to heat between the coleoptile

(2 cm) and 6-leaf (50 cm) stages. Heat tolerance is greatest at the 0–2-cm stage, when the apical meristem is protected. However, when the first leaves emerge, exposure to flaming may damage the plant. At the 6–7-leaf stage (50 cm), the foliage is tall enough for the flame to be directed at the base of the stem, which spares the foliage from the effects of heat (California Weed Conference 1985).

3.3 Effects of Flaming on Weeds

The experiment was carried out on four weed species common in Quebec cornfields: redroot pigweed (*Amaranthus retroflexus* L.), wild mustard [*Sinapis arvensis* L.], lamb's-quarters (*Chenopodium album* L.) and green foxtail [*Setaria viridis* (L.) Beauv.]. These weeds were treated at three different growth stages (0–2 leaves, 4–6 leaves and 8 or more leaves) at temperatures between 110 and 390 °C. At the 0–2-leaf stage, all four weed species were destroyed by temperatures at or above 110 °C (Fig. 3). At the 4–6-leaf stage, however, temperatures of at least 175 °C were required to reduce biomass by 80%. Lastly, at the 8-or more leaf stage, weeds were very difficult to control and, the level of control was below 85% even at temperatures of 350 °C (data not shown). Similar studies done in Sweden by Rahkonen and Vanhala (1993) and Ascard (1995) produced the same results.

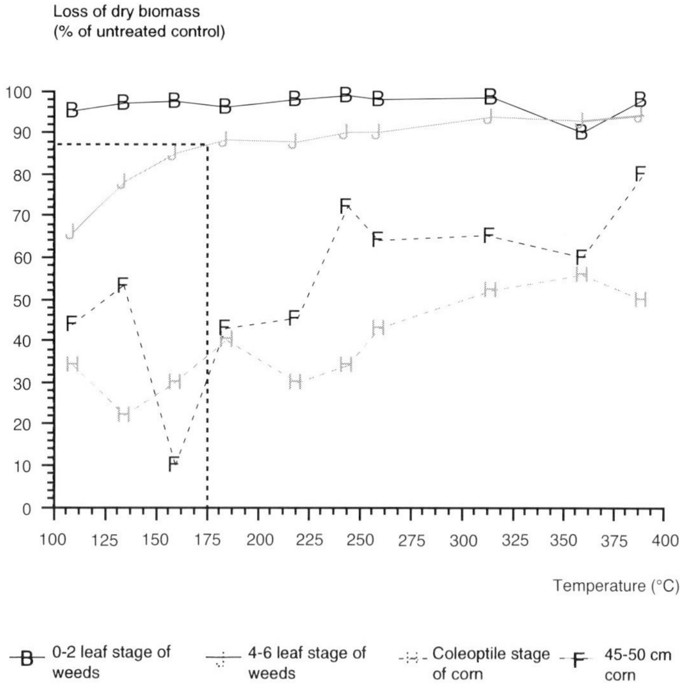

Fig. 3. Mean effect of temperature on dry biomass of weeds treated at the 0-2-leaf and 4-6-leaf stages, compared to that of corn at the coleoptile and 45-50-cm stages.

4 Field Methodology

The laboratory studies allowed us to propose the following hypotheses to be tested in the field:

- The best treatment time for corn is the coleoptile stage. The 6-leaf stage (45-50 cm) is also fairly tolerant to flaming since the foliage is not exposed to the flames to any great degree.
- Young weed seedlings (at the 0–2-leaf stage) are very sensitive to heat and even the lowest temperatures will kill weeds at a rate of over 90%.
- In older weeds, the temperatures required to obtain an 85% decrease in growth (the minimum level of effectiveness required to register a herbicide) also cause a significant decrease in growth of corn (over 200 °C for the 4–6-leaf stage and over 350 °C for the 8 leaf or more stage).

Flame weeding is therefore only feasible for controlling weeds no later than the 4-leaf stage. As the work of Geier (cited by Daar 1987) suggests, treatment must be done as early as possible, when the corn is most resistant to flaming and weeds are the most sensitive. This limits competition from weeds during the most critical period, from the 2–4-leaf stage to the 6–10-leaf stage, when corn is very sensitive to competition. Flaming can be used again once the corn has reached the 45–50-cm stage, as long as the weeds are still at a point where they can be controlled by flaming. To achieve this, weed growth has to be checked at the beginning of the season, to keep them at the 4-leaf stage when the corn is at the 6–7-leaf stage (45-50 cm).

The laboratory work showed that flame weeding can be used without risk for corn at the pre-emergence/early post-emergence stage or at the late post-emergence stage. However, the effectiveness of the treatment must be verified under field conditions, particularly regarding the maximum number of passes with the flame weeder corn plants can tolerate. Lastly, it is also essential to study the effects of flaming on corn yield and to compare the effectiveness of flame cultivation with that of mechanical or chemical treatments.

Our method of non-chemical weed control requires two separate treatments: one at the pre-emergence/early post-emergence stage and a later one when the corn is at the 6-7-leaf stage (45-50 cm). The purpose of the first treatment is to delay the establishment of weeds to ensure they are still amenable to treatment when the corn has reached the point (6–7-leaf stage or 45–50 cm) where it can tolerate flaming. The pre-emergence treatment can be mechanical or thermal. At the post-emergence stage, flame weeding is the only non-chemical method that can be used to remove weeds from the row; however, at this point, flame weeding can be combined with mechanical cultivation in the inter-row area.

4.1 Tools Used

4.1.1 Mechanical Cultivation

A John Deere rotary hoe with flex-discs was chosen (see Chap. 14) for the broadcast cultivation of corn at the pre-emergence/early post-emergence stage (Fig. 4). This machinery is used when the corn is no older than the 3-leaf stage, at a speed of at least 15 kmh^{-1}. It is most effective when weeds seedlings are sprouted but not yet emerged.

4.1.2 Flame Cultivation

Flat vapour burners were used for broadcast treatments. The burners were mounted and equally spaced on a tool bar, producing a uniform temperature rise along a 5-m-wide swath. This provides complete broadcast weed control without mechanical cultivation.
The flame cultivator used consisted of two burners per row mounted on a Kongskilde cultivator (adapted by Bervac, Thetford Mines, Quebec). Five goose-foot sweeps provided inter-row cultivation (Fig. 2). Round vapour burners (Flame Engineering Inc., LaCrosse, Kansas, USA) were used in the same configuration as the laboratory model. Gas was supplied from four 45.7 kg propane tanks mounted in front of the tractor. The cultivator was also equipped with an automatic guidance system manufactured by Sukup Manufacturing Company (Sheffield, Iowa, USA). This system uses two crop-sensing wands (free-floating metal rods) to align

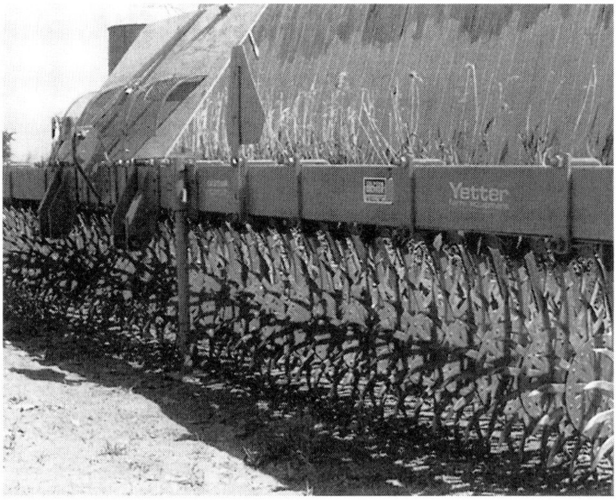

Fig. 4. Front view of rotary hoe.

the sweeps and burners with the crop rows. The rods determine the position of the corn row and control a hydraulic cylinder that directs the moves of the cultivator laterally.

4.2 Effectiveness of Pre-Emergence Weeding

We compared three methods of pre-emergence weeding with an unweeded control plot. They were: (1) standard herbicide treatment; (2) two passes with a rotary hoe (pre-emergence and early post-emergence of corn) and, (3) a broadcast thermal treatment (Table 1). The non-chemical treatments, which have no residual effects, were supplemented by a post-emergence treatment. The results will be discussed in the next section.

The three pre-emergence treatments all reduced weed density, with similar results in reducing the density of dicot weeds (Fig. 5). However, the broadcast thermal treatment left a greater number of annual grasses, probably because of the timing of the treatment, which was done early in the season when few grasses were present. Since thermal treatment only affects the above-ground portions of weeds, those there are not yet emerged at time of flaming remain protected. At this stage, none of the pre-emergence treatments was found to cause damage to the corn (data not shown).

Table 1. Treatments used in the field.

	Treatment	Timing
1	Control (weeds left untreated)	
2	Broadcast herbicide (standard treatment)	
	Metolachlor (2 kg-1ha) + Atrazine (1.125 kg-1ha)	PRE
3	Rotary hoe[a]	PRE + coleoptile
	+ flame cultivation[b]	POST (corn: 45 cm)
4	Broadcast flame weeding[c]	PRE
	+ flame cultivation[b]	POST (corn: 45 cm)

PRE: pre-emergence; POST: post-emergence.

[a] Rotary hoe at 15 kmh^{-1}.

[b] Flame cultivation: two round burners at 175 °C, 6 kmh^{-1}, gas operating pressure: 367 kPa. The burners are used in the corn row and the treatment is supplemented by sweep cultivation in the inter-row area.

[c] Broadcast flame weeding: flat burners at 175 °C, 6 kmh^{-1}, gas operating pressure: 367 kPa.

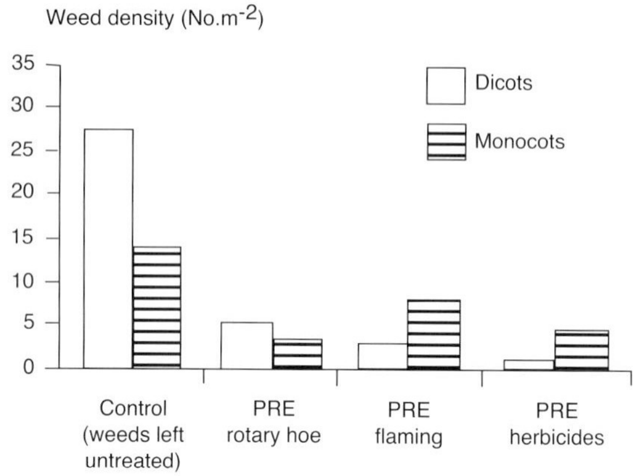

Fig. 5. Effect of pre-emergence weed control treatments on density of dicot and monocot (grasses) weeds, measured 21 days after corn emergence.

Although pre-emergence flaming does not control late-germinating grasses, it does have a significant advantage over cultivation since it can be used when the soil is wet. The rotary hoe is much less effective in wet soil because the exposed roots of the uprooted plants do not dry out. Another advantage of flame weeding is that, unlike tillage (which stirs up the soil, exposing weed seeds to the light), it does not stimulate the germination of weed seeds. Since the seeds of many weed species are photosensitive (they need a brief exposure to light to initiate germination), the rotary hoe may trigger the germination of an entire new cohort of weeds. However, the rotary hoe does have some advantages over thermal weed control since it can be operated at a faster speed (15–20 kmh^{-1} compared to 6 kmh^{-1}): the slow speed of the flamer can be a real handicap in spring, a very busy time of year, particularly if the window of opportunity for treatment is very small.

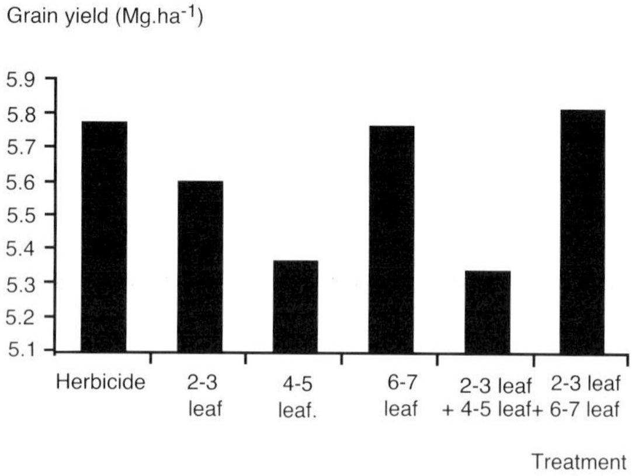

Fig. 6. Effect of various post-emergence fame weeding treatments on corn grain yields (T ha^{-1}) (at 15% moisture content).

4.3 Effectiveness of Post-Emergence Weeding

The performance of corn subjected to post-emergence thermal treatments, in both the absence and presence of competition from weeds was also monitored. Post-emergence flame weeding was done once, twice or three times at the following growth stages of the corn: 2–3-leaf stage (5–8 cm), 4–5-leaf stage (20–25 cm) and 6–7-leaf stage (45–50 cm). The burners were calibrated to generate a temperature rise of 175 °C.

As in the laboratory, corn treated at the 2–3-leaf stage (5–8 cm) and 4–5-leaf stage (20–25 cm) was sensitive to heat. However, despite significant foliage damage, kernel yields were reduced only in corn treated at the 4–5-leaf stage (20–25 cm), indicating that any slowdown in growth caused by flaming at the 2–3-leaf stage (5–8 cm) was made up for during the rest of the growing season. The 6–7-leaf stage (45–50 cm) of corn was more heat-resistant. With two post-emergence treatments, yield was most affected when the treatments were done at the 2–3-leaf and 4–5-leaf stages (5–8 and 20–25 cm) consecutively. When the two treatments were at the 2–3-leaf stage (5–8 cm) and the 6–7-leaf stage (45-50 cm), yields were not reduced (Fig. 6).

Although a single early post-emergence treatment (2–3-leaf stage or 5–8 cm) effectively controlled weeds, it did not cover the entire critical period for corn and yields suffered as a result. A single pass at the intermediate stage of 4–5 leaves

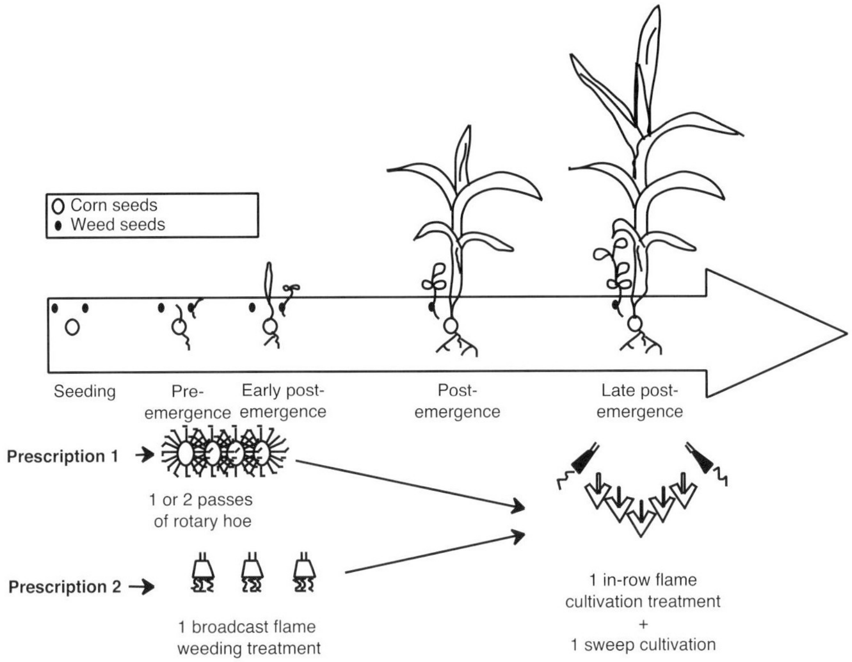

Fig. 7. Illustration of prescriptions for non-chemical weed control in corn.

(20–25 cm) controlled weeds effectively but is not advisable since corn is very sensitive at this stage. At the later stage of 6–7 leaves (45–50 cm), the weeds are too well established to be controlled effectively. The use of two passes, one at an early stage and the other at a late stage, is very effective and does not have a negative effect on yield. There is no advantage to using three flaming treatments. In presence of high weed populations, farmers forced to carry out flaming at the intermediate stage will suffer yield losses.

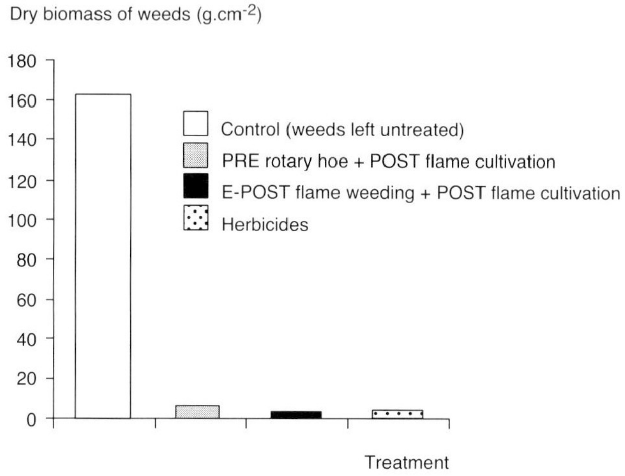

Fig. 8. Effect of prescriptions on dry weed biomass, measured 63 days after corn emergence (see Table 1 for description of treatments).

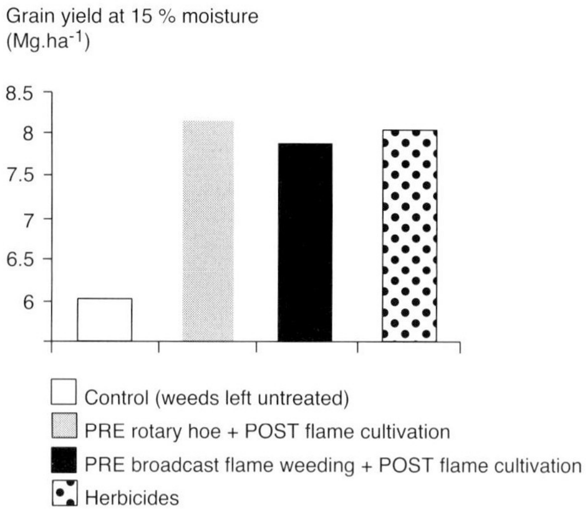

Fig. 9. Effect of weed control prescriptions on corn grain yield at 15% moisture content (see Table 1 for description of treatments).

4.4 Weeding Prescriptions

Based on these results, the best combinations of cultivation and flaming for effective weed control in corn can be determined. Weed control must be begun early in the season, either with a single pass (pre-emergence) or double pass (pre-emergence and early post-emergence) of the rotary hoe (prescription 1), or by broadcast flame weeding at the early post-emergence stage (prescription 2) (Fig. 7). At this time, the main weed cohort for the season is either in the process of germinating or has already emerged. Weather conditions will affect the choice of device used. However, if the soil is dry, the rotary hoe provides excellent results and can be operated at faster speeds than the flame weeder. If the soil is very wet or weeds have already emerged, flame weeding yields better results. These early treatments create a growth differential between the corn and weeds by destroying the first weed cohort. This approach allows the crop to reach the 6–7-leaf stage (40–50 cm) without being subjected to much competition. At this stage, weeds are at the four-leaf stage, at most, and are still susceptible to high temperatures. The corn can tolerate these temperatures, thus allowing selective flaming on the row.

Figure 7 shows the performance of both prescriptions compared to that of pre-emergence herbicides and a control group. Figure 8 shows weed biomass in plots 63 days after the emergence of corn. The treatments all resulted in a reduction of over 85% of weed biomass with respect to the control plot. There is no difference between the two prescriptions, which are also similar to the herbicide treatment. The non-chemical prescriptions resulted in identical yields to those obtained with chemical herbicides, an average of 8 $Mg.ha^{-1}$ (Fig. 9).

5 Economic Analysis

The figures in this study are based on those provided by the Comité de références économiques en agriculture du Québec (Quebec Committee of Economic Authorities in Agriculture). These costs include expenses for the tractor and depreciation, but not labour.

In the case of flame weeding treatments, the costs associated with the burner apparatus and gas consumption are included in the costs of using the machinery and tractor. Purchase costs, depreciation and the operating cost of the rig for broadcast flame weeding are not available since this machinery is not yet available commercially. It has been assumed that the cost is close to that of a post-emergence cultivator, 20.40 \$CAN ha^{-1} (85 FF ha^{-1}). Similarly, the cost of the burner apparatus (burners, piping and propane bottles) mounted on the cultivator for post-emergence flame weeding in the row is difficult to evaluate and cannot be factored in at the present time. We can estimate, however, the costs of propane consumption for both pre-emergence and post-emergence treatments at 14.30 \$CAN ha^{-1} (60 FF ha^{-1}) (Table 2). It should be noted that flame weeding treatments would probably be contracted out on a lump-sum basis since the use of propane gas requires special training, which could add significantly to the cost.

The cost of the proposed non-chemical treatments can be broken down as follows. For prescription 1, two passes of the rotary hoe (15.60 \$CAN ha^{-1} or 65 FF ha^{-1}), one pass of the cultivator (20.40 \$CAN ha^{-1} or 85 FF ha^{-1}) and propane (14.30 \$CAN ha^{-1} or 60 FF ha^{-1}), for a total of 50.30 \$CAN ha^{-1} or 210 FF ha^{-1}, to which must be added the cost of the contract work. Prescription 2, with two flame weeding treatments, costs 69.40 \$CAN ha^{-1} (290 FF ha-1) for operating costs and propane (Table 2). The costs of the burner apparatus and contract work must be added (Table 2).

Herbicides cost 60.40 \$CAN ha^{-1} (253 FF ha^{-1}) and spraying 22.40 \$CAN ha^{-1} (94 FF ha^{-1}) for a single broadcast treatment, with a total cost of 82.80 \$CAN ha^{-1} (347 FF ha^{-1}).

Table 2. Equipment and operating costs of proposed weed control prescriptions.

Treatment	Operating costs		Cost of prescriptions (FFha $^{-1}$)		
1 pass covers 6 rows	FF ha $^{-1}$ Prescription	\$CAN ha $^{-1}$	Prescription 1	Prescription 2	Herbicide
Equipment					
- Rotary hoe	32.50	7.80	32.50 x 2		
- Broadcast flame weeding	85	20.40		85	
- Cultivator	85	20.40	85	85	
- Burner apparatus	Not available		NA	NA	
- Sprayer	94	22.40			94
Products					
- Herbicides	253	60.40			253
-Propane	60	14.30	60	60 x 2	
		Total	210	290	347

6 Conclusion

The use of thermal weed control in corn is quite feasible. The difference in heat tolerance between weeds and corn, based on phenology (the stage of growth at which each is treated), is the basis for treatment selectivity. In corn, the apical meristem is located below the soil surface between the coleoptile stage and the 2–3-leaf stage, allowing the plants to resist temperatures of 175 °C. At these temperatures, weeds, in their 0–2-leaf stage, are destroyed, with over 90% efficacy. Later in the season, at the 6–7-leaf stage (45–50 cm high), corn is again tolerant to high temperatures. At this later stage, flaming can be successful in fields where early treatments have delayed weed growth such that weeds have not grown past their 4-leaf stage before this later treatment.

A broadcast treatment is therefore required in the pre-emergence or early post-emergence stage to slow the growth of weeds, with a rotary hoe or flame weeder. One of the advantages of flame weeding is that it can be done on wet soil. After emergence, a combination of mechanical cultivation between the rows and flame weeding in the row can be done at the 6–7-leaf stage of the corn (45–50 cm). Treatments at the 4–5-leaf stage should be avoided, because corn is very sensitive to heat at this stage. Both prescriptions result in weed control and crop yields comparable to pre-emergence treatment with herbicides. Our results confirm those already obtained by Liljedahl et al. (1964), Reece et al. (1964, 1966), Lalor and Buchele (1966, 1970), Parish (1989) and Balsari et al. (1991).

To maintain the medium-term effectiveness of the proposed prescriptions, they should be integrated with a crop rotation system to prevent massive weed infestations. Physical weed control methods have no residual activity. Although they limit competition between corn and weeds during the critical period of corn is susceptibility further on, nothing can stop weeds from getting established, growing and setting seeds. Thus, the weed seedbank may still increase. This is why these techniques must be combined with cultural methods that ensure crop competitiveness. Another important point is that non-chemical treatments require more time than do herbicides. The window of application is narrow, necessitating increased monitoring of fields. In addition, two or three passes are required, which is more work than a single pre-emergence application of a residual herbicide.

The greatest advantage of the methods proposed here is that they provide an alternative to herbicides and are relatively non-polluting (see Chap. 2). Although the advantages are presently difficult to quantify in terms of low-input farming and minimising input, they probably far outweigh the additional costs of the burner apparatus. Lastly, the appearance of biotypes resistant to thermal treatments is unlikely (which is not the case with herbicides), although it must be recognized that flame weeding may promote the growth of annual grasses in fields.

References

Ascard J., (1995). Effects of flame weeding on weed species at different developmental stages. Weed Res. 35: 397–411.

Ayotte P., Larue M., (1990). Micropolluants organiques. Campagne d'échantillonnage printemps/été 1987 et hiver 1988. Ministère de l'Environnement du Québec, Direction des écosystèmes urbains, Québec, Canada, 178 p.

Balsari P., Ferrero A., Airoldi G., (1991). Weed control in maize by flaming. Medelingen van de faculteit Landbouwetenschappen, Rijksuniversiteit Gent. 56(3a): 681–689

California Weed Conference, (1985). Principles of weed control in California, pp. 46-49, in Fresno, Ca.: Thomson Publications (eds.). Proceedings of California weed conference, El Macero, USA, 474 p.

Daar S., (1987). Flame weeding on European farms. The IPM practitioner 9(3): 1–5.

Economic Commission for Europe, (1993). The Environment in Europe and North America 1992,

United Nations, New York, 368 p.
Ellwanger T.C., Bingham S.W., Chappel W.E., (1973). Physiological and cytological effects of ultra high temperatures on corn. Weed Sci. 21: 296-303.
Gill J., Lagüe C., Lehoux N., (1995). Propane burner characterization for thermal weeding. ASHS Poster No. 878, Hortscience 30: 819.
Giroux I., Morin C., (1992). Contamination du milieu aquatique et des eaux souterraines par les pesticides au Québec. Revue des différentes activités d'échantillonnage réalisées de 1980 à 1991. Sainte-Foy, Québec, Canada: Ministère de l'Environnement du Québec, Direction du milieu agricole et du contrôle des pesticides, 74 p.
Laguë C., Gill J., Lehoux N., Péloquin G., (1997). Engineering performances of propane flamers used for weed, insect pest, and plant disease control. Applied Engi. Agric. 13: 7–16.
Lalor F.W., Buchele F.W., (1966). Chemical, mechanical, and flame weed control experiment in corn and soybeans, pp 27–32 in Proceedings of third annual symposium on thermal agriculture, Sponsored by Natural Gas Processor Association and National LP-Gas Association, Phoenix, Arizona.
Lalor F.W., Buchele F.W., (1970). Effects of thermal exposure on the foliage of young corn and soybean plants. Transactions of the ASAE, 13(4): 534–537.
Leroux G.D., Laganière M., Vanasse A., (1990). Méthodes alternatives de répression des mauvaises herbes dans les cultures fourragères, céréalières, le maïs et le soja. Ministère de l'Agriculture, des Pêcheries et de l'Alimentation du Québec, Ste Foy, Québec, 135 p.
Lijledahl J.B., Williams J.L., Albrecht K.J., (1964). Flame weeding research progress report, pp. 25-28 in Proceedings of first annual symposium, Research on flame weed control. Memphis, Tennessee: Natural Gas Processor Association.
Morez, (1985). Le désherbage thermique. Nature et progrès, Automne 1985: 9-12.
Parish S. (1989). Investigations into thermal techniques for weed control, pp. 2151-2156 in Dodd, V. A. and Geace P.M. (eds.). Proceedings of the 11th international congress on agricultural engineering (CIGR), Dublin, Balkema, Rotterdam.
Rahkonen J., Vanhala P., (1993). Response of a mixed weed stand to flaming and use of temperature measurements in predicting weed control efficiency, pp.167–171 in Thomas, J.-M. (ed.). Non chemical weed control, Papers from the 4th international conference of the IFOAM. Dijon, France.
Reece F., Hurst H., Russ O., (1966). Comparison of flame, mechanical and chemical weed control in corn and grain sorghum, pp. 34–35 in Proceedings third annual symposium on thermal agriculture, Sponsored by Natural Gas Processor Association and National LP-Gas Association, Phoenix, Arizona.
Reece F.N., Fitzgerald L.W., Anderson L.E., Larson G.H., (1964). Flame cultivation of corn and grain sorghum, pp. 22–25 in Proceedings of first annual symposium, Research on flame weed control. Memphis, Tennessee: Natural Gas Processor Association.
Vester J., (1986). Flame cultivation for weed control, two year's results, pp. 153-167 in Cavalloro, R. et El Titi, A. (eds.). Proceedings of a meeting of the EC experts group/Stuttgart, 28-31 October, 1986. Weed control in vegetable production, A.A. Balkema, Rotterdam, Brookfield.

Thermal Control of Colorado Potato Beetle

R.-M. DUCHESNE, C. LAGUË, M. KHELIFI and J. GILL

1 Introduction

The Colorado potato beetle (CPB) constitutes an interesting "target" insect for physical control, because several chemical treatments are otherwise required to control it, and since it is of major economic importance in potato production. Many research studies and various recent publications have led to the development of a number of non chemical means of control (Duchesne and Boiteau 1996), including thermal control. This work has stimulated the interest of potato growers in both the United States and Canada, where the problems associated with CPB are similar.

2 Recent Developments

Heat is a relatively old physical control tool. Laguë et al. (see Chap. 2) presented the background of plant protection by thermal control. The thermal control of CPB is associated with the use of propane equipment. This usage is relatively recent and its effectiveness in controlling spring CPB adults was demonstrated for the first time in 1989 in the United States. One grower, Ray Kuyawski (Riverhead, NY), obtained a mortality rate of 80% with a simple four-row prototype (Moyer 1991). Since then, many research and development studies were conducted in North America, in particular in the years 1991 to 1995. Two groups were particularly active in this area, one in the state of New York, USA (Moyer, 1991; Moyer et al. 1992), the other in Quebec, Canada, headed by Dr. Raymond-Marie Duchesne, entomologist at MAPAQ (Gill et al. 1994, 1995a,b; Laguë et al. 1994). The development of CPB thermal control equipment has also evolved during this period (Laguë et al. 1995).

3 Impact of Heat Treatments on the Target

The impact of heat treatments on the various targets varies according to the nature of the targets, the intensity of the thermal treatment, the time of exposure, and the equipment used (Gill et al. 1995a). For CPB, thermal control can either result in rapid death following a violent thermal shock, or in sub-lethal injuries affecting the insect's behavior. Temperatures of about 70 °C generate sufficient muscular injuries to reduce mobility and reproductive ability, ultimately causing the death of the insect (Pelletier et al. 1995).

Box 1. Thermosensitivity Threshold

The thermal sensitivity of live organisms represents an evaluation parameter of the susceptibility of the organism to support injuries directly caused by an increase or a decrease of its body temperature. In thermal control, temperature of exposure is used as an indicator of the thermal sensitivity of considered live organisms. The lower the temperature of exposure at which injuries are caused to an organism, the higher is its thermal sensitivity. In order to consider non-localized thermal treatments (e.g. combined weeding and control of CPB at the beginning of the season in the presence of potato plants), it is imperative that enough injuries be caused to undesirable organisms (weeds, CPB) at temperatures of exposure less than the thermal sensitivity thresholds of the organisms to be protected (potato plants). If these conditions are not met, it is possible for thermal treatments to induce important damages to the crops to be protected.

Laguë et al. (1994) and Pelletier et al. (1995) described similar injuries to CPB adults following the use of thermal equipment. According to Pelletier et al. (1995), surviving CPB adults suffer injuries to the antennae and legs. Segments, having the smallest diameters, are damaged first, followed by tarsi, antennae, femur-tibia joints, and coxa-body joints. Similar injuries have been demonstrated in the laboratory by dipping CPB adults in hot water at 70 °C (Pelletier et al 1995). According to Morelle (1993), the thermal shock starts by the coagulation of membrane proteins when the cellular tissue reaches a temperature between 50 and 70 °C. Mortality of CPB adults from exposure to flames would be a consequence of: (1) an impairment to their mobility affecting their ability to feed and leading to dehydration; (2) internal injuries; or (3) both types of impairment simultaneously. After thermal treatments, CPB adults are hardly moving due to non-functional tarsi, being supported only by the extremity of the tibia. This significant impact on locomotion has to be taken into account. If the insect is unable to climb normally on the potato plants, feeding is impaired and egg layoff will occur (Pelletier et al. 1995). This inevitably entails a significant impact on the seasonal dynamics of the CPB populations, mainly as a decrease in the larval densities during the first summer generation.

In addition, exposure to high temperatures totally or partially damages egg masses. During the first stages of embryogenesis, eggs are particularly sensitive. They become opaque immediately after the treatment. This can be attributed to the agglutination of the proteins within the egg. The effects on the larval stages have not been well documented because larval stages have never been targeted to thermal treatments. However, injuries are probably similar to those inflicted to adults. Immediate mortality of insect pests following thermal treatments should not be the only criterion for success. Indirect effects, including behavioral disorders, have a real impact on the population dynamics.

4 Thermal Control: Previous Results

From 1991 to 1993, many researchers reported results of the use of thermal control in potato crops. Data from Long Island, New York (USA) and New Brunswick (Canada), only dealt with the control of CPB adults in the spring, whereas in Quebec (Canada), the impacts of heat on weeds and for top killing of plants at the end of the season were also studied. In the United States, Moyer (1991), Moyer et al. (1992), and Olkowski et al. (1992) reported mortality rates from 70 to 90% for spring adults, 25 to 50% for eggs, and 30 to 88% for young larvae. Also, Bernard (1992) obtained rates of 50% for adults and 67% for eggs in New Brunswick and reported a global impact of about 80% on CPB adults, eggs, and larvae and a comparable mortality rate for flea beetles. During the same period, research in Quebec revealed mortality rates from 30 to 80% and 45 to 93% for spring and summer (during top killing) CPB adults, respectively. Regarding eggs and larvae (during top killing), mortality rates were 45% and 75 to 96%, respectively (Martel and Hamel 1993). Thermal top killing of plants was found to be as efficient as the chemical defoliator, REGLONE, and the impact on large leaf weeds in spring was very significant.
These favourable results should stimulate further projects to identifiy the advantages and disadvantages of this technology, taking into account its potential for weeding and top killing and its harmonisation with the integrated production of potato. Along these lines, an important research program was carried out in Quebec in 1994, by a multidisciplinary team including experts in entomology, engineering, weed science, pathology, and physiology (Ministry of Agriculture, Fisheries and Food, Université Laval, ICG Propane Inc.). For laboratory testing, a test bench designed for thermal control research was used (Gill et al. 1994). Results supporting the development of secure, efficient, and versatile equipment for thermal control will be discussed.

4.1 CPB Thermosensitivity

Thermal sensitivity thresholds of the different life stages of the CPB were first determined in the laboratory under controlled conditions and then validated in the field.

4.1.1 Laboratory Tests

Samples of CPB adults, larvae, and egg masses were exposed to thermal treatments that would generate temperature rises of 125, 150, 175, and 200 °C at the target level. Samples (adults and larvae) were set on Pyrex trays placed on the mobile carriage of the test bench. Egg masses were placed in two different manners, on the upper side of a leaf (exposed eggs) or under a leaf (protected eggs), in order to simulate the conditions encountered in the field. Eggs are more sensitive to heat than adults (Fig. 1). Moreover, a mortality of 100% was obtained in all larval stages at temperatures of 75 to 200 °C. For adults, temperatures beyond 100 °C resulted in mortality rates of 57% or higher. At temperatures exceeding 150 °C, more than 75% of the adults died within 2 days, whereas surviving individuals showed highly reduced mobility and feeding capacity. These results are comparable to those obtained by Pelletier et al. (1995). Treatments leading to adequate control of adults (severe mortality and injuries) also resulted in a satisfactory destruction of eggs and larvae present on the plants.

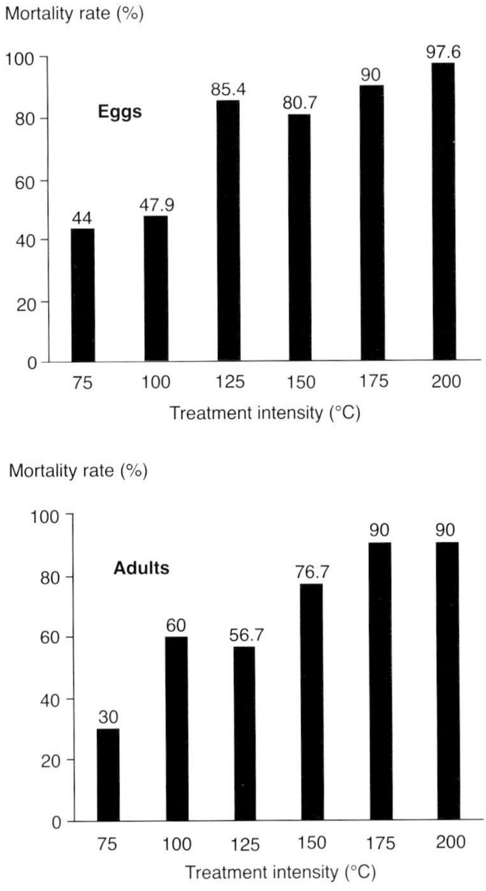

Fig. 1. Mortality of CPB eggs and adults during laboratory tests, four days after treatments.

4.1.2 Field Trials

For these tests, spring CPB adults placed on young plants (cv. Kennebec) of different heights (0 to 5, 5 to 10, and 10 to 15 cm) were exposed to two thermal treatments (150 and 200 °C). The mortality rates of spring adults ranged from 77 to 90% and were highest for insects located on young plants (0 to 5 cm). Thus, if the temperature near the plant is about 175 °C, the mortality rate could reach 80% for adults and 100% for larvae and eggs. In the laboratory, it was demonstrated that the mortality of eggs that were protected by leaves was 5 to 20% lower than that of those directly exposed. Similar results were obtained in the field. The effect of thermal shielding by foliage, which is also valid for other insect life stages, is more important when the control of CPB is combined with top killing because the foliage is at its maximum density.
In 1995, mortality rates of 83 to 90% and 85 to 92% for spring and summer (during top killing) CPB adults were obtained with thermal control equipment configured according to the laboratory and field results.

4.2 Sensitivity of Young Potato Plants

Thermal control can involve the treatment of the crop against CPB at the beginning of the season. Therefore, it is important to consider the resistance of young potato plants to these treatments. For this purpose the thermal sensitivity thresholds of young potato plants were established. Trials were first conducted in the laboratory under controlled conditions, then in experimental plots, and finally at the commercial scale.

4.2.1 Laboratory Tests

Work was conducted under controlled conditions on a test bench (Gill et al. 1994) to determine plant resistance to thermal treatments at stages that correspond to early CPB invasions. The tests were conducted on young greenhouse-grown potato plants (cv. Superior and Kennebec). The growth stages were 0 to 5, 5 to 10, and 10 to 15 cm. Plants were exposed to treatments generating temperature elevations of 125, 150, 175, and 200 °C.
Results showed that the young potato plants were quite tolerant to exposure at 175 °C, a temperature required for an efficient control of adult CPBs. Based on the plant height or on the presence of external damage, it was found that younger plants were more resistant to heat and recovered more rapidly.

4.2.2 Experimental Plot Trials

Only young plants (0 to 5, 5 to 10, and 10 to 15 cm) of the variety Kennebec were used in these studies. They were exposed to treatments generating temperature elevations of 125, 150, 175, 200, and 250 °C at the ground surface (Fig. 2). Treatments were applied using thermal control equipment having individual burners producing a uniform temperature elevation at the ground surface level over the whole operating width (Laguë et al. 1994).

Results confirmed the trends previously observed on plants grown in greenhouses. Twenty days after the treatments, the differences in height between treated and untreated plants were 3.2, 6.7, and 9.3 cm on average for the treatments at 175 °C applied at the stages of 0 to 5, 5 to 10, and 10 to 15 cm, respectively. The same pattern was observed for the visual damage index, which spans from 0 when there is no visual damage to 10 when 91 to 100% of the plants are burned (Laguë et al.

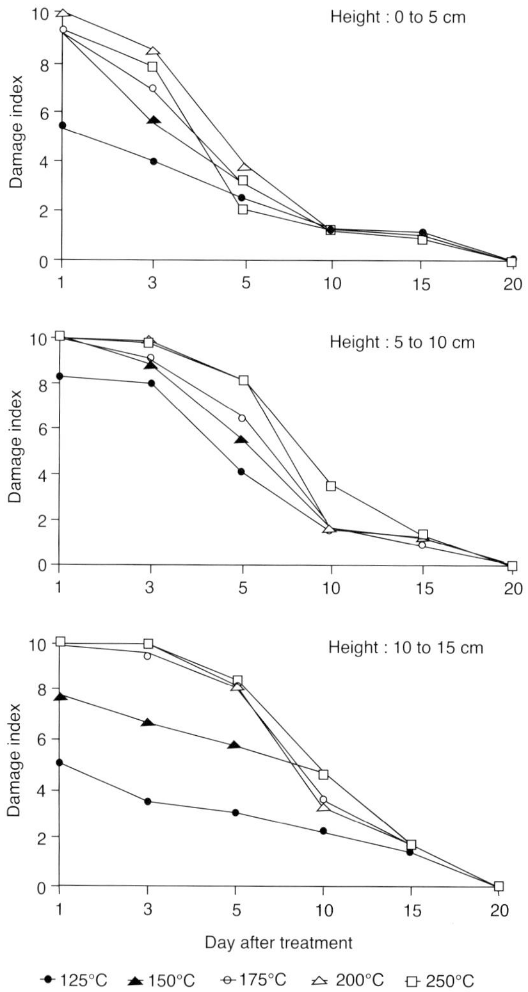

Fig. 2. Evolution of the damage index over a 20-day period for young potato plants (cv. Kennebec) treated in the field with heat using a propane thermal control equipment.

1994). Ten days after the treatments, values of this index were 1.5, 1.5, and 3.5, respectively, for growth stages of 0 to 5, 5 to 10, and 10 to 15 cm. All traces of damage disappeared and the indices were null for all stages and treatments 20 days after application.

4.2.3 Trials Under Commercial Production Conditions

In order to validate the results obtained in the laboratory and in experimental plots, spring treatments were applied in 1995 on parts of cropped fields located in two regions of Quebec. These treatments were aimed at the simultaneous thermal control of CPB and weeds using pre-commercial prototypes of thermal control units capable of producing a uniform heat application at the ground surface level. The trials were conducted on plants less than 10 cm tall.

Results on plant recovery confirm those observed during the previous experimental tests. In addition, yield evaluation confirmed that spring treatments on young plants had no negative impact on yields.

4.3 Control of CPB and Top Killing

In this part of the study, the objective was to determine the thermal treatment parameters leading to adequate defoliation for a second control strategy, which combines thermal top killing and control of CPB. This also involved measuring the resulting temperature profile over the plants. Trials were conducted in the laboratory on greenhouse plants, in experimental plots, and finally in grower's commercial fields.

4.3.1 Laboratory Tests

Laboratory tests were conducted on greenhouse plants on three potato varieties (Chieftain, Kennebec, and Superior). These three varieties were selected because they are the most widely cultivated in Quebec and because their foliage densities are different. Three gas supply pressures and three travel speeds were combined to obtain nine thermal treatments. Temperature elevations inside the plant foliage were measured at ten locations for each treatment, and the evolution of the plant damage index was monitored.

Results of the temperature monitoring inside the foliage showed a large variability and did not reveal a clear relationship between the nominal intensity of the treatments and the measurements inside the foliage (i.e. high pressure + low speed = stronger intensity). Most of the treatments resulted in a defoliation comparable to that obtained chemically.

4.3.2 Tests in Experimental Plots

Further to the laboratory trials, top killing tests in commercial plots were conducted on mature potato cultivars (Chieftain, Kennebec, and Superior). Treatments were applied using pre-commercial thermal control equipment. Temperature was monitored inside the foliage during the application of the treatments and plant defoliation estimated 20 days after treatments.
Temperatures inside the foliage exhibited large variability. However, temperature fluctuations associated with each treatment accurately followed those expected based on the general equation of treatment intensity (high pressure + low speed = stronger intensity). These measurements also indicated that mean temperatures for treatments resulting in an efficient top killing, corresponded to the heat intensities required for the control of adult CPBs at a mortality rate of 75% (150 to 200 °C). It is therefore possible to reduce the CPB population densities at the end of the season using the thermal method to defoliate potato plants.

4.3.3 Tests Under Commercial Conditions

In 1995, CPB mortality rates of 87 and 92% were obtained in commercial fields of Snowden and Atlantic cultivars, respectively. The quality of the thermal top killing was similar to and even better than chemical defoliation with REGLONE at all sites, even on varieties as turgescent as Kennebec. Sugar contents and chip color tests indicated that the quality of Atlantic, Snowden, and Niska tubers intended for chips market was excellent. Tuber emergence rates of cv. Kennebec, stored during the whole winter season, were not affected by thermal defoliation. In general, these results confirm the potential of using the thermal method to control CPB adults and to provide efficient top killing without reducing tuber quality at harvest.

5 Limitations and Constraints

5.1 Biology of the Plant and the Insect

Thermal control is a non-selective and non-specific technology. Indeed, it can destroy natural enemies present during treatments. Some lady beetles (for example *Coleomegilla maculata lengi* Timberlake) are very vulnerable. However, if used early in the spring, the impact on this predator can be minimized, particularly in the absence of CPB egg masses. The impact could be greater during top killing, especially when early or mid-season cultivars are involved. This is because lady beetle densities may be very high at this time of the season due to the presence of large numbers of aphids on the plants.

The effectiveness of the CPB thermal control and of the top killing depends on the cultivar grown. Mortality rates of existing insects are lower in cultivars with denser foliage. The cultivar, Kennebec, is therefore more difficult (dense foliage, long and thick stems) to defoliate than the cultivar Superior (slight dense foliage, short and small stems). Early spring treatments, when the plants are 0 to 10 cm tall, have no impact on the yield or on the quality of tubers at harvest. Plants recover over a period of about 14 days. The recovery of very young plants, 0 to 5 cm tall, takes 10 to 12 days. At other periods, growth and yield are affected.
Finally, the CPB biological cycle and its interrelation with potato growth and development restrict the use of this technology and make it necessary to consider other strategies. The technology is currently intended to control CPB adults at two particular periods of the season (Fig. 3). In the spring, the time frame for application is restricted. The emergence of spring adults may be spread over 4 to 6 weeks. The rapid colonisation of fields at high densities is irregular and the distribution of populations is heterogeneous from one field to the other and from one region to another. Interventions in the spring and at the end of the season are optional and not necessary (Boiteau et al. 1996). In the spring, they are only justified when the adult densities are high and have a real impact on the emergence of potato plants. This situation is hard to predict and is dependent on climatic conditions. Thermal control in the spring is therefore of limited value. A preventive program, based on crop rotations, reduces the need to intervene against adults in spring. In addition, other means could reduce adult densities on plants early in the spring (Boiteau et al. 1996; see Chap. 19). Therefore, if thermal control is only intended to control emerging spring adults, the technology is not likely to be implemented.

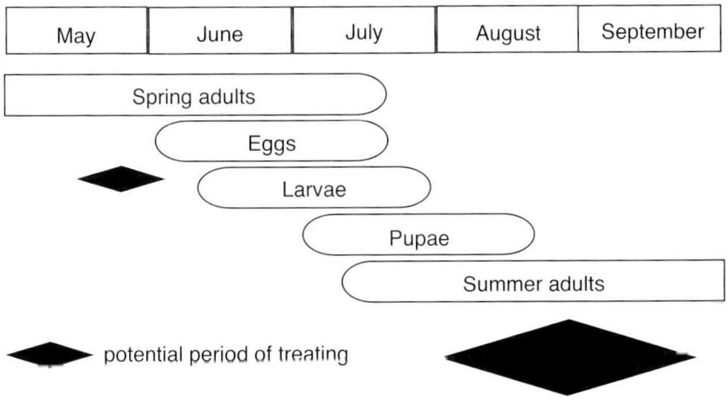

Fig. 3. Potential usage periods (in spring and at the end of the season) of the thermal control method within the seasonal biological cycle of CPB.

5.2 Effect of Meteorological Conditions

To be effective, thermal control should be applied under specific meteorological conditions. First of all, treatments have to be carried out on calm, sunny days that stimulate the activity of adult CPB feeding on the upper part of the plants where they are more exposed to the effect of the thermal treatments. Treatments must not be carried out early in the morning, late in the evening, or during cold and cloudy days because under such conditions, adults are on the bottom part of the plants or on the ground. Since strong winds could result in heat drift, treatments must be carried out on calm days, under light wind conditions, or under the protection of heat shields. Moyer (1992) observed that at wind speeds of 24 km h^{-1} or greater, mortality decreased by over 30% when treatments are applied in the spring travelling upwind as opposed to applications performed travelling downwind.

5.3 Proper Use of Equipment

Thermal control stresses potato plants. Only one treatment can be done in the spring when potato plants are less than 10 cm tall. Irreversible effects on plants and decreases in yield were observed when more than one treatment was applied. Plant height was reduced by 30% when thermal control was applied two or three times a week. This would be more critical if the plants are already stressed by unfavorable conditions such as drought. During top killing, more than one pass can be made without affecting tuber quality.
According to some studies carried out at Université Laval, the work rate of tested prototypes remains the major weakness of the thermal method. Indeed, travel speeds of 6 km h^{-1} for spring applications and of 3.5 km h^{-1} for top killing operations are in general slower than those of conventional sprayers (8 to 10 km h^{-1}). In addition, tested prototypes have smaller effective widths (about 6 m), which greatly reduces their work rate when compared to sprayers (20 to 30 m and more).

5.4 Costs

According to Moyer (1992), building a machine costs 10 030 to 15 930 FF (2380 to 3780 $CAN). The need to comply with the standards and security rules compulsory for safe use of propane gas would have an impact on the costs. On the other hand, operational costs of the thermal control in the spring (CPBs and weeds: 222.09 to 298.79 FFha^{-1}, i.e. 52.70 to 70.90 CANha^{-1}$) compare favorably with conventional chemical methods when heat is applied over the entire ground surface. However, when top killing cultivars of high density foliage, the thermal control is more expensive (207.76 to 444.60 FFha^{-1} / 49.30 to $105.50 CANha^{-1}$) (Laguë et al. 1994).

6. Environmental Impacts

Environmental impacts of the thermal method are discussed in Chap. 2. It should be stressed that the major impact of the thermal method is on air quality. Impacts on soil and water resources are negligible compared to those of chemical control. The impact of thermal method on air would be more severe when propane combustion is performed at high pressure supply without sufficient oxygen input. This can happen during thermal top killing on varieties having a high density foliage. Spring treatments against colonizing populations of CPB adults are performed at such pressure that negative impacts on the environment are lower. In this period of the season, thermal control is performed under optimal combustion conditions. However, the combustion of propane involves the use of a pure active ingredient and therefore consumes more energy than other methods (Jolliet 1994). The high fuel consumption needed for top killing makes this technique less energy efficient and less ecological. Finally, the impact of a thermal treatment is limited to the ground surface only and does not affect the microorganisms living in the soil.

7 Effectiveness and Strategies of Use for Thermal Control

The thermal technique could be used to simultaneously control weeds and CPBs (adults and eggs) in the spring. However, it is only efficient against weed and pests present during the treatments. For top killing, this technique is affected by the maturity of the plants (green or senescent plants), the volume of foliage, and the shape of the plants. For similar temperatures (150 to 200 °C), the success of thermal top killing relies on a good penetration and distribution of heat inside the foliage.

For potato production, the thermal control could represent an alternative to chemical pesticides. It contributes to the reduction of the problems related to the contamination of underground and surface water by pesticides. In addition, it does has no residual effect on the crop. The use of this technique also reduces the risk of developing resistant populations of CPBs.

However, since the impact is mainly on CPB adults and egg masses, thermal control technique cannot be used as the sole technique for the control of CPB populations. It should be integrated with other control strategies in several ways. In the spring, thermal control could be applied to the whole field or strictly over sections (border rows, crop traps) that are rapidly colonized by CPB adults. At the end of the season, during top killing, the thermal control could be useful in association with a chemical defoliator like REGLONE or alone, to defoliate rows within fields at different dates. This approach favors the accumulation of important densities of CPB adults on non-defoliated plants and increases the impact on the adults that are present.

8 Conclusion and Recommendations

The use of chemical insecticides for protecting potato crops has been reinforced with the introduction of the insecticide ADMIRE. However, based on the past 15 years' experience, it is likely that resistance will develop against this chemical. Many methods of control of the CPB are now available. The variety of methods for this pest is more extensive than for those on other crops. This offers flexibility for choosing an appropriate method in the context of sustainable agriculture. The potential of thermal control of CPB in North America is now well established from the agronomic, economic, and technical point of view, as are the operational parameters of this technique for weeding and top killing. Its potential for controlling CPB, weeding, and top killing facilitates its integration within a global approach.

In order to benefit from this technique in an optimal manner, some avenues have to be explored. A first improvement consists in enlarging the intervention window at the beginning of the season. In this manner, the required equipment could be combined on the frame of a cultivator or a hiller to control at the same time CPBs and weeds, thus reducing operational costs. Thermal control could also be combined with another mechanical or pneumatic system that would dislodge insects from the plants and deposit them on the ground surface between the rows where a burner could destroy them. One may also consider the control of larvae by making passages during the appearance of floral buds and at blooming.

References

Bernard G., (1992). Évaluation et démonstration en 1992 de brûleurs au propane utilisés pour la lutte contre les doryphores de la pomme de terre qui hivernent dans les champs, Technical reports on potatoes, New-Brunswick: Coordination committee of potato manufactoring ''de l'industrie'', Ministry of Agriculture of New-Brunswick. Vol 5, 24-28.

Boiteau G., Duchesne R.-M., Ferro D.N., (1996). Use and significance of traditional and alternative insect control technologies for potato protection in a sustainable approach pp. 169-188, in Duchesne, R.-M. and G. Boiteau (eds). Symposium 1995, Potato insect pest control. Sustainable approach, Quebec (Canada). 204 p.

Duchesne R.-M., Boiteau G., (1996). Symposium 1995: Potato insect pest control. Sustainable approach, Quebec (Canada). 204 p.

Gill J., Laguë C., Lehoux N., Péloquin G., (1994). Test bench for thermal weed and pest insect control, ASAE paper 94-8512. American Society of Agricultural Engineers, St-Joseph, MI, USA.

Gill J., Laguë C., Lehoux N., Péloquin G., (1995a.) Propane burner characterization for thermal weeding. ASHS poster 878. Hortscience. 30 : 819.

Gill J., Laguë C., Lehoux N., Duchesne R.-M., (1995b). Use of propane flamer in potato production. ASHS poster 899. Hortscience. 30 : 828.

Jolliet O., (1994). Bilan écologique de procédés thermique, mécanique et chimique pour le défanage des pommes de terre. Revue Suisse Agric. 26 : 83-90.

Laguë C., Bernier D., Duchesne R.-M., (1994). Use of propane flamers for weeds and Colorado potato beetle control and for top killing in potato crops, Centre de recherche en horticulture, Université Laval, Sainte-Foy (Canada). 83 p.

Laguë C., Gill J., Lehoux N., Péloquin G., Leroux G., Yelle S., (1995). Recherche et développement sur le contrôle thermique des parasites en agriculture à l'Université Laval. 17ème Colloque de génie rural, Département de génie rural, FSAA, Université Laval, Québec : 87-122.

Martel B., Hamel G., (1993). Influence du brûleur au propane sur la densité de population du doryphore en fin de saison. Technical report: Stratégie phytosanitaire, Ministry of Agriculture, Fisheries and Food of Quebec. 16 p.

Morelle B., (1993). Le désherbage thermique et ses applications en agriculture et en horticulture, pp. 109-115, in J.M. Thomas (eds). Proceedings of the fourth IFOAM International Conference. 177 p.

Moyer D.D., (1991). Development of a propane flamer for Colorado potato beetle control. Riverhead, New-York: Cornell Cooperative Extension, 7 p.

Moyer D.D., (1992). Fabrication and operation of a propane flamer for Colorado potato beetle control. Valley Potato Grower. pp. 14-20.

Moyer D.D., Derksen R.C., Mcleod M.J., (1992). Development of a propane flamer for Colorado potato beetle. Am. Potato J. 69 : 599-600.

Olkowski W., Saiki N., Daar S., (1992). IPM options for Colorado potato beetle. The IPM Practitioner 14 : 1-21.

Pelletier Y., McLeod C.D., Bernard G., (1995). Description of sub-lethal injuries caused to the Colorado potato beetle (Coleoptera: Chrysomelidae) by propane flamer treatment. J. Econ. Entomol. 88 : 1203 - 1205.

Control of Insects in Post-Harvest: High Temperature and Inert Atmospheres

Francis FLEURAT-LESSARD and Jean-Marc LE TORC'H

1 Introduction

The methods used in the physical control of stored-product insects are drawn from several different sources. They have been adapted to the many different pest problems that arise during storage, transportation, and processing of cereals and their processed products. Food manufacturers want to be sure that their product will remain free of infestation throughout the distribution and retail system before being consumed. Among the physical methods of control, hermetic packaging with modified atmospheres is one way to insure pest-free delivery of the product. However, this technique is too costly to be used for primary products such as; cereals, dried fruits and dried plants. An alternative to chemical control methods for these products is extreme temperatures, the physical control method of choice (Fleurat-Lessard 1987). In certain situations, mechanical control methods are used to disinfest flour and semolina through impact devices (see Fields et al., Chap. 17). Physical control in stored products integrates many diverse methods, and it requires an in-depth knowledge of their various characteristics to be able to use them effectively in commercial operations. The operations concerned range from the initial storage of a raw product to the delivery of a finished product free of infestation.

2 High Temperature

2.1 Mode of Action

2.1.1 Biological Effects of Heat (Thermobiological Scale)

The biological effects of temperature on insects are described by a diagram called the thermobiological scale (Fig. 1). Vannier (1987) defined an optimum temperatu-

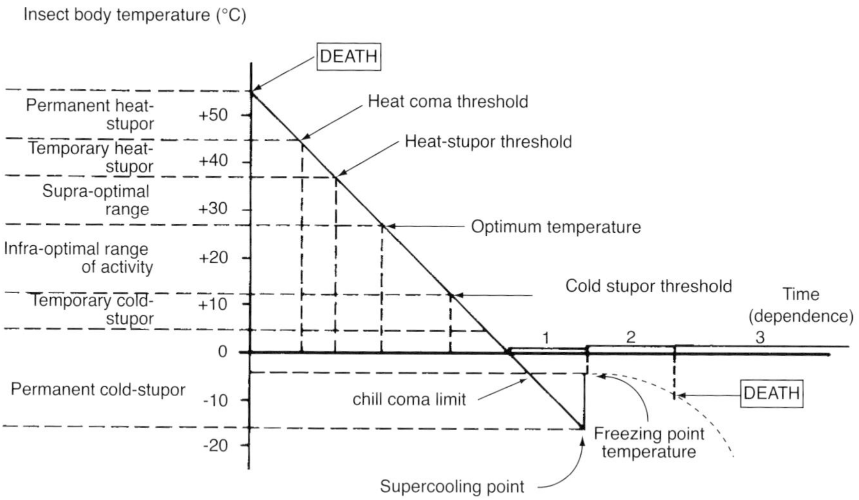

Fig. 1. Thermobiological scale of an insect (After Vannier 1987).

re that corresponds to the range of temperatures at which an insect functions normally. When the temperature goes above this optimum, the insect functions at a reduced level. If the temperature continues to rise, the insect becomes very active, and various stress mechanisms become activated. At a certain temperature, the insects enter a phase called heat stupor, where there is a rapid reduction in movement and breathing (Chauvin and Vannier 1991; Fig. 2). The upper lethal temperature is only a few °C after heat stu-

Table 1. The permanent heat stupor points for different mobile stages of stored-product insect pests (Chauvin and Vannier 1991).

Species	Stage	Heat stupor (°C) ± SD
Sitophilus granarius	Adult	53.1 ± 1.0
Oryzaephilus surinamensis	Larval (stage IV)	54.2 ± 1.3
	Adult	54.4 ± 0.9
Anthrenus verbasci	Larval (last stage)	54.0 ± 1.0
	Adult	53.1 ± 0.5
Galleria mellonella	Larval	55.1 ± 0.6
Tineola bisseliella	Larval (stage II)	48.9 ± 0.3
	Larval (last stage)	50.7 ± 0.7
	Adult	49.2 ± 0.6

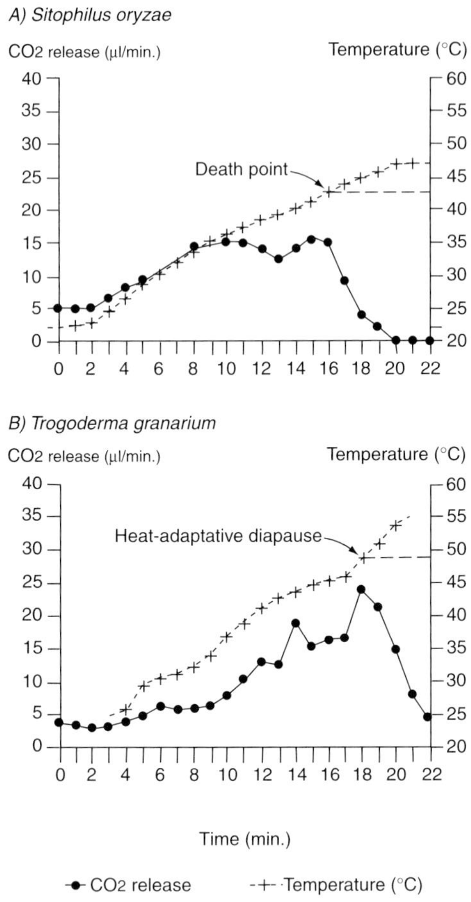

Fig. 2. The production of carbon dioxide during progressive heating beyond the point of heat stupor. **A** *Sitophilus oryzae*. **B** *Trogoderma granarium* (Fleurat-Lessard unpublished).

por is induced. The temperature at which heat stupor is induced is dependent on the species, stage of development, and physiological state of the insect. The heat stupor point, like the supercooling point (see Fields, Chap. 6), is essential for using effectively extreme temperatures to control pests. Chauvin and Vannier (1991) measured the temperature level of heat stupor point for several stored-product insects (Table 1), that is situated in the range 49-55 °C. The upper lethal temperature for insects is approximately 4 to 6 °C above the heat stupor level, i.e. in the range of 53-60 °C. It is thought that high temperature causes death because of damage to the phospholipid membranes and destruction of intracellular proteins. Other processes involved as possible physiological mechanisms of death due to high temperature are: ionic or osmotic imbalances, enzyme rate imbalances, and desiccation (Fields 1992).

Box 1. Thermobiological Limits of an Insect (Ectotherm)

The effect of temperature on insects is described by a graph called the temperature response curve. It describes the succession of physiological states that an insect undergoes, from the low temperatures at which insects freeze to the high temperatures that cause coagulation of the most heat-sensitive proteins. The best graphic representation of this is a straight line where temperature is on the y axis and time is on the x axis (Fig. 1). In the middle of the graph there is a temperature where most physiological functions are optimal. Above and below this temperature is a range of temperatures that are sub-optimum for the insects, but they can still reproduce and complete their life cycle (Vannier 1987). Nevertheless, death can occur before freezing if a slightly positive temperature level is maintained during several weeks or months. The time scale at temperatures below 0 °C and above 40 °C is not precise because there are several variables species, stage, acclimation – that affect the duration – needed before mortality occurs. When temperatures rise above the upper sub-optimum zone, the insect stops moving and enters into a state called heat stupor. A few degrees above this is the upper lethal limit of the insect and it dies after a few minutes.

At the other end of the scale, at sub-zero temperatures, ice can form inside the insect. There are two types of insects: insects that cannot survive the formation of ice inside their bodies, freeze-intolerant and those that can, freeze-tolerant. The lower limit of the thermobiological limits for freeze-intolerant insects is the supercooling point, or the point at which ice forms inside the insect. Although the supercooling point represents the lowest temperature a freezing-intolerant insect can survive, insects often die before they freeze. Insects can acclimate to low temperatures. This acclimation can greatly increase their tolerance to low temperatures. Many insects that overwinter in the field have a diapausing stage that is induced for a given stage (obligatory diapause) or depending upon day length and temperature and other cues (facultative diapause). Diapause often increases the insect's tolerance to cold and sometimes to high temperature. There are only a few stored-product insects that have a diapause, *Plodia interpunctella* for example. There are some insects that are freeze-tolerant, they usually freeze around –10 °C but survive the freezing and temperatures below their supercooling point. None of the stored product insects is freeze-tolerant.

2.1.2 Damage to Phospholipid Membranes

Changes in the properties of the phospholipid membrane are thought to be the prime reason for death at high temperatures (Fig. 3). The phospholipid membrane becomes more fluid at high temperature. As the nervous system is dependent upon the integrity of the cell membrane, it stands to reason that it is very sensitive to high temperature. As with low temperatures, insects can acclimate to a certain extent to high temperature. That is, exposure to a moderately high temperature (35 °C) will increase tolerance to higher temperatures (48 °C). It is thought that the basis of this acclimation is due to higher melting points in the synaptic membranes (Cossins and Prosser 1978). Insects reared at high temperatures have cell membranes with higher melting points (Fraenkel and Hopf 1940).

2.1.3 Damage to Proteins

Secondarily, high temperatures from the heat stupor point change the structure and the functional properties of insect proteins. There is a close link between the temperature at which pyruvate kinase is inactivated (a few minutes at 56 to 60 °C) and the upper lethal temperature of most insects. The function of enzymes can also be adversely affected at lower temperature levels if these levels are maintained during several hours. Since the rate at which an enzyme functions is affected by temperature, and different enzymes have different temperature coefficients, high temperatures can cause a decoupling of metabolic reactions, resulting in a breakdown of the biochemical equilibriums necessary to maintain important physiological systems such as energy metabolism, nervous, sensory, and respiration. We have noted that there is a wide variation in the response to high temperature of different stored-product insect pests. For example, at 45 °C the larvae of *Tenebrio molitor* L. has an erratic rate of respiration, with sometimes very high levels of gas exchange

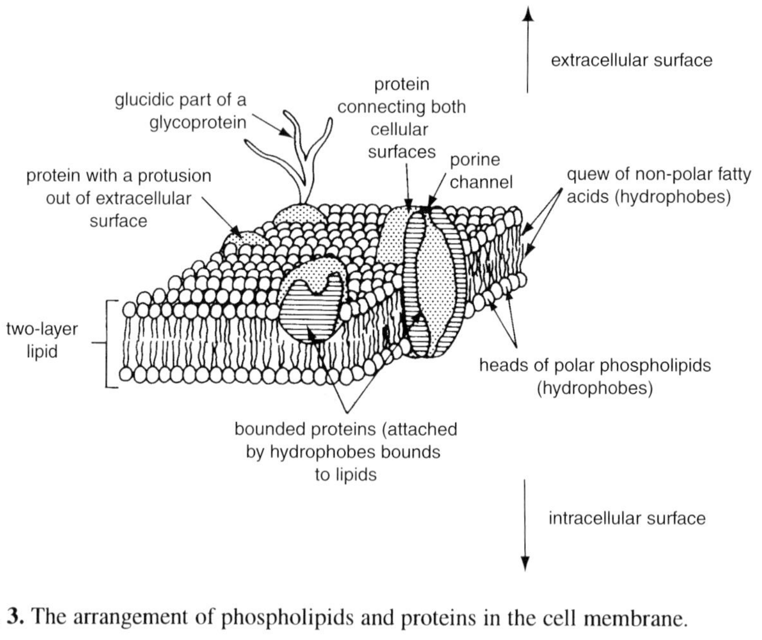

Fig. 3. The arrangement of phospholipids and proteins in the cell membrane.

(Fig. 4). In contrast, the last larval instar of the khapra beetle, *Trogoderma granarium* Everts, has a regular rate of respiration as the temperature rises to 50 °C, at which point it enters heat stupor. This physiological process of resistance to heat stress is considered to be a protective diapause, as we have observed larvae are able to survive for several minutes at 53 °C (Fig. 2). Under the same rate of temperature increase, the adults of *Sitophilus granarius* (L.) died at 45 °C.

2.1.4 Ionic Imbalances

Adequate pH is important for the proper functioning of many enzymes. The pH may change with temperature, causing acid-base imbalances. An example of this is that desert insects have a more acid haemolymph just before dying of heat stress than do healthy insects. Desiccation stress is often coupled with heat stress and, in general, insects that are not adapted to xeric habitats are also sensitive to heat (Fields 1992).

2.1.5 Heat Shock Proteins

Proteins are produced in response to short exposure to heat (Petersen and Mitchell 1985). The majority of studies on heat shock proteins have used fruit flies from the genus *Drosophila*. Heat shock proteins are produced between 30 and 40 °C, and are correlated with increased tolerance to temperatures above 40 °C. These proteins attach to RNA, and are thought to aid in the repair of proteins and lipids damaged by heat stress (Nagao et al. 1990). Further work is needed to verify this hypothesis, and to elucidate the other roles heat shock proteins may play in adaptation to high temperature.

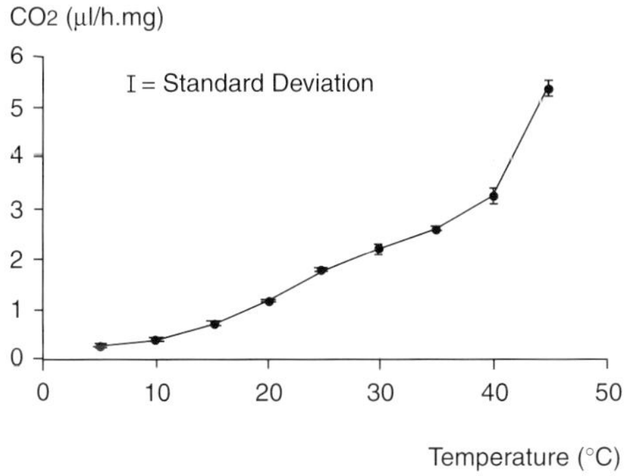

Fig. 4. The production of carbon dioxide during progressive cooling beyond the point of cold stupor of the yellow mealworm *T. molitor* (Fleurat-Lessard unpublished).

2.2 Practical Applications

2.1.1 Traditional Use of Heat Disinfestation of Stored Foodstuffs

Heat treatments have been used since the beginning of the 20th century to disinfest non-perishable foodstuffs (Goodwin1922; Grossman 1931) or even perishable foodstuffs (Baker 1939). For non-perishable foodstuffs, the thermal shock is obtained by using dry heat at temperatures above 50 °C, while heat treatment with perishable foods, such as fruits, uses hot water baths or hot water mist at temperatures below 50 °C.

The use of high temperature to control insects in post harvest has a long history. It has often been used to control insects in flour mills infested with the flour moth *Ephestia kuehniella* (Zeller), that at the beginning of the 20th century was still considered a dangerous pest of flour mills (Goodwin 1922; Dean 1913, cited in Sheppard 1992). Heat treatments were a common practice in late 19th century in France. Dean (1913) discusses the work carried out by Webster in 1883, which gave precise guidelines for heat disinfestation of *Sitotroga cerealella* (Ol.)(the Angoumois grain moth), an insect that feeds on wheat kernels. Winterbottom (1922) describes the heat disinfestation methods used during World War I (1915-1919) for export wheat produced in Australia to satisfy sanitary requirements of the customers. Pepper and Strand (1935), citing work by Goodwin (1922), state that 48.8 to 54.4 °C for a duration of 10 to 12 h is sufficient to control all stored-product insect pests without reducing grain quality. Nevertheless, the method of hea-

Table 2. The decreasing ranking of resistance to heat of stored-product insects (After Fields 1992).
Upper lethal temperature (T in °C)

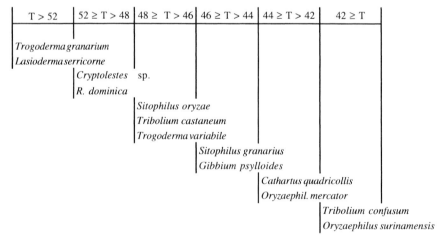

T > 52	52 ≥ T > 48	48 ≥ T > 46	46 ≥ T > 44	44 ≥ T > 42	42 ≥ T
Trogoderma granarium					
Lasioderma serricorne					
	Cryptolestes sp.				
	R. dominica				
		Sitophilus oryzae			
		Tribolium castaneum			
		Trogoderma variabile			
			Sitophilus granarius		
			Gibbium psylloides		
				Cathartus quadricollis	
				Oryzaephil. mercator	
					Tribolium confusum
					Oryzaephilus surinamensis

ting and the inevitable heat gradients required relatively long durations to completely control insects. Before the discovery of synthetic insecticides, heat disinfestation was the method of choice for disinfestation of flour mills in North America (Dean et al. 1936).

The relative ability of insects to survive at or around 49 °C, the critical temperature for heat tolerance, has recently been reviewed (Fields 1992) and is presented in Table 2. This figure is in agreement with the rankings by Oosthuizen (1935). With *Rhyzopertha dominica* (F.), the lesser grain borer, the experimental setup and the method of heat production can have an enormous effect on survival during heat treatment, and this probably holds true for other species. Also, it is often more difficult to control the juvenile stages of primary pests that are found inside the kernel, such as for the lesser grain borer, *R. dominica*, or the grain weevils because the internal stages are thermally buffered hence more difficult to control (Evans and Dermott 1981; Fields 1992). For ease of experimentation, the majority of the studies use adult insects, even though they are not necessarily the most heat tolerant stage. For example, at 49 °C, the adults and eggs of *Tribolium castaneum* (Herbst) survive only 30 min, whereas the larvae can survive 45 and the pupae 180 min (Grossman 1931).

Heat tolerance of stored-product insects is studied today in view of using heat disinfestation as an alternative to chemical insecticides (Forbes and Ebeling 1987). At high temperatures, there is little variation between species, and 7 min at 54 °C or 16 min at 51 °C is sufficient for complete control (Forbes and Ebeling 1987). Fields (1992) analyzed the survival data of postharvest insects from heat treatments from several publications. This author summarized the conditions necessary for a successful heat treatment:

1. avoid cold spots created from leaky windows, roofs and doors,
2. avoid closed spaces that heat slowly due to poor air circulation,
3. open machinery before the heat treatment,
4. add extra heat in the basement, and direct heat at the floor,
5. clean thoroughly before a heat treatment to prevent insects from being insulated from heat by dust and to reduce reinfestation,
6. provide good air circulation for an even distribution of heat,
7. treat doors and walls with residual insecticides around the heat-treated area to reduce the problem of immigration into the heat treated area.

2.2.2 More Recent Renewal of Interest of Heat Disinfestation

After the Second World War, heat treatment disinfestation was replaced in large part by chemical insecticides, such as the organo-chloride residual sprays and the fumigant methyl bromide. Chemical insect control had fewer operational inconveniences, and it was less expensive in time and money than heat treatments. However, during the past decade, there has been a renewed interest in heat disinfestation because of the concerns over chemical insecticide residues in food, worker safety, resistant pest populations, and environmental impact. Methyl bromide, the control method of choice for most cereal food processors, is slated to be phased out in most industrial countries by 2005 because it is an ozone depleting substance. Heat treatment has been improved since it was first used to control insects in flour mills (Heaps 1988). The main improvement has been the reduction in time needed to complete a treatment, 2 to 3 days, which is similar to the time needed to complete a fumigation with methyl bromide. Even though the killing effect of temperatures of 50 to 55 °C is obtained after 12 h, additional time is needed to reach the efficient temperature level in all locations and to wait for the building to cool. Heat treatments are the main method of insect control for a few companies such as Quaker Oats and Pillsbury, in the USA and Canada (Sheppard 1992; Ebeling 1994). It is also used on a limited scale in Scandinavia (P. Fields pers. comm).

2.2.3 Rapid Heat Treatments

Industrial disinfestation of grain and milled products (durum semolina in particular) is best done as a high throughput procedure. The first improvements in this type of treatment used infra-red (IR) ovens (Kirkpatrick 1975). However, the shallow penetration of IR into the foodstuffs necessitated that the foodstuffs, have a low throughput with a thin layer to ensure that insects were controlled in the entire product. Disinfestation and pasteurization of flour using heat was first studied by Chapman (1921), and Cotton et al. (1945) further refined the technique that treated flour in-line in 13 min (Fig. 5).

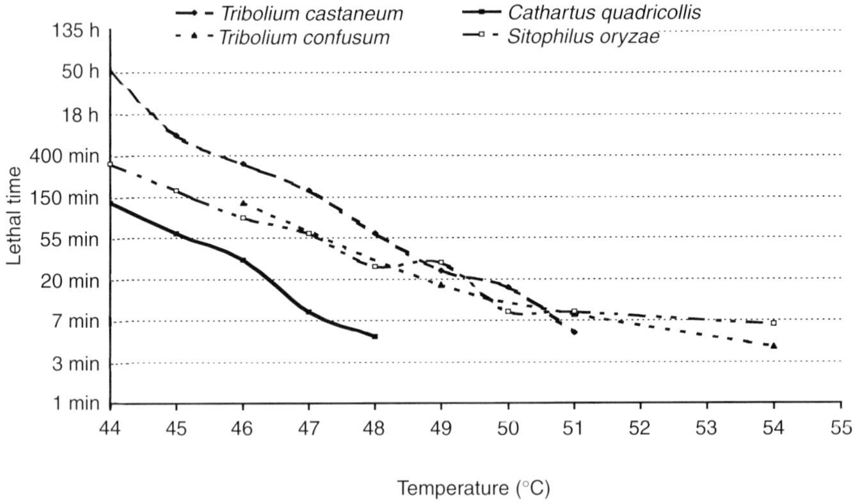

Fig. 5. The duration of survival before death of four insect pests common in food-processing facilities between 44 and 55 °C (After Grossman 1931; Forbes and Ebeling 1987).

Heat disinfestation of cereals and cereal products is a by-product of heat treatments in ovens mentioned above. Unlike drying, it is desirable for economic reasons (within certain limits, grain is sold on a wet-weight, not a dry-weight basis) that the moisture lost is keep to a minimum, during heat treatments for disinfestation. Since the heating is rapid, the moisture migration has only just begun, when the grain is rapidly cooled ending the migration of water and preventing the quality deterioration.

Two techniques are now used commercially for rapid heat disinfestation of foodstuffs: (1) heated fluidized-bed (or pneumatic conveyance), with hot air being the heat transfer medium, (2) dielectric heating, without a transfer medium by using microwaves or high frequency waves that heat from the inside out. (see Fleurat-Lessard, Chap. 11).

2.2.4 High-Temperature Fluidized Beds

Grain is a dielectric material with low thermal conductivity and hence is difficult to heat. One solution to this problem is to heat the grain in a fluidized bed (Botteril 1975) or a spouted bed (Becker and Sallans 1960), or a centrifugal fluidized bed (Brown et al. 1972). The rapid heating must be immediately followed by rapid cooling to avoid condensation problems and product quality deterioration. The time-temperature window that describes the conditions necessary to control insects by heat is separate but very close to the conditions that cause quality deterioration. Therefore, it is important to be able to tightly control the grain temperature at every step of the heating and cooling cycle. The effectiveness of the treatment depends on a number of factors, the most important of which are: grain moisture content, air temperature, air flow and insect pest to be controlled (the immature stages being the most difficult to control) (Dermott and Evans 1978; Evans and Dermott 1981; Fleurat-Lessard 1987; Table 3; Fig. 6). Australia has zero tolerance for insects, but has a climate that is ideal for stored-product insects and the insects have developed resistance to residual insecticides. In response to these problems, they have developed a pilot scale hot air fluidized bed at the grain terminal port of Dunolly, Victoria, that can treat grain at 150 th^{-1}. Grain is heated to about 60 °C in 2 to 4.5 min before being rapidly cooled in a second fluidized-bed with ambient air and water mist. The operating costs are similar to those of a residual insecticide treatment, about $0.80 Australian, and the capital costs for the unit was $500 000 Australian.

France has also developed heat treatments for disinfestation. To control insects in semolina, the product is heated during pneumatic conveyance, with air at 156 to 200 °C for 6 to 7 s. The semolina heats to a maximum of 70 °C, before being cooled with ambient air and stored (Fig. 7). Only 3-4% of the water is lost, which causes a total loss of 0.5% for the entire product. The heat treatment also increases the yellow pigment, a desired quality when the semolina is made into pasta. Pneumatic conveyance, has an advantage over fluidized-beds, because it kills 85% of insects just from mechanical damage to the insect. The operational costs of this system are ($1.6 US) per tonne of semolina treated.

Table 3. Durations (s) needed to control different insect species living in durum or semolina using a high-temperature fluidized bed in the laboratory (exiting air speed, 2.5 ms^{-1}).

Product treated	Pest	Stage	Air temperature (°C)				
			60	90	120	150	180
Durum	*S. oryzae*	All	66	16	12	9.5	
Soft wheat	*S. granarius*	All		18.5	9.2	8.5	
Fine	*A. kuehniella*	Late instar larva			11.5		
Semolina	*T. molitor*	Late instar larva			15		
	T. granarium	Late instar larva			14.2	8.5	5.3
Medium	*T. granarium*	Late instar larva			16.1	9.8	5.7
Semolina	*O. surinamensis*	Adult			10.5		

2.2.5 Hot Vapor or Baths for Fruit

Quarantine treatments of fruit using hot vapor or baths were common in the 1930s and 1940s; however, these were progressively replaced by fumigations with ethylene dibromide and methyl bromide in the 1950s. There has been a renewed inter-

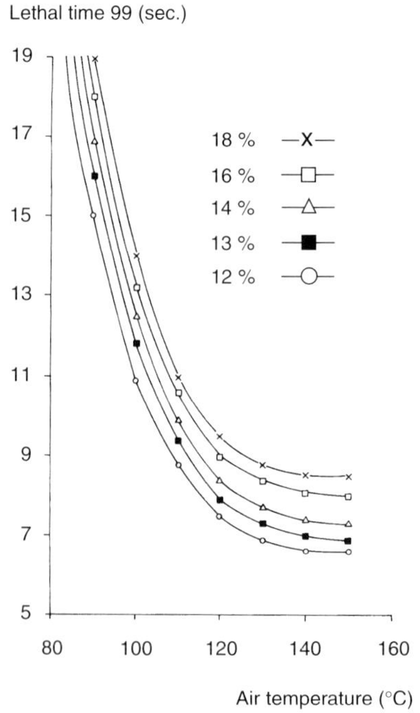

Fig. 6. The influence of moisture content of durum on the 99% isomortality lines for the internal stages of *S. granarius* in a heated fluidized bed at 90, 120 or 150 °C.

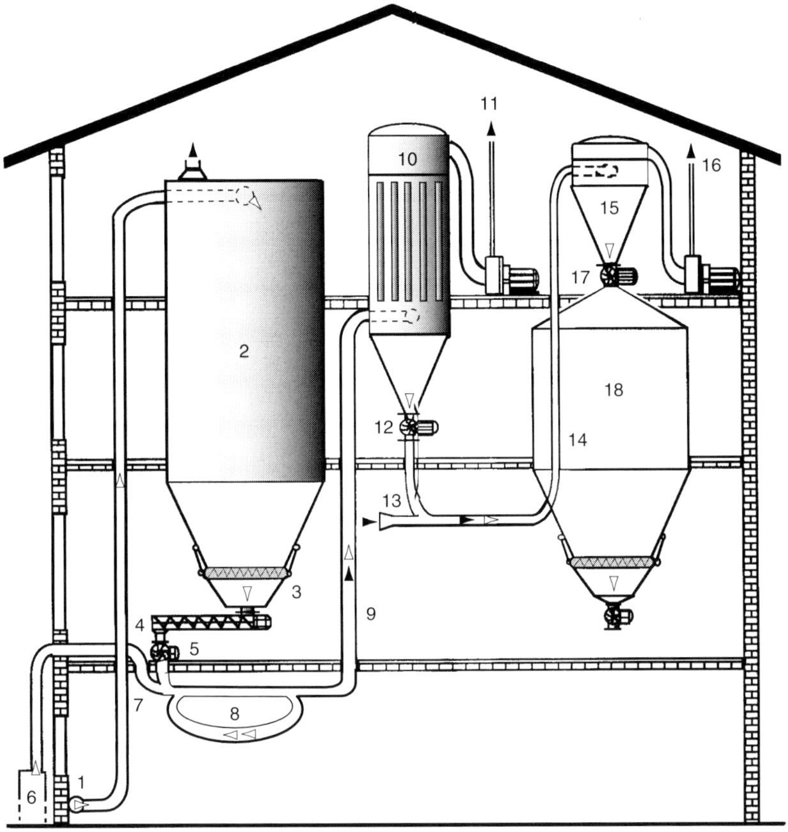

1 : Semolina or grain feeder
2 : storage bin
3 : vibrating bottom
4 : product flow-rate regulator
5 : rotating valves
6 : heat generator
7 : hot air pipe
8 : by-pass
9 : hot fluid-lift conveyance circuit (hot air + product)
10 : separation air/product (filter)
11 : hot air exhaust
12 : rotating valves
13 : fresh air inlet
14 : fresh air fluid-lift circuit (fresh air+ product)
15 : separation air/product (filter)
16 : fresh air exhaust
17 : rotating valves
18 : Buffer bin for the heat-treated product

Fig. 7. A pilot-scale facility for the heat curing and disinfestation of semolina or wheat during pneumatic conveyance in hot air.

est in heat treatments because of concerns of residues left by ethylene dibromide and the environmental impact of methyl bromide (Heather 1994; Paull and Armstrong 1994). Baker (1939) was able to control the fruit fly [*Ceratitis capitata* (Wiedeman)] in lemons using vapor heat at 44 °C for 8h. Modern techniques use lower temperatures for a few minutes to several hours, which are precisely controlled to minimize the damage to the fruit yet still ensure the control of the suite of possible insect pests (Sugimoto et al. 1983; Armstrong et al. 1989; Corcoran et al. 1993; Paull and Armstrong 1994). As in a fumigation, the time and temperature (dose) determine the mortality of the insects. The relationship between mortality and temperature can be modelled by an asymptotic regression curve, and the equation parameters are very different from one insect to another (Heather 1994). Quarantine treatments will often require that Probit 9, or 99.9968% mortality of the population will be achieved. For example, Japan requires that imported fruit from areas that are endemic to the fruit flies *C. capitata* or *Bactrocera tryonii* (Froggatt) undergo treatments that have been shown to obtain more than 99.9968% mortality. There are some systems that have been able to obtain this level of control and are currently used by the fruit and vegetable industries (Sharp 1990; Miller et al. 1990). The product [mango, tomatoes, carambola (star fruit) or lemons] is heated to 44-47 °C for a few minutes to several hours, depending upon the fruit or vegetable. In Australia, there are quarantine systems that can treat 20 t of fruit in 2 to 4 h using saturated-vapor heat (Corcoran et al. 1993). Vapor heat is destined to replace the older hot water bath treatments for quarantine treatment of fruits (Hallman and Sharp 1990).

3 Confined and Controlled Atmospheres

The use of low oxygen atmospheres has been a method of insect control since the 1920s (Winterbottom 1922; Cotton 1930). The first large-scale trials were begun at about the same time around the world: Australia (Bailey and Banks 1980), United Kingdom (Hyde et al. 1973), USA (Jay and d'Orazio 1984), Italy (Shejbal et al. 1973), Israel (Navarro 1978), Japan (Mitsuda et al. 1971), and France (Cahagnier and Poisson 1973). These pioneers in modified atmosphere storage were present for the first international symposium on the topic held in 1980 in Castelgandolfo, Italy (Shejbal 1980). The wide variation in the physiological reactions of stored-product insects to low oxygen or high carbon dioxide has necessitated the undertaking of fundamental studies to understand the effects of these modified atmospheres on insect respiration and to determine the conditions needed to effectively use these techniques to control insects in commercial grain stores.

3.1 Mode of Action of Hypoxia (Low Oxygen)

Almost all living organisms access the energy present in food through oxidation. Electrons or hydrogen ions are transferred to an acceptor molecule or to the electron transfer chain, with the last acceptor accumulating oxidation products in a reduced state. Oxygen is the ideal final acceptor because it combines with the reduced products to form water, which can be released without danger to the organism.

There are several other advantages of having oxygen as the final acceptor: (1) oxygen allows for the metabolism of any energy source whether it is lipid, protein or carbohydrate, whereas only glycolysis is possible via the anaerobic fermentation pathway; (2) aerobic metabolism is much more efficient than anaerobic metabolism; (3) lipids and sugars can be catabolized completely and the residues excreted as carbon dioxide and water, whereas anaerobic metabolism increases the acidity of the cell, which interferes with the cell biochemistry and makes the waste products much more difficult to excrete (Fleurat-Lessard 1990).
Lowering oxygen concentrations causes a series of physiological responses in anaerobic organisms: (1) to conserve energy there is a reduction in respiration and an increase in the conversion of food into energy; (2) anaerobic glycolysis is begun with the reduction of pyruvate to lactate, which uses NADH instead of oxygen as the final acceptor. Three moles of ATP are produced per second during anaerobic catabolism, whereas 36 moles are produced in aerobic catabolism of pyruvate to malate in the citric acid cycle. Simple hypoxia, the replacement of oxygen with nitrogen, causes a reduction of respiration, and a conversion to anaerobic metabolism when there is less than 3% oxygen in nitrogen (Adler 1994). Anaerobic metabolism causes an accumulation of the waste products from glycolysis and leads to a lack of metabolic water. Navarro (1978) noted that desiccation could be an important mortality factor under these anaerobic conditions. However, there are other critical systems that could be adversely affected. The permeability of the cell membrane is affected by the acidification of the cytosol and the build up of K+ ions outside the membrane. This is especially critical for the axons and causes a breakdown in the selective permeability of the ion channels and translates into the beginning of narcosis (Miller 1966).

3.2 Modified Atmospheres Using Carbon Dioxide (CO_2)

The increase in carbon dioxide in the air, with or without residual oxygen, is called hypercarbia. This causes different effects than the simple deprivation of oxygen. The physiological mechanisms, energy metabolism and neurotoxicity, underlining the toxic effects of high carbon dioxide atmospheres have been reviewed by Fleurat-Lessard (1990). The first measurable effect is the reduction of the energy charge (E):

$$E = \frac{[ATP] + 1/2\,[ADP]}{[ATP] + [ADP] + [AMP]},$$

which translates into a reduction in ATP, the basic fuel for the cell. Under these conditions, there is a marked reduction in the synthesis of protein, which is particularly important for larvae and pupae. Glutathion is not regenerated and its role in detoxification or biosynthesis is greatly reduced. This allows the toxic reduced free radicals to go unchecked, causing an accumulation of these products in metabolic

dead-ends. Changes in the oxygen-reduction pathways cause changes to the pyruvate/lactate ratio. In the nervous system, there is a reduction in the regeneration of acetylcholine from acetate and choline, mainly due to the reduction in the ATP levels. Another factor is that NADPH levels in the mitochondria drop because the cytosol has been transformed to the reducing state due to the lack of oxygen. Initially, insects are anesthetized, but if the exposure to carbon dioxide is prolonged there is damage to their central nervous system.
Also, the reversible decarboxylation of the 6-phosphogluconic-acid is inhibited by 66% when the insect has been exposed to a 10% CO_2 atmosphere. The malic enzyme is 100% inhibited at 40% CO_2, which stops the production of NADPH and involves a loss in reducing power. As CO_2 levels rise above 40%, there is a corresponding rapid decline in insect survival (Annis 1986; Adler 1994). At elevated CO_2 levels, the development times increase significantly. Given that carbon dioxide dissolves easily in water, increased levels of carbon dioxide cause an acidification of the cytosol. Glycolysis can be rapidly inhibited in high CO_2 atmospheres, but can still function in an anaerobic atmosphere (low oxygen, normal carbon dioxide, the balance in nitrogen), even though it is much more productive under aerobic conditions. This can explain why modified atmospheres with 40% or greater CO_2 are much more effective in killing most stored-product insect species than simply replacing the oxygen with nitrogen.

3.3 Practical Applications of Modified Atmospheres (MA)

Modified atmospheres have been used in post-harvest protection since antiquity in the majority of countries with hot and dry climates (Spain, Morocco, Egypt, India, Somalia, etc.). In these countries, there is a long history of underground storage, with sufficient sealing to increase the levels of CO_2 through the natural respiration of grain, insects and microorganisms until all O_2 is consumed (Sigaut 1980).

3.3.1 MA Fumigation at Atmospheric Pressure

Since the 1970s, research in the principal cereal producing or consuming countries (Australia, Israel, USA, Canada, France, Morocco, Argentina, etc.) has focused on accelerating the natural loss of O_2 by injecting inert gases into sealed granaries. Even though CO_2 is preferred for reasons discussed above, as well as increasing the effectiveness of modified atmospheres at temperatures close to the insect's development threshold (Calderon et Navarro 1980; Jay 1980; Annis 1986; Banks and Annis 1990), modified atmospheres using N_2 have also been used (Tranchino et al. 1980; Reichmuth 1986). A set of guidelines presented as flow diagrams was even proposed to determine the best solution according to storage conditions and gas availability (Banks et al. 1980; Jay 1980; Jay and d'Orazio 1984; Banks and Annis 1990). However, because the mode of action of anoxia is different from hypercarbia, N_2 fumigations are longer and hence more difficult to achieve than fumigation with CO_2. One of the intermediate solutions between pure N_2 and CO_2 at more than 40%, are the exothermic inert gas generators (Fleurat-Lessard 1987; White and Jayas 1993). If hydrocarbon gases are burned completely, the exhaust is an inert atmosphere made up of less than 1% O_2 and between 12 and 14% CO_2, with N_2 making up most of the balance. Under these conditions, insects are controlled in cool grain (11 to 14 °C) after a 45-to 60-exposure day only (Fleurat-Lessard 1987). At present, this system is only suited to high value products that are to be stored for a long time, since it is approximately eight to ten times more expensive than a phosphine fumigation.

Fumigation with CO_2 is facilitated by the physical properties of the gas. When CO_2 is released from compressed gas cylinders, it is cold and dense. Injected into the base of the granary, equipped with a gas vent on the roof, it acts as an "immiscible liquid" and displaces the air upward without any appreciable loss of CO_2. Nitrogen does not benefit from this "piston effect", making the displacement of intergranular air more difficult. Carbon dioxide fumigation in bulk commodity storages began to be used on a regular basis in Australia in the 1970s and 1980s (Delmenico 1993) and then in Israel and Canada (White et al. 1993). It has been used in several other countries on a demonstration basis to evaluate the costs and technical difficulties compared to phosphine or methyl bromide fumigations (Jay and d'Orazio 1984; Reichmuth 1986; Fleurat-Lessard 1987; White and Jayas 1999).

From the many studies on the effectiveness of the different modified atmospheres, there have emerged a few general principles. Higher temperatures increase effectiveness. At temperatures below their development threshold, approximately 15 °C, insects are more tolerant of modified atmospheres (Adler 1994). Carbon dioxide fumigations of structures such as churches, museums and flour mills, have been attempted in Germany. To increase the effectiveness, the temperature is generally increased in fumigated buildings sealed and tarped (Meeus 1998). Even with these modifications, the fumigations are longer than the traditional fumigations with methyl bromide and require several tonnes of CO_2 for a modest building of about 10 000 m^3. Yet it is not typical to fumigate structures with CO_2 in the USA. When it is possible, PH_3 is used. Otherwise (because of corrosion risks), CO_2 and heat are used. Insects in quiescence almost always have their stigmata closed. This barrier to the respiratory system is one reason that these insects are tolerant to modified atmospheres. Several laboratories are now investigating ways to overcome closing of the stigmata to allow for a more effective fumigation with carbon dioxide.

3.3.2 High Pressure CO_2

The effectiveness of combining CO_2 at high levels (80 to 90%) with elevated temperatures or pressure, confirms the hypothesis that the stigmata are an essential part of an insect's defence system against carbon dioxide fumigation. Laboratory studies have been followed through to the industrial use of high-pressure CO_2 fumigation in Germany and France. The first laboratory experiments showed that by increasing the pressure of CO_2 from 1 to 20 atmospheres (2 MPa), the time needed to control *S. granarius* larvae decreased from 18 days to 4 h (Le Torc'h and Fleurat-Lessard 1991; Reichmuth and Wohlgemuth 1994). Given the success of our initial experiments, our studies expanded to insects commonly found in processed cereal foodstuffs. The duration and CO_2 pressure needed to control the eggs, larvae and adults of *T. castaneum*, *Dermestes maculatus* (de Geer) and *Plodia interpunctella* (Huebner) were determined. Two hours at 16 atmospheres (1.6 MPa) was sufficient to control all stages (Table 4). The eggs of *T. castaneum* were the most resistant, but an exposure of 1 h at 19 atm was sufficient to control them. The adults of *D. maculatus* and *P. interpunctella* are controlled after 2 h at 19 atm, but the eggs are unaffected under these conditions. Further experiments were carried out with a 0.75-m^3 autoclave that allowed for cycling of the pressure, as would occur in an industrial facility. The result of this work was that there was 99% control of insects placed in the centre of a 3- or 15-kg bag of pet food after 1.5 h at 19 atm. This work also underlined the need to purge the chamber with CO_2 before increasing the pressure. Without purging, there is a slight chance that the residual air (1/19th of the total) could remain trapped at the top of the chamber, producing a possible refuge for insects. Testing at this stage also demonstrated that packaging must have a certain level of porosity or it will burst during the rapid degassing. This work led to the building of a high pressure fumigation chamber by a pet food manufacture in the south of France that is 80 m^3 capacity and enables the treatment of 32 standard palettes. Each complete fumigation cycle takes less than 4 h. The loading is done automatically and the fumigation is controlled remotely for added safety. As a preventive measure, all product destined for export is fumigated. The system is used around the clock and completes five to six fumiga-

Table 4. Duration of exposure (min) with carbon dioxide at various high pressures necessary to reach at least 99% kill of a given stage of three insect pests of processed foods.

Insect	Stage	CO_2 pressure (atmosphere)			
		10	13	16	19
Tribolium castaneum	Egg	> 240	≈240	120	60
	Larvae	240	120	120	45
	Adult	240	90	90	15
Dermestes maculatus	Egg	> 240	60	< 60	30
	Larvae	240	90	30	45
	Adult	240	90	60	45
Plodia interpunctella	Egg	> 240	45	< 60	< 45
	Larvae	120	60	30	15
	Adult	120	45	30	15

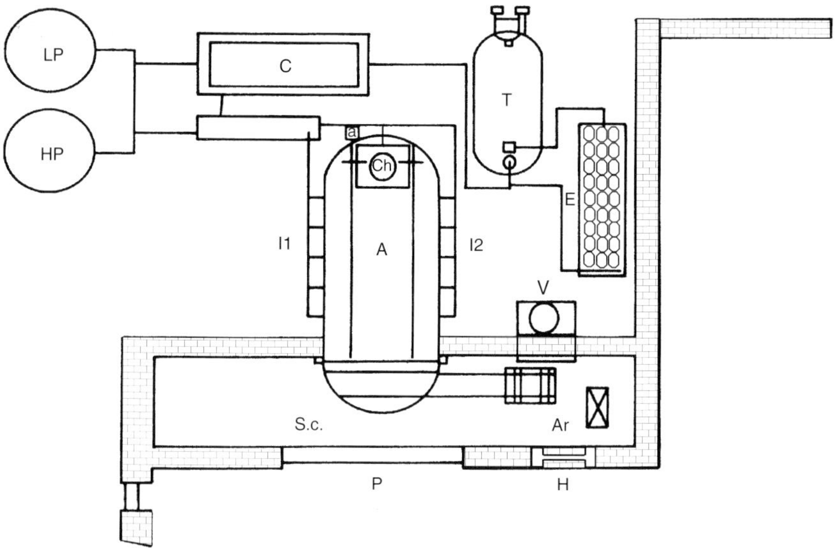

A : Gas autoclave
a : analyzer
Ar : remote control
C : compressor
Ch : mixing chamber
H : control window

Ii : CO_2 injection pipe
HP : high pressure tank
LP : low pressure tank
P : safety closing
S.c.: loading room
T : CO_2 reserve tank

Fig. 8. An industrial-scale facility for the high pressure CO_2 fumigation of pet foods. Schematic drawing (manufacturer MG-SIAC, France).

tions a day. Seventy-five percent of the carbon dioxide is recaptured, stored in a separate tank and reused in the next fumigation. Some carbon dioxide is lost because it is necessary to finish the cycle at atmospheric pressure to open the door of the chamber (Fig. 8). This system is suited for any dry food in a package porous enough to allow for the rapid penetration and escape of gases during treatment, and offers an alternative to fumigation with methyl bromide or phosphine.

4 Conclusions

Physical control is an indispensable tool for post-harvest integrated pest management (IPM). During the different phases of storage, transportation and processing of agricultural products the control of stored-product insects must be approached very differently for:

1. Storage of commodities (grain, fodder, fruit, dried vegetable, etc.) is facilitated by the use of residual insecticides and fumigation (mainly with phosphine for grain, and the preventive application of aerosols in warehouses.

2. In facilities for the initial processing of commodities there is a growing desire to purchase raw product and to deliver processed products that are free of insecticide residues. Once a commodity has been processed, the only method of control allowed is fumigation. However, the growing list of constraints on fumigants has increased the use of physical methods of control, such as thermal treatments.
3. Finally, packaging with modified atmospheres is the only way to ensure that the finished product remains insect-free in the distribution and retail system.
However, it should be underlined that the majority of physical control methods do not protect products after treatment, as do residual insecticides. The counter measure to this minor inconvenience is to promote insect-proof packaging at the same time as the physical control measures are introduced, so that reinfestation of uninfested products does not occur.

References

Adler C., (1994a). Carbon dioxide - more rapidly impairing the glycolytic energy production than nitrogen ? pp. 7-10 in E. Highley, E.J. Wright, H.J. Banks and B.R. Champ (eds.) Stored Product Protection, CAB International, Wallingford, (UK), 1274 p.

Annis P.C., (1986). Towards rational controlled atmosphere dosage schedules : a review of current knowledge, pp. 128-148 in E. Donahaye and S. Navarro (eds.) Proc. 4th Int. Working Conf. Stored Product Protection, Tel Aviv, 668 p.

Armstrong J.W., Hansen J.D., Hu B.K.S., Brown S., (1989). High-temperature forced-air quarantine treatment for papayas infested with Tephritid fruit flies. J. Econ. Entomol. 82:1667-1674.

Bailey S.W., Banks H.J., (1980). A review of recent studies of the effects of controlled atmospheres on stored product pests, pp. 101-118 in J. Shejbal (ed.) Controlled Atmosphere Storage of Grains, Elsevier, Amsterdam, 608 p.

Baker A.C., (1939). The bans for treatment of products where fruit flies are involved as a condition for entry into the United States. Washington D.C., USDA Circular No. 551, 7 p.

Banks H.J., Annis P.C., (1990). Comparative advantages of high CO_2 and low O_2 types of controlled atmospheres, pp. 94-122 in M. Calderon and R. Barkai-Golan (eds.) Food preservation by modified atmospheres, CRC Press, Boca Raton, 402 p.

Becker H.A., Sallans H.R., (1960). Drying wheat in a spouted bed. Chem. Eng. Sci. 13:97-112.

Botteril J.S.M., (1975). Fluid-bed heat transfer. Academic Press, London, 230 p.

Brown G.E., Farkas D.F., de Marchena E.S., (1972). Centrifugal fluidized bed - blanches, dries, and puffs piece-form foods. Food Technol. 26:23-30.

Cahagnier B., Poisson J., (1973). La microflore des grains de maïs humide. Composition et évolution en fonction de divers modes de stockage. Rev. Mycol. 38:23-43.

Calderon M., Navarro S., (1980). Synergistic effect of CO_2 and O_2 mixtures on two stored grain pests, pp. 79-84 in J. Shejbal (ed.) Controlled Atmosphere Storage of Grains, Elsevier, Amsterdam, 650 p.

Chapman R.N., (1921). Insect infesting stored food products. Minn. Agric. Expt. Sta. Bull. 198:1-76.

Chauvin G., Vannier G., (1991). La résistance au froid et à la chaleur : deux données fondamentales dans le contrôle des insectes des produits entreposés, pp.1157-1165 in F. Fleurat-Lessard and P. Ducom (eds.) Proc. 5th Int. Working Conf. Stored Product Protection, INRA, Bordeaux, 2066 p.

Corcoran R.J., Heather N.W., Heard T.A., (1993). Vapour-heat treatment for zucchini infested with *Bactrocera cucumis* (Diptera : Tephritidae). J. Econ. Entomol. 86:66-69.

Cossins A.R., Prosser C.L., (1978). Evolutionary adaptation of membranes to temperature. Proc. Natn. Acad. Sci. USA 75:2040-2043.

Cotton R.T., (1930). Carbon dioxide as an aid in the fumigation of certain highly adsorptive commodities. J. Econ. Entomol. 25:1088-1103.

Cotton R.T., Frankenfeld J.C., Dean C.A., (1945). Controlling insects in flour mills. USDA circular No. 720, 75 p.

Dean G.A., (1913). Further data on heat as a means of controlling mill insects. J. Econ. Entomol. 6:40-53.

Dean G.A., Cotton R.T., Wagner G.B., (1936). Flour-mill insects and their control. Circular No. 390, USDA Washington D.C., 34 p.

Dermott T., Evans D.E., (1978). An evaluation of fluidized-bed heating as a means of disinfesting wheat. J. Stored Prod. Res. 14:1-12.

Ebeling W., (1994). The thermal pest eradication system for structural pest-control. IPM practitioner 16:1-7.

Evans D.E., Dermott T., (1981). Dosage-mortality relationships for *Rhyzopertha dominica* (F.) (Coleoptera : Bostrychidae) exposed to heat in a fluidized bed. J. Stored Prod. Res. 17:53-64.

Fields P.G., (1992). The control of stored-product insects and mites with extreme temperatures. J. Stored Prod. Res. 28:89-118.

Fleurat-Lessard F., (1980). Lutte physique par l'air chaud ou les hautes fréquences contre les insectes des grains et des produits céréaliers. Bull. Techn. Info. Minist. Agric. Paris, 349:345-352.

Fleurat-Lessard F., (1987). Control of storage insects by physical means and modified environmental conditions: Feasability and applications, pp. 209-218 in T.J. Lawson (éd.) , Stored Product Pest Control, BCPC monograph 37, Thornton Heath, 277 p.

Fleurat-Lessard F. (1990). Effect of modified atmospheres on insects and mites infesting stored products, pp. 21-38 in M. Calderon and R. Barkai-Golan (eds.) Food preservation by modified atmospheres. CRC Press, Boca Raton, 402 p.

Forbes C.F., Ebeling W., (1987). Update : use of heat for elimination of structural pests. IPM practitioner. 9:1-5.

Fraenkel G.S., Hopf H.S., (1940). The physiological action of abnormally high temperatures on poikilothermic animals. 1 - Temperature adaptation and the degree of saturation of the phosphatides. Biochem. J. 34:1085-1092.

Goodvin W.H., (1922). Heat treatment of cereal insects. Ohio Agr. Stat. Bull. 354:1-18.

Grossman E.F., (1931). Heat treatment for controlling the insect pests of stored corn. Fla. Agr. Expt. Sta. Bull. 239:3-24.

Hallman G.J., Sharp J.L., (1990). Hot-water immersion quarantine treatment for carambolas infested with caribbean fruit fly (Diptera : Tephritidae). J. Econ. Entomol. 83:1471-1474.

Heaps J.W., (1988). Turn on the heat to control insects. Dairy Food Sanit. 8:416-418.

Heather N.W., (1994). Commodity disinfestation treatments with heat, pp. 1199-1200 in E. Highley, E.J. Wright, H.J. Banks, and B.R. Champ (eds.) Stored Product Protection CAB International, Wallingford, U.K., 1274 p.

Hyde M.B., Baker A.A., Ross A.C., Lopez C.O., (1973). Airtight grain storage. Agricultural Services Bull. FAO Rome, 71 p.

Jay E.G., (1980). Methods of applying carbon dioxide for insect control in stored grain, pp. 225-234 in J. Shejbal (ed.) Controlled Atmosphere Storage of Grains, Elsevier, Amsterdam, 608 p.

Jay E.G., D'Orazio R., (1984). Progress in the use of controlled atmosphere in actual field situations in the United States, pp. 3-13 in B.E. Ripp et al. (eds.) Controlled atmosphere and fumigation in grain storages. Elsevier, Amsterdam, 800 p.

Le Torc'h J.M., Fleurat-Lessard F., (1991). Effet des fortes pressions sur l'efficacité insecticide des atmosphères modifiées par CO_2 contre *Sitophilus granarius* (L.) et *S. oryzae* (L.) (Coleoptera: Curculionidae), pp. 847-856 in F. Fleurat-Lessard and P. Ducom (eds) Proc. 5th Int. Working Conf. Stored Product Protection. INRA Bordeaux (France), 2066 p.

Kirkpatrick R.L., (1975). Infrared radiation for control of lesser grain borers and rice weevils in bulk wheat. J. Kansas ent. Soc. 48:100-104.

Miller P.L., (1966). The regulation of breathing in insects. Adv. Insect Physiol. 3:279-286.

Meeus P., (1998). Various methods to use heat for enhancing fumigation results. Proceedings of 7th Int. Workink Conf. On Stored-Product Prot., Beijing, Oct. 1998, Abstract p. 70.

Miller W.R., Mc Donald R.E., Sharp J.L., (1990). Condition of Florida carambolas after preliminary tests of forced warm air treatment and storage. Proc. Fla. State Hort. Soc. 103:238-241.

Nagao R.T., Kimpel J.A., Key J.L., (1990). Molecular and cellular biology of the heat-shock response. Adv. Genet. 28:235-275.

Navarro S., (1978). The effects of low oxygen tensions on three stored-product insect pests. Phytoparasitica 6:51-58.

Oosthuizen M.J., (1935). The effect of high temperature on the confused flour beetle. Minn. Agr. Expt. Sta. Bull. 107:1-45.

Paull R.E., Armstrong J.W., (1994). Insect Pests and Fresh Horticultural Products: Treatments and Responses. CABI, Wallingford, U.K., 360 pp.

Pepper J.H., Strand A.L., (1935). Superheating as a control for cereal-mill insects. Bull. Mont. Agr. Expt Stat. 297:1-26.

Petersen N.S., Mitchell H.K., (1985). Heat shock proteins, pp. 347-365 in G.A. Kerkut and L.I. Gilbert (eds.) Comparative Insect Physiology, Biochemistry and Pharmacology, Pergamon Press, New York.

Prozell, S. and C. Reichmuth 1990. Response of the granary weevil *Sitophilus granarius* (L.) (Col.: Curculionidae) to controlled atmospheres under high pressure, vol. II, pp. 911-918. In F. Fleurat-Lessard and P. Ducom (eds) Proc. 5th Int. Working Conf. Stored Product Protection. INRA Bordeaux (France) 2066 p.

Reichmuth C., (1986). Low oxygen content to control stored product insects, pp. 194-207 in E. Donahaye and S. Navarro (eds.) Proc. 4th Int. Working Conf. Stored Product Protection, Tel Aviv, 668 p.

Reichmuth C., Wohlgemuth R., (1994). Carbon dioxide under high pressure of 15 bar and 20 bar to control the eggs of the Indian meal moth *Plodia interpunctella* (Hübner) (Lepidoptera: Pyralidae) as the most tolerant stage at 25°C, pp. 163-172 in E. Highley, E.J. Wright, H.J. Banks and B.R. Champ (eds.) Stored Product Protection, CAB International, Wallingford, Oxon, 1274 p.

Sharp J.L., (1990). Effectiveness of conventional commodity treatments (heat, refrigeration, chemical, others) to satisfy quarantine regulations. Proc. Fla. State Hortic. Soc. 103:175-186.

Shejbal J., (1980). Controlled atmosphere storage of grains, Elsevier, Amsterdam, 608 p.

Shejbal J., Tonolo N., Careri G., (1973). Conservation of wheat in silos under nitrogen. Ann. Technol. Agric. 22:773-785.

Sheppard K.O., (1992). Heat sterilisation (superheating) as a control for stored-grain pests in a food plant, pp. 194-200 in F.J. Baur (ed.) Insect management for food storage and processing, AACC, St Paul MN, 384 p.

Sigaut F., (1980). Significance of underground storage in traditional systems of grain production, pp. 3-14 in J. Shejbal (ed.) Controlled atmosphere storage of grains, Elsevier, Amsterdam, 608 p.

Sugimoto T., Furazawa K., Mizobuchi M., (1983). The effectiveness of vapour heat treatment against the Oriental fruit fly, *Dacus dorsalis* Hendel in green peppers and fruit tolerance to treatment. Res. Bull. Plant. Prot. Serv. of Japan 19:81-88.

Thorpe G.R., (1987). The thermodynamic performance of a continuous-flow fluidized bed grain disinfestor and drier. J. agric. Eng. Res. 37:27-41.

Tranchino L., Agostinelli P., Costantini A., Shejbal J., (1980). The first Italian large scale facilities for the storage of cereal grain in nitrogen, pp. 445-459 in J. Shejbal (ed.) Controlled Atmosphere Storage of Grains, Elsevier, Amsterdam, 608 p.

Vannier G., (1987). Mesure de la thermotorpeur chez les insectes. Bull. Soc. Ecophysiol. 12:165-186.

White N.D.G., Jayas D.S., (1993). Effectiveness of carbon dioxide in compressed gas or solid formulation for the control of insects and mites in stored wheat and barley. Phytoprotection 74:101-111.

White N.D.G., Jayas D.S., 1999. Controlled atmosphere use during the storage of grain. In Handbook of Postharvest Technology A. Chakraverty, A.S. Mujumidar, G.S.V. Raghavan and H.S. Ramaswamy, (eds), Marcel Dekker, New York, in press.

Winterbottom D.C., (1922). Weevils in wheat and storage of grain in bags: a record of Australian experience during the war period (1915 to 1919). Austr. Govt. Printing. Adelaide, 122 p.

Control of Insects in Post-Harvest: Low Temperature

Paul G. FIELDS

1 Introduction

Stored-product insects have been controlled using physical means for thousands of years. The golden rule of sound seed storage, i.e. keep seed cool and dry, was employed in neolithic times in the Nile delta by placing clay jars underground (Levinson and Levinson 1989). Nowadays, physical control of stored-product insects manipulates environmental factors (e.g. temperature, relative humidity, moisture content) in structures containing commodities (silos, elevators, bags, packaging) or forces on a commodity (compression, impaction) and irradiation to reduce or eliminate insect populations (Fields 1992; Banks and Fields 1995; Fields and Muir 1995; Mason and Strait 1998).

For a number of years, the use of extreme temperatures, particularly low temperatures, has been extensively used to control stored-product insects. The advantages of physical control methods are that : (1) there are no residues are left on the product after treatment, (2) they are effective against insecticide resistant strains, and (3) there are few risks for operators. Inconveniences that prevent the widespread use of low temperatures are : (1) the treatment is dependent upon cold ambient air or refrigeration equipment and (2) durations of treatment needed for control can be long depending upon temperatures achieved. There are a number of variables that influence the success of a low temperature treatment, e.g. species, stage of target insect, level of cold acclimation, temperature attained, duration and relative humidity. The goal of this chapter is to introduce some basic concepts related to insect cold tolerance, and to demonstrate how low temperature may be used to prevent or to control stored-product insect infestations.

2. Adaptations of Insects to Low Temperature

The response of insects to temperature can be divided into three zones: optimal, sub optimal and lethal (Table 1). The differences in cold susceptibility observed during laboratory experiments (Table 2) or under storage conditions may depend on behavioral, physiological, or biochemical adaptations to cold. A better unders-

Table 1. The response of stored-product insects to temperature[a].

Zone	Temperature (°C)	Effect
Lethal	50 to 60	Death in minutes
	45	Death in hours
Sub-optimum	35	Development stops
	33 to 35	Development slows
Optimum	25 to 33	Maximum rate of development
Sub-optimum	13 to 25	Development slows
	13 to 20	Development stops
Lethal	5	Death in days (unacclimated), movement stops
	-10 to 5	Death in weeks to months (acclimated)
	-25 to -15	Death in minutes, insects freeze

[a] Species, stage of development and moisture content of food will influence the response to temperature.

tanding of these mechanisms should lead to more effective use of low temperature to control stored-product insects.

Most insects that overwinter in temperate regions find overwintering habitats that are protected from the low air temperatures. Such habitats include under leaf litter, under bark and the south side of trees. The grain stored in silos has a much higher temperature than the ambient winter temperatures due the insulating properties of grain. There are steep temperature and moisture content gradients in winter since the outermost grain cools faster than the core (Hagstrum 1987). Intuitively, one would think that the insects would follow a temperature gradient to find the temperature that is closest to their optimum. Laboratory experiments have shown that *Cryptolestes ferrugineus* (Stephens) aggregate to the warmest temperatures and can detect differences of less than 1 °C (Flinn and Hagstrum 1998). *Oryzaephilus surinamensis* (L.) and *Tribolium castaneum* (Herbst), also exhibit a positive thermotaxis. However, not all insects have this behaviour. *Tribolium confusum* J. du Val aggregates at 15 °C, the lowest temperature available in the experiment (Graham 1958). As for the Lepidoptera that infest stored products, the larvae can only penetrate a few centimetres into the grain, and hence are exposed to some of the most extreme temperatures in the grain bulk. The Lepidoptera are also some of the most cold hardy insects that infest stored products (Table 2). Reduction of exposure to low temperatures by avoidance behavioral means is insufficient to prevent cold damage. Physiological and biochemical mechanisms may also contribute to protect insects from the damage caused by low temperature.

A number of mechanisms are involved in cold injury: "frozen" membranes, rate imbalances, changes in ionic activity, and ice formation (Storey and Storey 1989; Fields 1992). They are not mutually exclusive, and one or more of them may cause the final demise of the insect at low temperature. The "frozen" membrane theory is the one most often put forth as the reason for lethal damage, and it is discussed in detail by Fleurat-Lessard and Le Torc'h (Chap. 5). Cell membranes must remain homeoviscous

Table 2. The relative cold hardiness of some stored-product insects* (Fields 1992).

		Tolerance index			
Site[a]	Stage[b]	1 (most susceptible)	2	3	4 (most tolerant)
Field	l or a	*T. castaneum* *T. confusum*	*C. turcicus* *E. cautella* *L. serricorne* *R. dominica* *S. oryzae*	*O. surinamensis* *S. granarius*	*E. elutella* *E. kuehniella* *P. interpunctella* *T. granarium* *T. molitor*
Field[a]	e,l,p,a	*O. mercator*	*E. cautella* *S. oryzae* *S. granarius* *T. castaneum* *T. confusum*	*S. cerealella*	*C. ferrugineus* *C. turcicus* *E. elutella* *E. kuehniella* *O. surinamensis* *P. interpunctella* *T. granarium* *T. molitor*
Field and Lab	l or a	*E. cautella* *O. surinamensis* *R. dominica* *S. oryzae*	*C. ferrugineus* *E. elutella* *E. kuehniella* *T. granarium* *S. granarius*		
Lab	e,l,p,a	*R. dominica*	*S. oryzae*	*S. granarius*	
Field	l,a	*C. pusillus* *O. mercator* *R. dominica* *T. castaneum* *T. confusum*	*S. cerealella* *S. oryzae* *S. zeamais*	*C. turcicus* *C. ferrugineus* *O. surinamensis* *S. granarius*	
Lab	e	*E. cautella* *O. surinamensis* *T. castaneum*	*C. maculatus*	*L. serricorne*	
Lab	a	*T. castaneum*	*O. surinamensis* *C. ferrugineus* *S. oryzae* *R. dominica*	*S. granarius*	
Lab	l,p	*O. surinamensis* *T. castaneum*	*S. oryzae*	*S. granarius*	*R. dominica*
Field	l,a	*T. confusum*	*C. turcicus* *O. surinamensis*	*T. granarium*	*S. granarius*

[a] Only a partial list of species studied.

[b] e = egg, l = larva, p = pupa, a = adult.

* Within a given site, species in the same column had similar cold hardiness.

to carry out their various functions. As the temperature drops, membranes become less viscous, eventually becoming non-functional. One adaptation to low temperature is that membranes remain homeoviscous over a wide range of temperatures.
Temperature also affects the ionic activity of molecules. For example, a drop from 35 to 0 °C doubles the solubility of oxygen in water, increases the hydrogen ion content by 2.5 times and has no effect on the solubility of sodium chloride. Small changes in pH can radically change the performance of enzymes. For example, the activity of digestive enzymes is reduced 20% by a 20 °C drop in temperature solely due to pH changes (equivalent to a 0.34 unit increase in activity).
Freezing is lethal for all stored-product insect pests, as it is for most insects. The supercooling point varies between -10 and -25 °C for most species and is dependent upon a number of factors, e.g. species, stage, feeding state and acclimation. The supercooling point therefore represents the lower lethal temperature for stored-product insects. It is thought that ice causes death because ice crystals mechanically damage the cells, cause osmotic imbalances when salts are concentrated as the water is bound in ice and that ice prevents the movement of molecules within and between the cells (Storey and Storey 1989).
Usually, insects that are freeze-intolerant or freeze-susceptible have lower supercooling points in the winter than in the summer. With stored-product insects, there is a slight drop in supercooling points when insects are exposed to temperatures between 5 and 15 °C. Under these conditions there is a dramatic increase in the ability of the insects to tolerate low temperatures. This indicates that the supercooling point should not be used as the only indicator of cold-tolerance. If one species freezes at -20 °C, it is not necessarily more cold tolerant than one that freezes at -15 °C. Most stored-product insects die from chill injury before they freeze.

Box 1. Supercooling Point

The supercooling point is the temperature at which ice crystals start to form. Small quantities of water, cooled, rapidly supercool to -40 °C before freezing (Sømme 1982). However if ice crystals are present, they act as ice nucleators, water is unable to supercool and it freezes at 0 °C. There are other ice nucleators with varying degrees of effectiveness; silver iodine, ice-nucleating bacteria, dust, proteins, amino acids and alcohol. Other substances, such as glycerol, mannitol, sorbitol, trehalose and fructose, lower the supercooling points (Storey and Storey 1989). To measure supercooling points in insects, a thermocouple is attached to the insect, the temperature lowered at about 1°Cm^{-1}(standardized rate) and the supercooling point is signalled by the abrupt increase in temperature caused by the heat of crystallization. Insects can be divided into two groups, freeze-tolerant (can survive freezing) and the more common freeze-intolerant (cannot survive freezing of tissues). For freeze-intolerant insects the supercooling point is the lower lethal temperature; however, many insects die of chill injury before the freeze (Table 1).

Freeze-susceptible insects go through a number of changes during cold acclimation that increase their survival at low temperatures. Before overwintering, they void their gut or mask ice nucleators so that their supercooling points are decreased. There is often an increase in the concentrations of low molecular weight molecules such as glycerol, mannitol, sorbitol, trehalose, fructose or proline. In addition to lowering the supercooling point, these compounds are thought to have other roles such as stabilization of proteins and membranes (Sømme 1982; Storey and Storey 1989; Fields et al. 1998).

3. Control of Insects Using Low Temperature

3.1 Temperatures above 0 °C

Any reduction in temperature below the optimum will reduce the growth rate of insect populations (Table 1), mites and moulds. Howe (1965) summarized the temperatures and humidities at which stored-product insects would have only two generations per year. Grain held at 20 °C will prevent most stored-product insects from completing their development. The most important exception is *Sitophilus granarius* (L.), which is able to complete its life cycle at 15 °C. To prevent the development of mites in humid grain, temperatures must be below 2 °C, but in grain at 13% moisture content or lower, the development threshold temperature rises to 6 °C (Burges and Burrell 1964). Cold temperatures can also prevent the development of moulds in grain waiting to be dried or processed, but in one study, a temperature of -23 °C did not reduce survival of moulds (Brown and Hill 1984).

3.2 Lethal Low Temperatures

Below the developmental threshold, insects eventually die due to cold injury. The variables that are important in determining the susceptibility to low temperatures are the minimum temperature attained, the duration of exposure, the species, the stage, acclimation, sex and humidity. The relationship between the time necessary to kill a given percentage of the population (y-axis) and the temperature (x-axis) is usually a concave (j-shaped) curve (Fig. 1; Fields 1992). The most susceptible stored-product insect species are: *T. castaneum*, *T. confusum* and *Oryzaephilus mercator* (Fauvel). The most tolerant are *Trogoderma granarium* (Everts), *S. granarius* and the *Lepidoptera Ephestia elutella* (Hübner), *E. kuelniella* (Zeller) and *Plodia interpunctella* (Hübner) (Table 2).
Relative cold tolerance poorly explains the distribution of stored product pests. If cold tolerance were the quality for determining distribution, cold-tolerant species

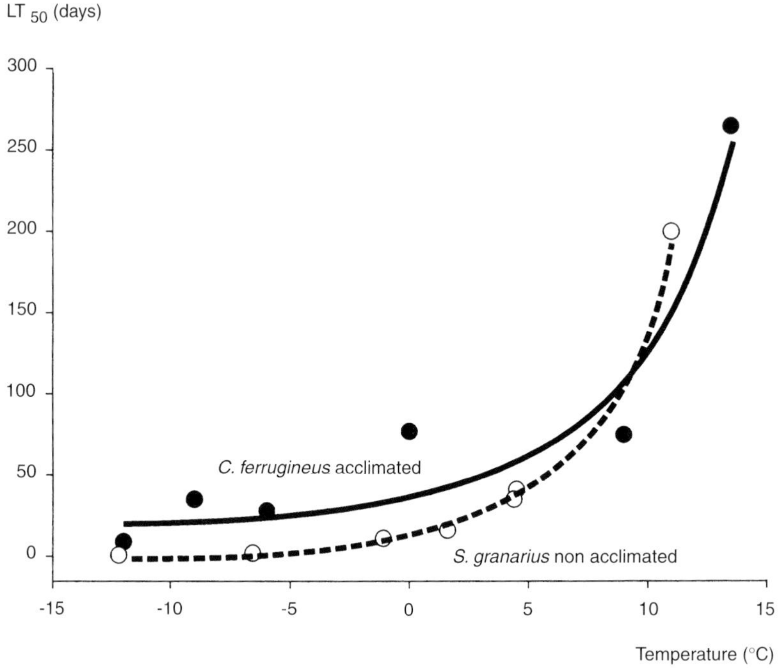

Fig. 1. The lethal time for 50% of the population for two stored-product insects at temperatures below their development threshold (Fields 1992).

would be expected to be predominant pests in cold countries such as Canada. Although *T. castaneum* is one of the least cold-tolerant species (Table 2), it is the second most common species in farm storages in Western Canada. The most common insect is *C. ferrugineus*, an insect with moderate cold tolerance. Species such as *T. granarium, S. granarius, P. interpunctella, E. elutella* and *E. kuehniella* are rarely found in Western Canada, but can be more cold-hardy, than *C. ferrugineus*. It could be that the intrinsic rate of increase is more important than cold tolerance for stored-product insects in this region. A high intrinsic rate of increase would allow for quick establishment of populations in a habitat (stored grain) that is warm

Box 2. Probit Analysis

Probit analysis is often used to predict the mortality of insects due to insecticides, extreme temperatures, or other factors. Mortality (y-axis) plotted versus concentration (insecticide, temperature or duration) is usually an S-shaped or sigmoidal curve. Transforming the mortality data using a probit function often gives a straight line, from which is calculated the concentrations needed to cause a given mortality with 50, 95 and 99% being the most often reported. Probit analysis also gives error estimates (fiducial limits) that are useful when comparing between treatments and studies.

for only a few months. The fact that *C. ferrugineus* and *T. castaneum* have some of the highest intrinsic rates of growth supports this hypothesis.

The stage of development is an important factor in determining the cold tolerance of insects (Fields 1992). Temperate insects usually overwinter in a diapausing stage, but very few stored-product insects have a diapause. Diapause is an arrested state of development that is stage-specific. It is induced either every generation (obligatory) or by conditions that signal the coming of unfavourable conditions (facultative), such as short days and cool temperatures that signal the coming of winter in temperate regions. The Lepidoptera in the family *Pyralidae* and *T. granarium* are the only stored-product insect pests that have a diapause. For the Pyralidae, diapause occurs in the prepupal stage and increases their cold hardiness. In other stored-product insects the egg is generally the most sensitive stage to cold. Within the egg stage there is significant variation in the cold hardiness. The young and old eggs are the least cold-hardy. The larvae are the most cold-tolerant stage of *Rhyzopertha dominica* (F.) and *Sitophilus oryzae* (L.). The adult is the most cold-hardy stage of *C. ferrugineus* (Smith 1970) and *T. confusum*. Certain larval stages of *S. granarius* are as cold-hardy as the adults.

Acclimation is the factor most often ignored when studying cold tolerance in stored-product insects. In general, the cold tolerance of insects exposed to temperatures between 10 and 20 °C before exposure to cold temperatures, is two to ten times greater than that of insects that are held at 25 to 35 °C. What is the best method to acclimate insects to cold? In Australia, Evans (1987) used temperatures that mimicked the temperatures that could be attained using chilled-air aeration. I have acclimated insects by placing them in cages in granaries for the winter. Smith (1970) chose 15 °C, which is the development threshold for many stored-product insect pests. He found that the longer he held the insects at 15 °C, the greater the acclimation. Many studies on cold tolerance do not acclimate or acclimate insects only for a very short period. In consequence, the cold tolerance of stored-product insects has often been underestimated. To be used in predicting the temperatures needed to control insects that infest stored products, acclimation regimes must reflect the cooling rates observed in storage and processing facilities. The drawback of this method is that it increases the time needed to conduct experiments, and cooling rates vary widely depending upon the commodity, weather and storage structure.

Box 3. Estimating Cold Tolerance

The criteria chosen for determining cold tolerance depend on the developmental stage tested. The criteria used most often for eggs and pupae are emergence to the next stage. For adults it is the capacity for coordinated movement. The ultimate measure of the effects of low temperature would be calculating the effects on the insects overall fitness tested under low temperature conditions that approximate those the pest insect is found. A few studies have shown that in addition to mortality there are non-lethal detrimental effects of low temperature (Fields 1992).

Table 3. The supercooling points of eight stored-product insects (Fields 1992).

Species	Stage	Acclimation	Supercooling point (°C) (Standard error of the mean)	
Cryptolestes ferrugineus	a	30 °C	-17.9	(-)
	a	30 °C, 15 °C/28 days	-20.2	(-)
	a	30 °C, ice-nucleating bacteria	-8.1	(0.5)
Cryptolestes pusillus	a	30 °C	-14	(1.0)
	a	30 °C, ice-nucleating bacteria	-12.0	(0.5)
Ephestia kuehniella	l	26 °C, 20 °C/2 days	-16.9	(0.5)
	l	26 °C, 0 °C/7 days	-20.3	(0.3)
	l	26 °C, 6 °C/14-42 days	-18.3	(0.7)
Oryzaephilus surinamensis	a	30 °C	-13.7	(1.9)
	a	30 °C, ice-nucleating bacteria	-11.0	(0.3)
Plodia interpunctella	l	23 °C	-10.3	(0.4)
	l	23 °C, ice-nucleating bacteria	-5.4	(0.5)
Rhyzopertha dominica	a	23 °C	-15.2	(0.8)
	a	23 °C, ice-nucleating bacteria	-3.3	(0.1)
Sitophilus granarius	a	30 °C	-14.3	(0.8)
	a	30 °C, ice-nucleating bacteria	-7.8	(0.5)
Tribolium castaneum	a	30 °C	-12.3	(1.0)
	a	30 °C, ice-nucleating bacteria	-5.8	(0.3)

a: adult, l: larva.

[a] data not available.

Relative humidity affects many aspects of the biology of stored-product insects, and cold tolerance is no exception. Low relative humidity can shorten the time needed to reach 100% mortality at low temperatures by 50% (Evans 1987; Fields 1992). The sex of an insect can have a small but significant effect on the cold tolerance of some species (Fields 1992; Fields et al. 1998).

Smith (1970) found that the supercooling point of *C. ferrugineus* dropped by only 2 °C during an acclimation of 4 weeks at 15 °C. Their LT_{50} (lethal time for 50% of the population) at -12 °C increased from 2 to 30 days. It is usual for pupae to have lower supercooling points than other stages, but there is usually no corresponding increase in cold tolerance (Sømme 1982), probably because the pupae have fewer ice nucleators (none from the gut) than larvae and adults. Sømme (1968) showed that injecting glycerol into unacclimated *E. kuehniella* increased their survival at -10 °C beyond that of cold-acclimated larvae. Subsequent work suggested that the glycerol aided in maintaining the balance of enzymes during cold acclimation. *Cryptolestes ferrugineus* and *S. granarius* do not increase their levels of glycerol during cold acclimation, but there is an increase in several amino acids, proline being the most important. There is also an increase in trehalose (Fields et al. 1998). One way to decrease an insect's cold tolerance is through the use of ice nucleating bacteria (Fields 1992). The supercooling point of *C. ferrugineus* reared on wheat dusted with ice nucleating bacteria was -8 °C, but was -17 °C for individuals reared on untreated wheat (Table 3). At -10 °C, there was 8% survival of the group treated with ice nucleating bacteria and 81% survival in the control group. Other stored-product insects responded similarly (Table 3).

4 Methods to Achieve Low Temperature

4.1 Turning Grain

Three methods are used to cool grain: turning, ambient air aeration and chilled air aeration. In winter, moving grain from one granary to another will break up "hot spots" but leave the mean temperature largely unchanged. Some grains such as maize are turned as little as possible to avoid breakage of kernels. Turning is not an effective way to cool grain.

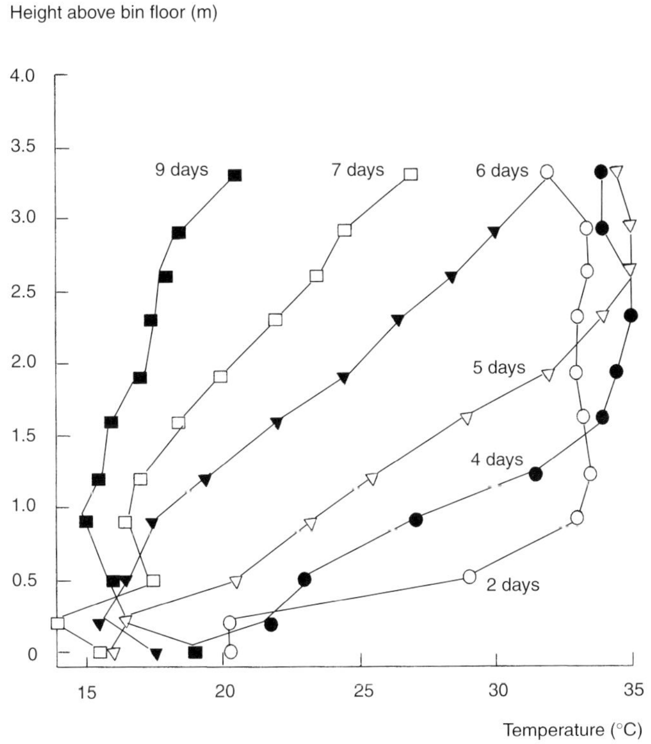

Fig. 2. Grain temperature during an initial aeration of a granary of wheat aerated continuously with upward airflow of 0.8 $(ls^{-1})m^{-3}$ of grain at various times after aeration was begun (After Sanderson 1988).

4.2 Ambient Air Aeration

Ambient air aeration is a common method of cooling vertical and horizontal storages. Although numerous systems exist, they are based either on pushing air up from the bottom or pulling air from the top. The grain temperature in a silo can be brought to the average air temperature in approximately 10 days with an air flow of 7 m^3 h^{-1} for each m^3 of grain (Lasseran 1994). A cooling front is established and moves slowly through the grain mass (Fig. 2; Sanderson 1986).
Automated aeration control systems have been developed to maximize the effectiveness of the cooling and to minimize the energy costs of aeration. The simplest control is a timed switch that turns the fans on at night and off during the day. A more sophisticated and expensive system monitors the temperature of air and the core grain temperature. The aeration fans are turned on when the ambient air temperature is cooler than the core grain temperature. This ensures that the cool night air can be used optimally and aeration can begin immediately after harvest (Armitage and Llewellin 1987; Fields and Muir 1995). The rewetting of grain due to high relative humidity is only a minor problem since the cooling front moves through the grain 30 times faster than the moisture migration front. It is recom-

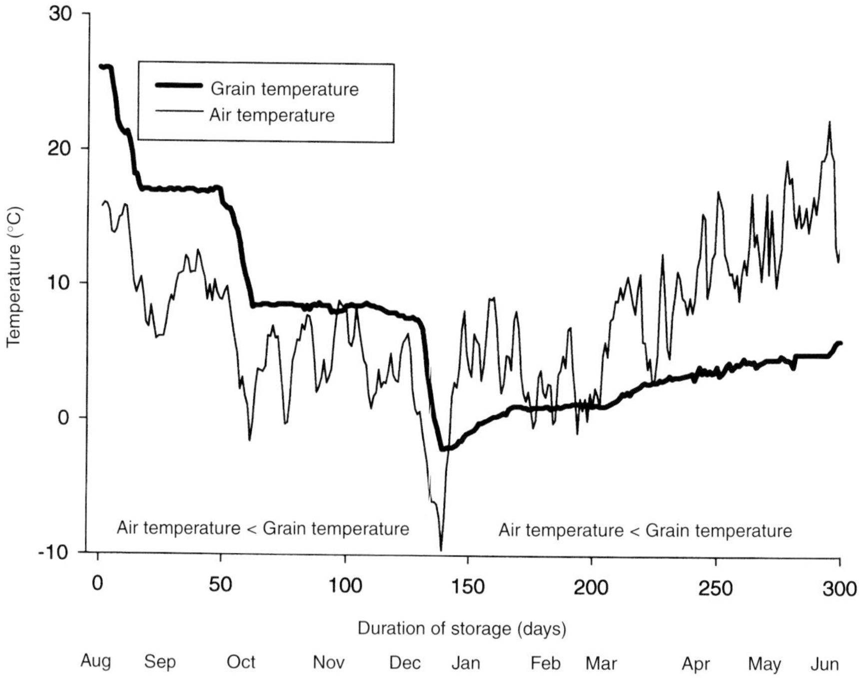

Fig. 3. The temperature at the centre of a granary of 500 t of wheat aerated with ambient air with an automatic controller that had fixed temperature thresholds. Air temperatures are the minimum daily temperatures during cooling and the average temperatures during reheating (After Lasseran et al.1994).

mended that the aeration continue even during rainy conditions, since the grain will gain only a small amount of water, which will be rapidly lost once the rain stops. This technique has been used in France to cool grain down in stages until the average grain temperature was 2 °C (Fig. 3). Cooling was sufficiently rapid to prevent the first generations of *S. granarius* and *S. oryzae* from developing (Lasseran 1994; Lasseran et al. 1994). In France, it is possible to cool grain to 18 °C in August. A second stage of cooling in October brings it to 12 °C and the final cooling in January can bring it down to 2-4 °C. After 100 to 120 days, laboratory tests that simulated the temperatures that insects would be exposed to during this type of cooling showed 95% mortality of *S. granarius*, a relatively cold-hardy species after 100 to 120 days. In the United Kingdom and France, grain is often cooled to only 7-10 °C so that cooling costs are comparable to chemical treatments. However, under these conditions there are a greater number of insects that survive the winter (Armitage and Stables 1984).

In certain very dry regions, such as Australia, grain can also be cooled using the heat of vaporization as the grain dries during ambient air aeration (Fields 1992). A unit that passes cool night air over a desiccant before passing the dried air through the grain has been designed and field tested. The desiccant is dried out the next day using solar energy (G. Thorpe, pers. comm.).

Ambient air cooling is also used to control insects found in food-processing facilities in Western Canada. During the winter, water is drained from the pipes, windows opened up and fans are used to bring in cold air. This method has largely been replaced by fumigations with methyl bromide (Fields 1992).

Mathematical models have been used to study and predict the rate and uniformity of cooling. They are also useful for comparing fan management policies (Lasseran et al. 1994; Fields and Muir 1995). A simplified method [Eq. (1)] for calculating the approximate time to cool a grain bulk by uniform aeration was presented by Navarro and Calderon (1982):

$$\theta = (M \times \Delta T \times c) / (Q \times \Delta H \times E), \tag{1}$$

where:

θ = cooling time (h)
M = mass of grain to be cooled (kg)
ΔT = difference between initial and final grain temperatures (°C)
c = specific heat of grain [kJ (kg °C^{-1})]
Q = mass flow rate or air (kg h^{-1})
ΔH = change in enthalpy of air between entering and leaving the grain bulk (kJ kg^{-1})
E = correction factor, dimensionless.

The correction factor is required because the change in enthalpy of the ventilating air, DH, decreases as the cooling front passes through the grain. Navarro and Calderon (1982) recommended a correction factor of 0.5. A simpler, but less accurate method to calculate the time needed to cool grain is to assume that 1000 m^3 of air is needed to cool 1 m^3 of grain (Lasseran 1994) and the time needed is a function of the airflow produced. Another rough estimate of cooling time can be based on the assumptions that an airflow rate of 1 s^{-1} m^{-3} is equivalent to a cooling time of about 10 days and that the rate of cooling is directly proportional to airflow rate. One problem with these methods is that it is difficult to measure the air flow through the grain. The accuracy of the predicted cooling time is also affected by changes in ambient air conditions and uniformity of airflow through the grain bulk. More accurate predictions can be obtained from computer simulation techniques (Sanderson 1986).

4.3 Refrigerated Aeration

Chilled-air aeration is another method of grain cooling. It is used when the ambient air is not cool enough, such as in tropical areas or in temperate regions during the summer (Hunter and Taylor 1980; Fields 1992; Navarro and Calderon 1982; Fields and Muir 1995). In the majority of cases, the grain is cooled only to the development threshold of the insect, approximately 15 °C, to minimize the cost of treatment. The cost of treatment is similar to that of a residual insecticide in insulated granaries. Reheating chilled air reduces the humidity of the air and ensures that the grain does not gain moisture as it is cooled. The cost of a chilled air treatment can vary from 3 kW ht^{-1} in a well-insulated granary to 12 kW ht^{-1} under unfavourable conditions.

5. Conclusions

Low temperatures have been used for thousands of years to control stored-product insects. Today, this method is widely used in granaries to limit the growth of populations or to eliminate infestations. There are a number of advantages, i.e. few risks to operators, no residues on the grain, and it is effective against insecticide-resistant populations. However, there are several limitations that prevent a more general adoption of cold treatments; the cost of equipment and energy, some systems are dependent on cool ambient air, limiting treatments to the fall and winter in temperate regions. A thorough understanding of insect cold tolerance can in part allow the use of low temperature as an insect control method. One possible solution is to block the cold acclimation of insects (blocking synthesis of cryoprotectants, deregulating the control systems or using ice-nucleating bacteria). This would reduce the cold tolerance of insects three to ten fold, making it easier to control insects with low temperature.

Acknowledgements: I thank Francis Fleurat-Lessard for reviewing the manuscript.

References

Armitage D.M., Llewellin B.E., (1987). The survival of *Oryzaephilus surinamensis* (L.) (Coleoptera: Silvanidae) and *Sitophilus granarius* (L.) (Coleoptera: Curculionidae) in aerated bins of wheat during British winters. Bull. Entomol. Res. 77:457-466.

Armitage D.M., Stables L.M., (1984). Effects of aeration on established insect infestations in bins of wheat. Protect. Ecol. 6:63-73.

Banks H.J., Fields P.G., (1995). Physical methods for insect control in stored grain ecosystem. pp. 353-410 in D.S. Jayas, N.D.G. White and W.E. Muir (Eds.) Stored-grain ecosystems, Marcel-Dekker Inc, New York, 757 p.

Brown C.W., Hill S.T., (1984). Survival of micro-organisms in deep-frozen barley and pig feed. J. stored Prod. Res. 20:145-150.

Burges H.D., Burrell N.J., (1964). Cooling bulk grain in the British climate to control storage insects and to improve keeping quality. J. Sci. Food Agr. 15:32-50.

Evans, D. 1987. The survival of immature grain beetles at low temperatures. J. stored Prod. Res 23:79-83.

Fields P.G., (1992). The control of stored-product insects and mites with extreme temperatures. J. stored Prod. Res. 28:89-118.

Fields P.G., Muir W. E., (1995). Physical control. pp 195-221. In Integrated management of insects in stored products. Eds B. Subramanyam and D. W. Hagstrum. Marcel Dekker Inc. New York. 426 p.

Fields P.G., Fleurat-Lessard F., Lavenseau L., Febvay G., Peypelut L., Bonnot G., (1998). The effect of cold acclimation and deacclimation on cold tolerance, trehalose and free amino acid levels in *Sitophilus granarius* and *Cryptolestes ferrugineus* (Coleoptera). J. Insect Physiol. 44: 955-965.

Flinn P.W., Hagstrum D. W., (1998). Distribution of *Crypotlestes ferrugineus* (Coleoptera: Cucujidae) in reponse to temperature gradients in stored wheat. J. stored Prod. Res. 34:107-112.

Graham W.M., (1958). Temperature preference determinations using *Tribolium*. Anim. Behav. 6: 231-237.

Hagstrum D.W., (1987). Seasonal variation of stored wheat environment and insect populations. Environ. Entomol. 16:77-83.

Howe R.W., (1965). A summary of estimates of optimal and minimal conditions for population increase of some stored products insects. J. stored Prod. Res. 1:177-184.

Hunter A.J., Taylor P.A., (1980). Refrigerated aeration for the preservation of bulk grain. J. stored Prod. Res. 16:123-131.

Lasseran J.C., (1994). L'optimisation du refroidissement par la ventilation. p. 31-40 in La maîtrise du stockage économie et qualité, ITCF-FFCAT-INAC (Eds.) Paris, 76 p.

Lasseran J.C., Niquet G., Fleurat-Lessard F., (1994). Quality enhancement of stored grain by improved design and management of aeration pp. 296-299 in E. Highley, E.J. Wright, H.J. Banks and B.R. Champ (Eds) Stored Product Protection, CAB Int., Wallingford, UK, 1274 p.

Levinson H.Z., Levinson A.R., (1989). Food storage and storage protection in ancient Egypt. Boletin De Sanidad Vegetal. 17:475-482.

Mason L.J., Strait C.A., (1998). Stored product integrated pest management with extreme temperatures. pp 141-177. In Temperature sensitivity in insects and application in integrated pest management. Eds G.J. Hallman and D.L. Denlinger. Westview Press, Boulder, Co, 311 p.

Navarro S., Calderon M., (1982). Aeration of grain in subtropical climates. FAO, Agriculture Services Bulletin 52, Rome, 119 p.

Sanderson D.B., (1986). Evaluation of stored-grain ecosystem ventilated with near-ambient air. Winnipeg, Canada, M.Sc. Thesis, University of Manitoba. 191 p.

Smith L.B., (1970). Effects of cold-acclimation on supercooling and survival of the rusty grain beetle, *Cryptolestes ferrugineus* (Stephens) (Coleoptera: Cucujidae), at sub-zero temperatures. Can. J. Zool. 48:853-858.

Sømme L., (1968). Acclimation to low temperatures in *Tribolium confusum* Duval (Col., Tenebrionidae). Norw. J. Entomol. 15:134-136.

Sømme L., (1982). Supercooling and winter survival in terrestrial arthropods. Comp. Biochem. Physiol. 73A:519-543.

Storey K.B., Storey J.M., (1989). Freeze tolerance and freeze avoidance in ectotherms, pp. 52-82 in L.C.H. Wang (Ed.) Advances in Comparative and Environmental Physiology, Springer-Verlag, Berlin, 441 p.

Electromagnetic Methods

Electromagnetic Radiation for Plant Protection

Jacques Lewandowski

1 Introduction

This chapter describes techniques of plant protection using electromagnetic (EM) waves. EM protection depends on the nature of the interaction between EM waves and the target plants, soil, devastating insects and diseases. The examination of each factor affecting this interaction is thus essential since it will set up the limits and possibilities of applications for these EM techniques. First, we will describe the main characteristics of EM waves; then we will analyse their interaction with plants, the soil and the insects.

2 Some Notions about the EM Wave-Matter Interaction

2.1 The EM Spectrum

All EM waves have the same structure (Fig. 1): they consist of an electric field E and a magnetic field H vibrating in phase and perpendicular to the direction of propagation of the wave (Bonn and Rochon, 1992). EM waves differ by their frequency (f) or their wavelength (λ) which are connected by the following expression:

$$\lambda = c / f, \tag{1}$$

where $c = 3 \times 10^8\ ms^{-1}$ is the speed of EM waves in vacuum. The EM spectrum is divided in eight spectral bands (Bonn and Rochon, 1992) illustrated in Fig. 2. Starting with the longer wavelengths, one finds audio waves, radio waves, microwaves, infrared (IR), visible light, ultraviolet (UV), X-rays and gamma rays. The wavelength range is very wide, extending from the very long audio waves (some thousands of km and more) to the very short gamma rays (10^{-12} m).

There is no single source able to emit the entire EM spectrum. Specific sources are available for each of the eight spectral bands. Audio and radio waves are obtained with a ferro or piezoelectric transducer. Microwaves are emitted by a magnetron or a klys-

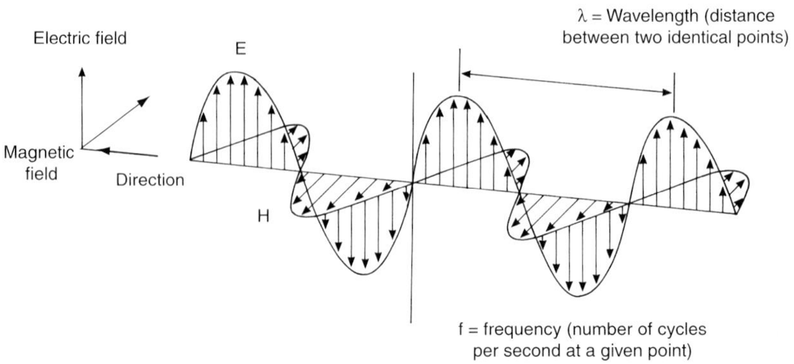

Fig. 1. The electromagnetic wave (After Bonn and Rochon 1992).

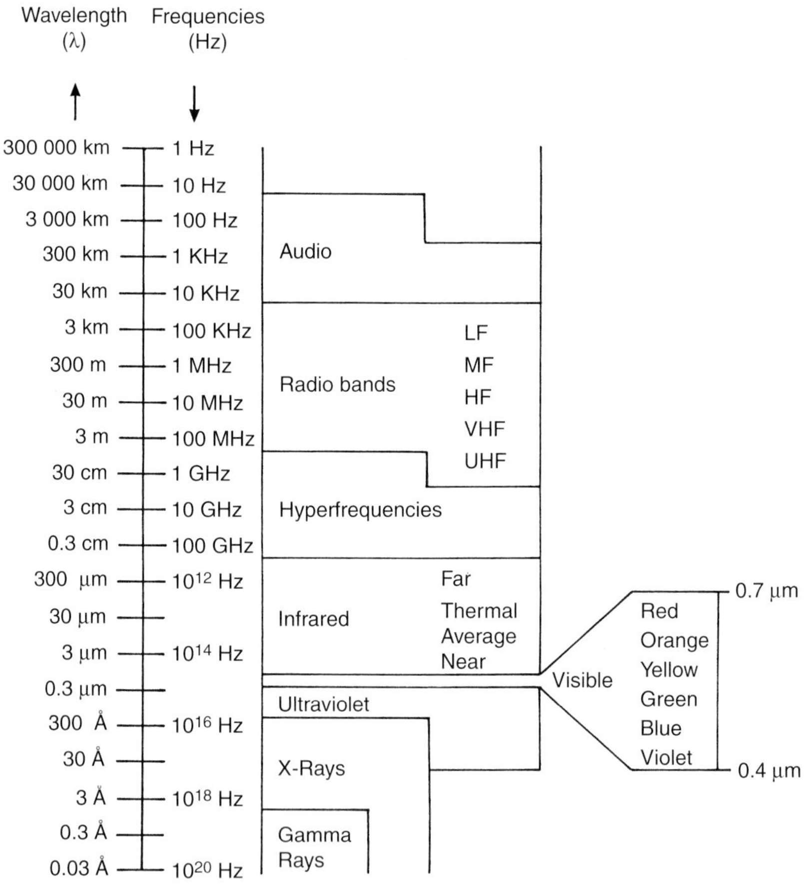

Fig. 2. The electromagnetic spectrum. (After Bonn and Rochon 1992). The eight spectral bands are illustrated. Note the great range in terms of frequency and wavelength.

tron. IR is emitted by an incandescent object. An electric light bulb emits visible radiation. UV is radiated with deuterium or mercury vapour lamps. X-rays are emitted when electrons collide on a metal plate. Gamma rays are emitted by radioactive elements. From a purely practical aspect, one can recognise that small wavelengths are in general more difficult to obtain, require complex hardware and are often more expensive. Therefore, gamma ray or X-ray equipment generally requires a higher investment than IR or radio wave experiments. This financial aspect will have to be considered when selecting an EM wave.

Each spectral band, from the audio waves to the gamma rays, may have a potential application in plant protection. Several modes of action can take place: some waves interact directly with the plant; others affect the insects selectively; others penetrate deeply into the ground and heat it. The type of interaction between the EM wave and the matter essentially relies on two factors: (1) the frequency of the wave and (2) the penetration depth of the wave in the medium. We will now describe the importance of these two factors in order to better determine the potential of EM waves in plant protection.

2.2 The Frequency or the Role of the Photon

An EM wave carries energy, but this energy is not emitted continuously. It is actually emitted in the form of small indivisible quantities (called photons) individually transporting an energy E_{ph} proportional to the frequency (Hecht 1987)

$$E_{ph} = h\,f = h\,c\,/\,\lambda\,, \qquad (2)$$

where $h = 0.66 \times 10^{-33}$ Js is Planck's constant. The concept of photons is fundamental because EM waves interact individually with matter through their photons. Because Planck's constant is very small, the energy of the photon is significant only for very small wavelengths. In other words, an EM wave with a small wavelength has a greater potential effect on matter.

For extremely small wavelengths — such as gamma rays, X-rays, UV and sometimes visible light — a photon has enough energy to extract an electron from the atom. In that case, the atom is ionised and the radiation is said to be ionising. If the extracted electron belongs to a molecule, ionisation will destroy a chemical bond. If the elec tron is part of a gene of a biological cell, the genetic code may be modified. One can thus predict that ionising radiations — small wavelengths like gamma rays, X-rays or UV — will interact deeply within the atomic or electronic structure of the matter components and, consequently, on the chemical or genetic integrity of the living organisms.

Photons of EM with longer wavelengths, i.e. visible light, IR, microwaves, radio and audio waves, do not have enough energy to extract an electron. These radiations are therefore known as non-ionising. In this case, the EM wave will nevertheless act on the matter through its electric field oscillating at the frequency of EM wave. The charged particles inside the matter will start to vibrate at the same frequency but will not be extracted from their chemical bonds. The resulting internal friction leads to a release of heat directly inside the matter. The matter is heated at the exact place where the EM is absorbed. Thus, non-ionising EM waves — long wavelengths like visible light, IR, microwaves, radio and audio waves — will produce vibrations of the matter components and heat the medium.

The EM wave frequency is therefore a fundamental characteristic which determines how the EM wave interacts with the medium, be it a plant, an insect or the ground. These interactions may range from generation of heat to breaking chemical links and possibly to atomic or genetic changes.

The energy delivered by the EM wave plays a different role since it determines the number of existing photons. This energy is proportional to the number of interactions involved: a larger energy level increases the number of interactions and allows a greater physical modification of the medium. On the other hand, if the EM wave frequency has no effect on a particular medium, increasing its energy will have no effect.

2.3 Penetration of EM Waves in Matter

For an EM wave to interact, it must penetrate and propagate inside the medium. The power P of an EM wave is defined as the energy carried per unit of time. When an EM wave of power P is incident on a medium, i.e. a plant, an insect or the soil, a fraction P_r is reflected at the interface and the remaining power P_t is transmitted inside the medium. The incident power can be expressed as follows:

$$P_r + P_t = P \tag{3}$$

Only the power P_t transmitted through the interface will interact with the medium since the reflected wave does not penetrate inside it. Introducing the reflectance $R = P_r / P$ and transmittance $T = P_t / P$, the preceding equation becomes

$$R + T = 1 \tag{4}$$

which is satisfied for any wavelength of the incident radiation. The coefficients R and T depend on the wavelength. Their variation with wavelength defines the spectral signature of the material. Figure 3 illustrates the spectral reflectance $R(\lambda)$ typical of lemon tree leaf.

The transmitted EM wave is gradually absorbed by the medium as soon as it penetrates. After a distance x of propagation, the remaining power $P_t(x)$ decreases according to the exponential law (Hecht 1989):

$$P_t(x) = P_t \exp(-\alpha x) \tag{5}$$

where α is the absorption coefficient of the medium expressed in m^{-1}. A more convenient measurement of the EM wave absorption in the medium is the penetration depth D defined by

$$D = 1/\alpha. \tag{6}$$

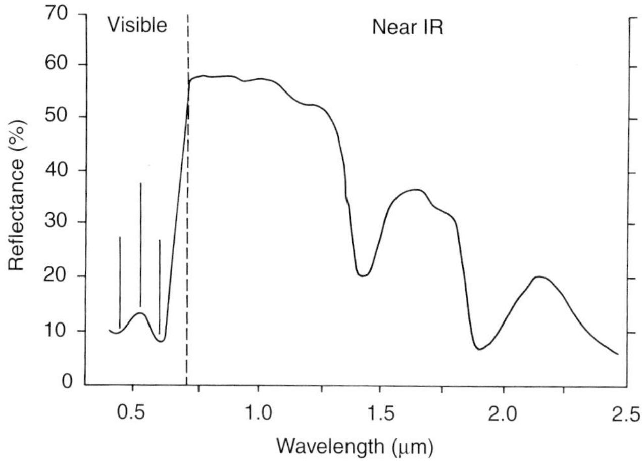

Fig. 3. Reflectance spectrum of a leaf of lemon tree. (After Gausman 1985) The higher reflectance of green (0.55 μm) explains the colour of the leaf. Leaves show a very high reflectance in near IR.

This penetration depth plays a significant role for EM techniques since it indicates the distance along which the EM wave/matter interaction takes place. If D is very small, the interaction will take place close to the surface or within the surface layers. If D is large, the EM wave will penetrate deeply and the interaction will not be efficient because of the extremely weak absorption. An EM wave can thus penetrate more or less deeply in the medium. The penetration depth depends on the dielectric properties of material and the frequency of the EM wave.

3 Use of Electromagnetic Waves

The interactions between EM energy and biological organisms are a very complex subject due to the great diversity of living organisms and the physical phenomena at play. Each living organism reacts differently to radiation. EM wave/matter interaction depends on the photon energy, ranging from atomic changes to the break of chemical or genetic links and possibly to the generation of heat. EM waves can also penetrate more or less deeply in the medium, going from surface interactions to very deep penetrations. An advantage of EM techniques over traditional chemical processes is that, in theory, they do not leave any toxic residue.

3.1 Combination of the Various Factors

For an EM wave to be effective in plant protection, several factors must be satisfied simultaneously: (1) the EM wave must eradicate the target without damaging the sur-

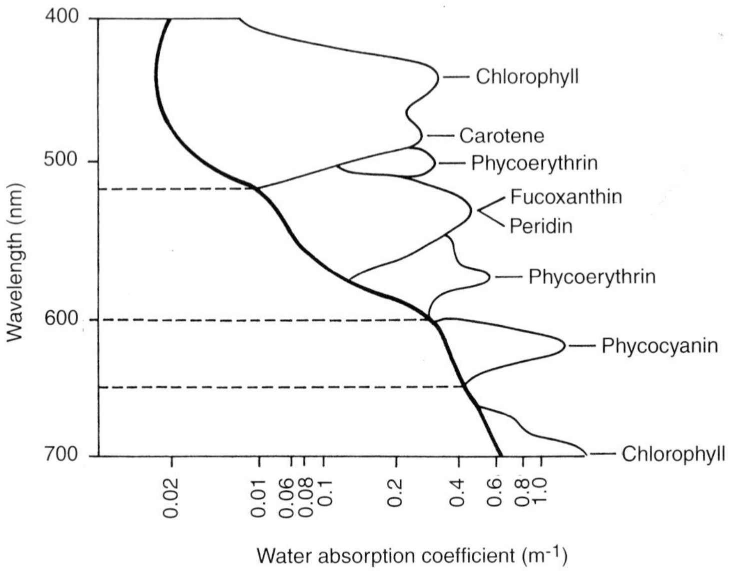

Fig. 4. Absorption coefficients of water and certain pigments associated with biological activity. (After Bonn and Rochon 1992) Minimum absorption of water is around 0.45 mm where the penetration depth is D = 1 / 0.02 = 50 m. Chlorophyll shows absorption peaks at 0.42 and 0.70 μm.

rounding medium, (2) it must penetrate the ground or the plant to reach the target, (3) the radiation source must be reasonable in terms of cost and ease of use, and (4) the energy should not have harmful side effects.

X-rays penetrate deeply in plants and are absorbed by insects. UV rays and IR are absorbed at the surface and microwaves have very different penetration depths and that depends on the nature of the material and its moisture content. For visible light, the absorption coefficients for the principal plant biological components and for water are illustrated in Fig. 4 (Bonn and Rochon 1992).

Figure 5 summarises the properties of EM waves in relation to the media found in plant protection, i.e. soils, plants and insects. The mode of interaction (ionising or non- ionising), the penetration depths in the medium (ground, plant or insect) and the facility of use of the radiation source are illustrated for each of the eight spectral bands. Figure 5 shows that radio waves penetrate a substrate very deeply and are thus practically ineffective in plant protection. In contrast, X-rays and gamma rays may kill insects inside a substrate like food, while UV rays will kill them only on the surface. On each side of visible light, the waves which offer the best compromise for penetration are X-rays and microwaves. The first acts at the chemical bond level, and the second overheats the medium.

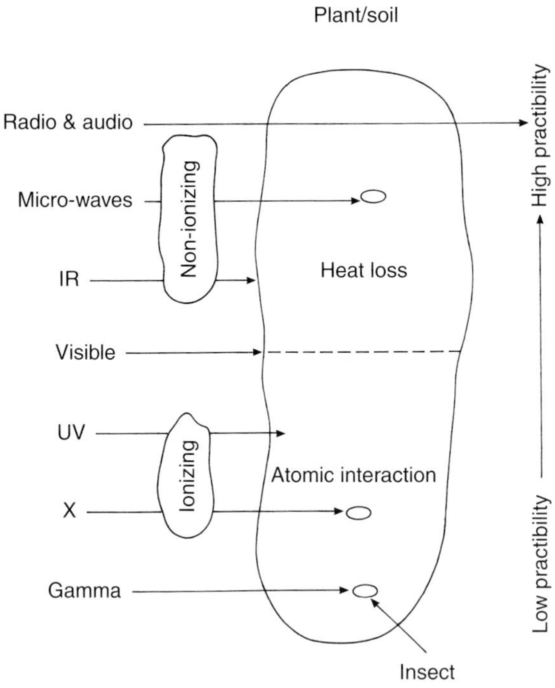

Fig. 5. Combination of the various factors for EM techniques in plant protection. EM waves are illustrated by their ionising or non-ionising behaviour, their penetration depth inside the plant or the ground, and their ease of use (not very practical to very practical). Notice that the extreme spectral bands (gamma rays, X-rays and UV, microwaves and radio waves) penetrate well inside the plant, whereas the intermediate bands (visible and IR) are absorbed at the surface. Ionising radiations interact at the atomic level, whereas non-ionising radiation tend to heat the medium. On a practical level, short wavelengths require expensive equipment that is difficult to set up.

3.2 The Terrestrial EM Environment

Plants, insects and soil are subjected daily to illumination by the sun. It is therefore important to be aware of the spectral characteristics of sun rays incident on Earth (Fig. 6). Solar radiation has the following properties:

- 44% of the emitted energy is in the visible band (0.4 to 0.8 μm). Visible photons have sufficient energy to initiate the phenomenon of photosynthesis, activate the retinal receivers or affect photographic films through the photochemical effect.
- UV-B rays (0.28 to 0.32 μm) are partially absorbed by the ozone layer. Many researchers simulated the impact of a ozone layer reduction on vegetation and agriculture (Flint et al. 1996) because UV-B rays have a major influence on the biochemical behaviour of plants.
- Solar radiation in the microwave band (1 mm to 10 cm) is negligible.

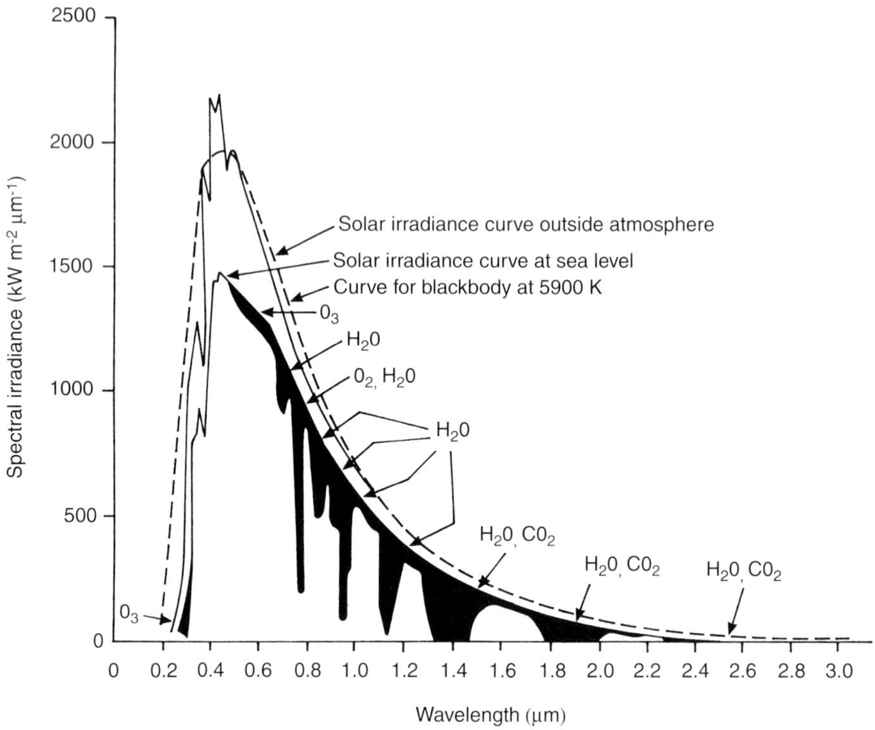

Fig. 6. Solar spectral illumination. (After Bonn and Rochon 1992). Solar illumination at sea level is significant in the visible band but not negligible in near-IR (0.70 to 2.50 µm) and in UV.

Box 1. Ionising Radiation

There is ionisation when an atom loses one or more electrons. An atom can be ionised by a radiation whose photon has sufficient energy, such as X-rays, UV or even visible light. The energy of ionisation required to extract an electron depends on the atom in question: it is different for each species. The effect of the ionising radiation on a living organism can be very complex. If the extracted electron belongs to a molecule, ionisation involves the destruction of a chemical bond. If this molecule belongs to a gene inside the core of a cell, its genetic code can be modified. Ionising radiation includes UV rays (bronzing and sunstroke), X-rays (radioscopy) and gamma rays (radioactive materials). UVs are used in microbiology to control and produce organisms with faded physiological characteristics. X-rays and gamma rays are used for microbiological sterilisation. These very penetrating radiations can produce ionisation in biological tissues, causing severe consequences if permitted doses are exceeded. In the food industry, ionising radiations are used to kill insects on fruits and many spices (black pepper, coriander, ginger and marjoram).

- Solar radiation is extremely variable (latitude, hour of the day, orientation of the leaves or plants) and depends strongly on the cloud cover (Schulzowà 1996).
We will describe separately each of the eight spectral bands, starting with short wavelength ionising radiation and mention some interesting EM applications for plant protection.

3.3 Gamma Rays (λ Lower than 0.3 Å)

Gamma ray photons are so energetic that they interact at the electronic and atomic level of matter. They offer the advantage of being selective. They are absorbed by the insects while penetrating deeply inside food material. The potential of application for gamma rays is therefore considerable in the food industry. Irradiation increases the storage time (Thayer 1985) because it destroys the harmful organisms inside the food. In 1963, the United States Department of Agriculture (USDA) authorised the use of gamma rays for irradiation of corn and corn flour (Van Kooij 1982). Gamma rays are also used to eliminate insects from spices such as black pepper, coriander, ginger and marjoram. However, some observers mentioned that the taste of these spices changed after irradiation (Thayer 1985). Nevertheless, gamma rays remain of little use because of the high cost of equipment and the precautions necessary for the protection of personnel. It should also be noted that gamma rays are dissipated over short distances in the atmosphere.

3.4 X-Rays (0.3 to 300 Å)

Like gamma rays, X-rays are used for post-harvest processing to selectively kill insects (or to make them sterile) without interacting with the surrounding medium. Banks (1976) reports a great variability of the results on the lethal doses of radiation for various insects. As for gamma rays, production of X-rays is not very effective and is expensive: their use remains therefore relatively limited.

3.5 UV Rays (300 Å with 0.4 μm)

UV rays are also ionising radiations, but their penetration depth is very small and the interaction with materials occurs at the surface. These radiations typically penetrate to depths of approximately 100 μm. Thus, the targets (e.g. insects and bacteria) must be exposed directly to the UV radiation. The UV band is divided into three spectral subbands:
- UV-A (0.32 to 0.40 μm) corresponds to energy photons having minor effects in plant protection. UV-A wavelength 0,365 μm was used, however, to kill some insects (Stuben 1973).

- UV-B (0.28 to 0.32 µm) is the most commonly band used because photons are sufficiently energetic to attack the plant morphology by causing damage to the cells and modifying their genetic constitution (Culotta 1994). This is due to the UV-B absorption by macromolecules such as nucleic acids, proteins and pigments. Even if UV-B rays constitute a small portion of solar radiation, they are sufficiently energetic to create a range of biological effects. For example, UV-B was used to sterilise insect eggs (Krieg 1975) and to decrease the impact of the post-harvest diseases of certain plants (Arul et al. 1996, see Chap. 10).
- UV-C (l < 0.28 µm) are the most energetic and their effects are even more detrimental on plants and microorganisms (Baden et al. 1996).

In laboratories, UV lamps reproduce solar radiation to grow plants (Olszyk et al. 1996). Fig. 7 shows that it is possible to rather accurately reproduce solar radiation between 0.29 and 0.30 µm, whereas radiation from 0.30 to 0.31 µm is inadequately reproduced in laboratories. Covers filtering the solar radiation can modify the spectral composition of the light incident on plants and gave interesting results in inhibiting grey rot in a greenhouse environment (Nicot et al. 1996, see Chap. 9).

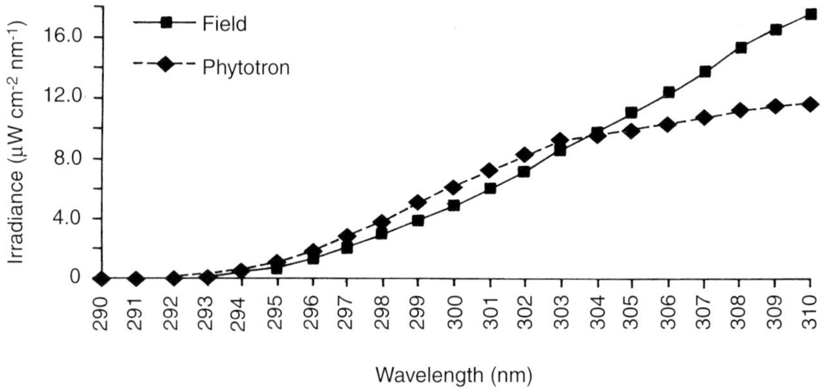

Fig. 7. Compared UV-B illuminations for sun and laboratory lamps (Olszyk et al. 1996). The solid line (field) corresponds to the UV solar spectral illumination (see Fig. 6). The reproduction of laboratory illumination (plant laboratory) is reliable up to 300 nm.

3.6 Visible and Near IR (0.4 to 2.5 µm)

Visible light (0.4 to 0.8 µm) and near IR (0.8 to 2.5 µm) have the advantage of offering a source of natural radiation available everyday. In this spectral band, the radiation can sometimes extract an electron, sometimes make the molecules of the medium vibrate. The non-ionising wavelengths are used to dry food, like fish or grapes. The ionising effect is the power house of plant photosynthesis (Fig. 4).

Box 2. Non-Ionising Radiation

Photons from low frequency EM radiation — visible light, IR, microwaves and radio waves — do not have enough energy to extract an electron or modify the chemical bond. For these non-ionising radiations, the electric field of the EM wave makes the charged particles vibrate within the matter, without dissociating them from their chemical bond. Internal friction, at the frequency of the EM wave, leads to energy dissipation in the matter. A very great number of photons of weak energy act together to heat the material. The principal effect of radio waves, microwaves and IR on living organisms is internal generation of heat. The choice of EM wave frequency depends on the object to be treated, the depth, the percentage of moisture, etc. The role of water is especially significant: microwave radiation agitates the molecules of water and fat of the insects, increasing the temperature to the lethal point. Thus, a heterogeneous mixture of matter, like insects and plants, subjected to a microwave radiation will experience a different heating for each component of the mixture.

IR waves, microwaves and radio waves produce a great quantity of heat whereas visible light is used to dry many foods, like fish or grapes. Microwaves are especially used in the food industry to thaw frozen products or for cooking.

3.7 Medium and Thermal IR (2.5 to 14 μm)

Medium and thermal IR possess photons that are only able to initiate vibrations and rotations of the molecules. IR rays are absorbed at the surface and are principally used for cooking. Thermal IR has been used to detect the presence of insects in food (Bruce et al. 1982) but failed to kill them. IR sources are relatively affordable and easy to set up. However, few applications of thermal IR have been reported in plant protection.

3.8 Microwaves (1 mm to 10 cm)

Microwaves cause molecules of the medium to oscillate and vibrate and part of the microwave radiation is thus transformed into heat. Thanks to their adequate penetration, microwaves are used in the food industry to thaw frozen products or for cooking. In EM applications, one can use microwaves to heat the soil and its organisms, increasing their temperature to lethal points. Each organism has a living temperature threshold (see Chap. 5). For human cells, destruction ocurs at about 40 °C, whereas a coleopteran of the rice flour dies at a final temperature ranging from 65 to 70 °C (Rosenberg and Bögl 1987).

Microwave frequencies cause the water molecule to vibrate within a very wide frequency band. Water plays a significant role because it is present everywhere in nature. The human body consists of approximately 70% of water; the majority of food

contains a significant proportion of water; and insects contain more water than the surrounding medium. For example, corn in the storehouse has a water content between 14 and 16%, whereas the insects consist of 50 to 65% water. Due to the particular absorption of water at certain frequencies, microwaves penetrate deeper in dry soils (Ulaby et al. 1982). Microwaves from 2 to 12 GHz penetrate to depths from a few centimetres to a few metres, depending on the type of soil and its percentage of moisture (Fig. 8). Ideally, one would like to heat the unwanted insect selectively without heating the surrounding medium. Frequencies from 10 to 100 MHz offer the best advantage for selective heating. Insects absorb 3 to 3.5 times more EM energy than the surrounding medium (Nelson 1985; see Fig. 2 of Chap. 11). Unfortunately, even if the insect is selectively heated, the conduction of heat into the medium decreases the selectivity of this technique.
Microwave sources, including oscillators and waveguides, must emit a large EM power, consume a lot of electric power and are expensive. The costs of microwave techniques for insect control are definitely higher than those of chemical processes.

3.9 Radio and Audio Waves (λ Higher than 10 cm)

Because of their excessive penetration in the soil (Figure 8), radio waves offer few applications in plant protection, except for de-infestation of dry or dehydrated food products (Fleurat-Lessard, Chap. 11). On the other hand, these waves are used in geo-

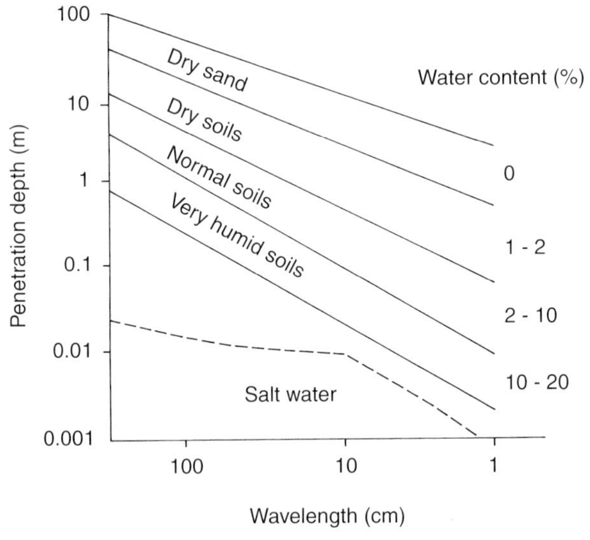

Fig. 8. Penetration depths of microwaves and radio waves in various types of soil (Ulaby et al. 1982). Microwaves and radio waves are very sensitive to soil moisture and penetrate deeper in dry soils. For example, an EM wave of 10 cm wavelength can reach a 10 m depth in dry soil, but penetrates only down to 5 cm in very wet soil.

logical prospecting because they allow the probing of deep underground layers (Fisher et al. 1994) and the detection of the presence of specific materials.

4. Conclusion

EM techniques are interesting alternatives to chemical methods because they offer the advantage of eliminating plant enemies without damaging the surrounding medium or leaving residues. Moreover, a study reported by the New York Times (1995) on chemical methods pointed out that insecticides and fungicides can spread thousands of kilometres from their place of origin and that some 10-year-old chemical pesticides still continue to affect the environment.

EM techniques have their limits and they may pose hazards. Indeed, it is rather easy to deteriorate the germinative faculty of seeds with an over-intense microwave exposure (Philbrick 1984) or to modify the immune system of certain mammals (Yang et al. 1983). So far there have been no reports of acquired tolerance to EM irradiation by plants or animals. Although adaptation to gamma rays seems very improbable, some form of tolerance to other EM waves is still possible. For practical applications, EM techniques seem more expensive than conventional methods. In spite of this additional cost, the use of EM waves for the control of crop pathogens does not rest solely on economic considerations, but also on the acceptance of these methods by society. Additional research on the interactions between EM waves and matter is necessary to discover all the potential applications and to enlarge the field of validity of EM techniques for plant protection.

References

Arul J., Mercier J., Charles M.T., Baka M., Benhamou N., (1996). Utilisation des UV pour réduire les maladies post-récolte chez les carottes et d'autres produits horticoles. Symposium La lutte physique en protection des plantes, SPPQ, Québec, 6-7 Juin 1996.

Baden H.P., Kollias N., Anderson R.R., Hopkins T., Raftery L., (1996). *Drosophila melanogaster* larvae detect low doses of UVC radiation as manifested by a writhing response. Arch. Insect Biochem. Physiol. 32: 187-196.

Banks H.J., (1976). Physical control of insects - recent developments. J. Aust. Entomol. Soc. 15: 89-100.

Bonn F., Rochon G., (1992). Précis de Télédétection, Vol. 1, Presses de l'Université du Québec, Québec, 485 p.

Bruce W.A., Street M.W., Semper A.R.C., Fulk D., (1982). Detection of hidden insect infestations in wheat by infrared carbon dioxide gas analysis. Agriculture Research Service, no. AAT-S-26, USDA, Washington DC.

Culotta E., (1994). UV-B effects: Bad for insect larvae means good for algae. Science, 265: 30.

Fisher E., McMechan G.A., Annan A.P., (1994). Acquisition and processing of wide-aperture ground-penetrating radar data. Geophysics 57: 495-504.

Flint S.D., Caldwell M.M., (1996). Scaling plant ultraviolet responses from laboratory action spectra to field spectral weighting factors. J. Plant Physiol. 148: 101-114.

Gausman H. W., (1985). Plant leaf optical properties in visible and near-infrared light, Lubbock, Texas: Texas Tech Press, p.78.
Hecht E., (1987). Optics, 2nd ed., Addison-Wesley, Mass., 640 p..
Krieg A., (1975). Photoprotection against inactivation of *Bacillus thuringiensis* spores by ultraviolet light. J. Invert. Pathol. 25: 267-268.
Nelson S.O., (1972). Insect-control possibilities of electromagnetic energy. Cereal Science Today 17; 377-387.
Nelson S.O., (1985). RF and microwave energy for potential agricultural applications. J. Microwave Power 20: 65-70.
New York Times, (1995). Analysis of tree bark shows global spread of insecticides. Oct. 10, 1995, Sec: C, p:4, col: 1.
Olszyk D., Dai Q., Teng P., Leung H., Luo Y., Peng S.B., (1996). UV-B Effects on Crop: Response of the Irrigated Rice Ecosystem. J. Plant Physiol. 148: 26-34.
Philbrick C.T., (1984). Comments on the use of microwave as a method of herbarium insect control: possible drawbacks, Taxon 33: 73-76.
Rosenberg U., Bögl W., (1987). Microwave Pasteurization, Sterilization, Blanching, and Pest Control in the Food Industry. Food Technology, June 1987, p. 93-97.
Schulzowà T., (1996). Photoinhibition in situ in Norway Spruce. J. Plant Physiol. 148: 129-134.
Stuben M., (1973). Studies on the influence of electronic flashes on the mortality and fertility of *Musca domestica*. Z. Ang. Entomol. 74: 35.
Thayer D.W., (1985). Application of Radiant Energy in Pest Management. Cereal Foods World, 30: 714-721.
Ulaby F., Moore K., Fung A., (1982). Microwave remote sensing - active and passive. Vol II, Artech.
Van Kooij J.G. (1982). International aspects of food irradiation, p. 84-97 in Food irradiation Now, M. Nijkoof et W. Junk Publishers, La Haye, Pays Bas, 157 p
Yang H.K., Lockwood J., Tompkins W.A..F. (1983). Effects of microwave exposure on the hamster immune system. Bioelectromagnetics 4: 123.

The Use of Microwaves for Insect Control

Yvan PELLETIER and Bruce G. COLPITTS

1 Introduction

Insects can maintain their metabolic activity only within a limited body temperature range. They are cold-blooded; that is, their body temperature is determined by climatic conditions in their immediate environment. This suggests that a possible approach to pest control might be to alter the physical conditions of an insect's environment so as to raise its body temperature to a lethal level.

Such a result requires a means of transferring energy efficiently into the body of the insect pest. This energy transfer may be brought about by increasing the temperature of the ambient air (Fields 1992; Duchesne and Boiteau 1991; Magan and Ingle 1992), by transferring heat in the form of steam (Hansen et al. 1992), or by using radiant energy (McFarlane 1989, see also Chap. 5 and 6). Insects themselves use the radiant energy of the sun to increase their body temperature, adopting behaviours that increase the surface area of their bodies that is exposed to the sun's rays. Some of their own physical characteristics, including absorptivity (see Chap. 7), mass and surface area, are factors affecting the extent to which their body temperature can be raised by exposure to solar radiation. However, it is the properties of the source of the radiation, including power, wavelength and polarization, that determine how efficiently energy is transferred to the insects.

Microwaves can be readily generated at a high power level. Consequently, they are potentially appropriate for quick, high-volume treatment of food products. However, the feasibility of microwave insect pest control will depend ultimately on compatibility between the physical characteristics of microwaves and those of insects. In this chapter, we shall briefly discuss the physical properties of microwaves and the transfer of microwave energy to insects. We shall then summarize the findings of research to date on the development of insect pest control methods in stored food products and in growing crops.

2 Microwave Radiation

2.1 The EM Spectrum

Microwaves are electromagnetic radiation in the 300 MHz-300 GHz frequency range (see Chap. 7). As electromagnetic waves propagate at a velocity effectively equal to that of light (3×10^8 ms^{-1}), the wavelength of microwave radiation ranges from 1 m to 1 mm. Microwaves are used mainly for telecommunications, but some frequencies have been set aside for industrial, scientific and medical applications (also known as ISM bands), in order to limit interference with communications. Safety standards applicable to radiation leakage from microwave appliances using the ISM bands are based on exposure hazards for human beings, whereas the standards used in telecommunications are concerned with protection from radio interference. Of the ISM frequencies, 915.0 MHz and 2.45 GHz are the most widely used for heating materials, and it is the second of these frequencies that is used in household microwave ovens.

A microwave electromagnetic field acts on charged particles of the matter traversed, producing a specific quantity of movement or particular orientation in them. The particles in question may be electrons, protons, atoms, ions or polar molecules like water. Water is an essential component of all living organisms. A water molecule is a dipole; that is, one side of it carries a positive charge and the other a negative charge. When subjected to an electromagnetic field, a water molecule pivots so as to align itself with the field. In an alternating electromagnetic field produced by microwaves, every water molecule is constantly in motion, oscillating back and forth. The rapid heating that results from the molecular friction caused by this oscillation is put to practical use in home and industrial microwave cooking ovens.

If the material exposed to the microwave electromagnetic field contains few charged particles to react to the electromagnetic field, it will be heated only slowly or not at all. In such cases, the material is said to be microwave-transparent. The strength of the electric current produced in the material depends on the shape and physical size of the exposed object. In any object subjected to an electromagnetic field, the charged particles move, creating an electric current. The more closely the longest axis of the object is aligned with the orientation of the electromagnetic field, the stronger the current will be and the greater the heating rate. This constant agitation of the particles is braked by the friction of the molecules within the exposed object. If the time required for the movement of the charged particles within the object is equal to one-half of the wave period, those particles will be oscillating at a frequency that is the same as that of the microwave radiation. Generally speaking, heat production is greatest when one of the dimensions of the exposed object is equal to or is a multiple of one-half of the wavelength of the microwave radiation. This principle is utilized in designing antennas for communication systems. The orientation of the object relative to the plane of polarization of the microwaves is also important. For example, a thin cylinder with dimensions approximating the wavelength will become hotter if its longitudinal axis is parallel to the microwave field. Although it is a simple matter to calculate the increase in temperature that will result from exposure to microwave radiation in the case of an object that is regular in shape and made of a material with known dielectric properties, this is not possible for an object of complex shape and heterogeneous composition, such as an insect or a plant. Modelling is not feasible, and only experimentation will tell us what effect microwave radiation will have on it (Olsen and Hammer 1982).

3 Generation and Application of Microwaves

Microwave radiation is usually generated by a magnetron. A magnetron converts high-voltage, low-frequency electric current into microwave energy. Where the electric current is 60 Hz AC, the conversion efficiency is of the order of 65% (Thuéry 1992).

Once generated, microwave radiation is transmitted through rectangular metal tubes known as waveguides. The dimensions of a waveguide are determined by the wavelength of the radiation to be carried by it. The width of the waveguide must be greater than one-half of the wavelength. In industrial applications, the part of the microwave energy that is transmitted to the material and the part that is reflected back toward the source are usually measured by devices mounted directly on the waveguide.

4 The Use of Microwave Radiation for Insect Pest Control

4.1 Laboratory Testing

Insect pest control methods must have maximal impact on the target insects but minimal harmful effects on the food product or crop. Investigators working on the development of methods involving microwaves have focused on the properties that influence the transfer of microwave energy to insects, food products and plants. Most of the work done to date has had to do with the effects of wavelength, target object orientation in the electromagnetic field, and radiation intensity on the temperature of the target.

The water content of cereal grains is ordinarily low, somewhere between 12 and 16%, whereas insects always consist of more than 50% water. This water content differential has direct repercussions on the dielectric losses sustained by an insect compared with those sustained by the substrate on which it lives (Table 1). The dielectric loss factor varies with the frequency. Because insects are so small in relation to the wavelengths of microwave radiation (12 cm at a frequency of 2.45 GHz), the smaller the wavelength, the more of the radiation they absorb. In the case of *Tenebrio molitor* L. nymphs exposed to microwave radiation, absorption increases from 8 Wkg^{-1} at 1.3 GHz to 17 Wkg^{-1} at 5.95 GHz and to 74 Wkg^{-1} at 10 GHz (Olsen and Hammer 1982). In some laboratory studies, insects have been placed inside the waveguide. In a standard waveguide for a frequency of 2.45 GHz, the body temperature reached by experimental *T. molitor* nymphs has been found to be the same regardless of whether the insects were placed longitudinally (with their longitudinal axes aligned with the long axis of the waveguide) (Fig. 1) or transversely (with their longitudinal axes aligned with the vertical axis of the waveguide) (Olsen and Hammer 1982). The heating

Table 1. Complex permittivity of insects, plants and stored products.

	Dielectric constant	Dielectric loss factor	Wavelength (GHz)	Reference
Tenebrio molitor	35.9	11.4	10.2	Nelson (1972)
Tenebrio molitor	31.5	12.7	9.4	Nelson (1976)
Sitophilus oryzae		0.4	1.10	Thuéry (1992)
Leptinotarsa (prothorax)	40.09	12.39	2.08	Colpitts et al. (1992)
Leptinotarsa (thorax)	39.98	12.2	2.08	Colpitts et al. (1992)
Leptinotarsa (female abdomen)	34.08	9.90	2.08	Colpitts et al. (1992)
Leptinotarsa (male abdomen)	40.92	11.74	2.08	Colpitts et al. (1992)
Wheat (11.5% water)	5.01	0.89	9.4	Nelson (1976)
Wheat (14.6% water)	5.56	1.38	9.4	Nelson (1976)
Wheat (17.8% water)	6.32	1.77	9.4	Nelson (1976)
Potato (leaf)	10.54	1.60	2.08	Colpitts et al. (1992)
Potato (petiole)	51.65	15.77	2.08	Colpitts et al. (1992)
Potato (stem)	49.67	14.57	2.08	Colpitts et al. (1992)
Tobacco		1.1	0.92	Hirose et al. (1975)
Tobacco		1.3	2.45	Hirose et al. (1975)
Tobacco		1.5	9.38	Hirose et al. (1975)

occurred mainly in the boundary region between thorax and abdomen. Fujiwara et al. (1983), showed that the power absorbed per unit volume in the various parts of the body of the nymphs was different at 2.45 GHz than at 6 GHz.

In some instances, it has proved feasible to measure the components of the complex dielectric constant of insect pest species and the substrates on which they live (Table 1). With the help of a coaxial probe for greater refinement, investigators have been able to measure the complex dielectric constant of various parts of the anatomy of the Colorado potato beetle (Fig. 2) and of the potato (Colpitts et al. 1992; Table 1). In general, the values of the dielectric constant and of the dielectric loss factor vary

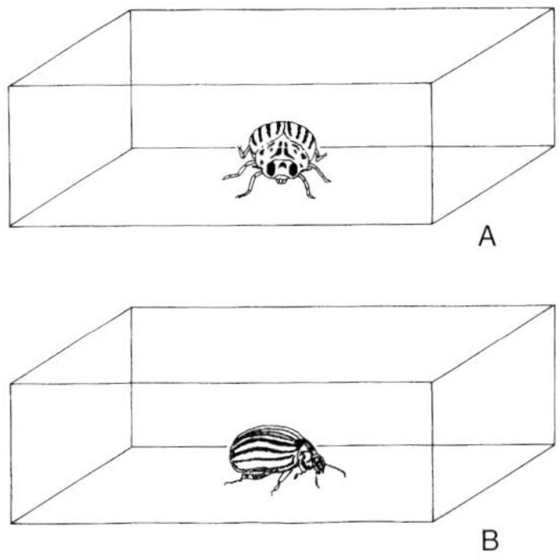

Fig. 1. Schematic representation of an insect placed transversely (A) and longitudinally (B) in a waveguide.

only slightly with frequency. The dielectric constant of the abdomen is slightly higher in the male potato beetle than in the female. In the case of potato plants, while small variations in the value of the dielectric constant have been observed, the values found for the leaves are clearly distinguishable from those observed for other parts of the plant. The leaves have a dielectric loss factor that is approximately one-tenth that of the other parts (Table 1). The energy densities that will kill Colorado potato beetles or damage potato plants have also been quantified through laboratory research (Colpitts et al. 1993). The microwave energy required to kill Colorado potato beetles is inversely proportional to their size. It takes 1230 J cm^{-2} to kill them in the first larval stage whereas 390 J cm^{-2} are enough to kill the adults (Fig. 3). The eggs are approximately the same size as the first larval stage, and they are deposited in clusters of 30 to 100 on the leaves. The clusters are held together with a sticky substance produced by the female. Egg clusters are more sensitive to microwave radiation than are individual eggs because of the larger size. The lethal energy density varies with the orientation of the egg cluster in the electric field, ranging

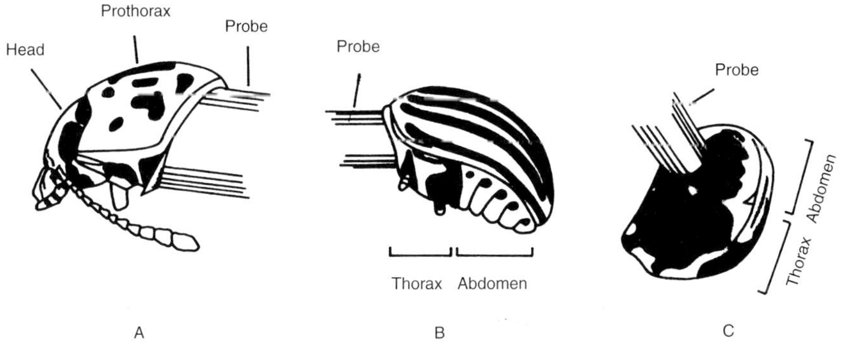

Fig. 2. Schematic representation of the equipment used to measure the dielectric constant of the prothorax, thorax and abdomen of the Colorado potato beetle (Colpitts et al. 1992).

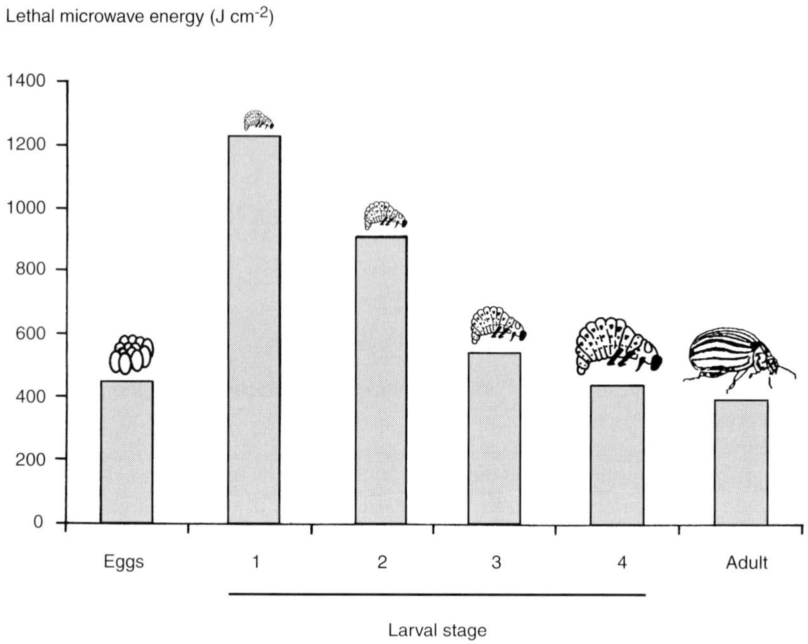

Fig. 3. Lethal microwave energy densities (J cm^{-2}) for the successive stages in the life cycle of the Colorado potato beetle (Colpitts et al. 1993).

from 60 J cm^{-2} for a parallel orientation, to 450 J cm^{-2} when it is perpendicular. Potato plants are quickly damaged by microwave radiation. Energy densities in the 10 to 160 J cm^{-2} range are enough to damage the leaves, peduncles or stems. The lethal energy density, i.e. that sufficient to cause irreversible damage to the stems and peduncles, depends on the orientation of the parts in question relative to the electric field. Maximum damage is observed when they are perpendicular to the field. These findings indicate that in the case of the Colorado potato beetle, the egg is the stage that is most sensitive to microwave radiation, but in practice the use of microwave radiation for control would risk damaging the plants. Consequently, this method is not usable with present microwave technology.
The possibility of using microwaves as a means of controlling underground insect pests has also been explored (Nelson 1996). The main problem with using microwave radiation to treat soil arises from the fact that microwaves cannot penetrate this type of substrate to any substantial depth. They weaken rapidly, at a rate of 2 db cm^{-1}, which means that the part of the energy penetrating to a depth of 5 cm represents only 10% of the energy initially developed (at the output of the generator or waveguide). At a depth of 10 cm, the value has dwindled to 1%, and at 15 cm it is no more than 0.1%. Treatment effectiveness thus decreases rapidly with depth. Heat penetrates moist soil by conduction more readily than dry soil. On the other hand, the greater the soil's water content, the lower the penetrating power of the radiation. It thus appears that the use of microwaves to treat soil would be advantageous only in the case of very dry

soils (Nelson 1996). Even so, some applications may be feasible in certain situations. Some investigators have reported that microwave radiation has other, non-thermal effects on insects, and that microwave treatment may be more effective than heat treatment (Thuéry 1992). Reports of effects at power levels which did not cause total mortality among the exposed insects have also been published (Fleurat-Lessard et al. 1979). These effects were caused by localized increases in body temperature or increases in the temperature of certain organs or tissues (Thuéry 1992; Olsen and Hammer 1982).

4.2 Tests for Commercial Applications

Under certain conditions, the prospect of using microwaves as a means of insect pest control has appeared economically advantageous, and various prototypes and commercial devices have been tested for effectiveness.

Microwaves have long been used for drying grain, some types of food products, and lumber. High-powered, large-volume microwave dryers suitable for industrial use have been developed. In these applications, the product is heated inside the dryer, so that the water migrates outward and evaporates at the surface. Any effect on insects would thus be an indirect result of the heating of the substrate. In terms of insect control, therefore, the effectiveness of a dryer of this type would depend on the tolerance of the insects in question to heat, the maximum temperature of the substrate and the exposure time (Burdette et al. 1975; Fanslow et al. 1975; Hirose et al. 1975; Tilton and Vardell 1982).

The fact that microwave treatment offers a potentially useful way of getting rid of insects in food products has long been acknowledged (Nelson 1974; Wilkin and Nelson 1987). The procedure leaves no residue and can readily be integrated with production-line methods. However, consumers have a very low tolerance level for insects in their food products, and this means that control methods must be highly effective. Insects are characterized by small size and a high surface-to-volume ratio, so that it is relatively difficult to heat them with microwaves. Furthermore, the temperature of the food product concerned must not be allowed to rise to a level at which its flavour or aroma would be affected. Even so, microwaves can be used for some applications, such as the disinfestation of grain (Nelson 1974), nuts (Wilkin and Nelson 1987), lumber (Burdette et al. 1975) and tobacco (Hirose et al. 1975).

A method of controlling Colorado potato beetles in field-grown potatoes was evaluated by the authors. Although lethal energy density values for the eggs of these insects are about the same as those for the most sensitive parts of the potato plant, it seemed likely that harmful effects on the plants could be minimized given adequate control of the plane of polarization of the microwaves. Unfortunately, in a system involving exposure of the entire plant, the depolarization resulting from scattering of the microwaves made the technique ineffective, dependent as it was on control of the polarization of the incident radiation. Owing to the size differen-

tial between the beetles and the potato plants, the microwaves affected the latter as well, and they were quickly heated to a lethal temperature.
Experiments designed to investigate the practicality of using microwaves to disinfest herbaria and botanical collections have been reported. With a dry substrate, the insects are directly affected by the microwaves. Here, too, however, the method has its limitations, owing to its negative impact on seed germination, fine structures such as trichomes, and the macromolecules used in taxonomy (Philbrick 1984).

5 Conclusion

In theory, the use of microwave radiation for insect control offers obvious advantages over chemical insecticides. Used on dry food products, it leaves no residue, is compatible with mechanized production systems, and can readily be automated. On the other hand, this insect control method has its limitations, which are inherent in the nature of microwaves and the dielectric properties of insects and the substrates on which they live. Insects are so small and have such a high surface-to-volume ratio that they are not very sensitive to microwaves. The insect-infested substrate is quickly heated along with the insects by the microwave energy. It is essential for substrate quality to be preserved, and this sets limits to the temperature conditions to which a substrate can be exposed. It thus appears likely that the use of microwaves on a commercial scale for insect control will be restricted to products that can tolerate heating (such as lumber) and dry products (such as grain and processed food products). Where these conditions can be met, microwave radiation raises the temperature of the target insects, directly or indirectly, to a lethal level. The method is still not practical for treating living plants, insects that inhabit the soil, or the many kinds of food products that are heat-sensitive. In time, progress in microwave technology may overcome some of these problems and make microwave radiation usable over a broader range of applications.

References

Burette E.C., Hightower N.C., Burns C.P, Cain F.L., (1975). Microwave energy for wood products insect control. pp. 276-281. In: East, T.W.R. (ed), Proceedings of the Microwave power symposium 1975: May 28-30. International Microwave Power Institute, Edmonton, Alberta.

Colpitts B.G., Pelletier Y., Cogswell S., (1992). Complex permittivity measurements of the Colorado potato beetle using coaxial probe techniques. J. Microwave Power & Electromagn. Energy 27: 175-182.

Colpitts B.G., Pelletier Y., Sleep D., (1993). Lethal energy densities of the Colorado potato beetle and potato plant at 2450 MHz. . J. Microwave Power & Electromagn. Energy 28: 132-139.

Duchesne R.-M., Boiteau G., (1991). Nouvelles perspectives sur les moyens de lutte mécaniques, physiques et culturaux contre le doryphore de la pomme de terre. pp. 33-54. Proceeding of the thirty-eight annual meeting of the Canadian Pest Management Society, Fredericton, New Brunswick.

Fields P.G., (1992). The control of stored-product insects and mites with extreme temperatures. J. Stored Prod. Res. 28: 89-118.

Fleurat-Lessard F., Lesbats M., Lavenseau L., Cangardel H., Moreau R., Lamy M, Anglade P., (1979). Effects biologiques des micro-ondes sur deux insectes *Tenebrio molitor* L. (Col. : Tenebrionidae) et *Pieris brassicae* L. (Lep. : Pieridae). Ann. Zool. Ecol. Anim. 11 : 457-478.

Fujiwara O., Goto Y., Amemiya Y., (1983). Characteristics of microwave power absorption in an insect exposed to standing-wave fields. Electronics and Communications in Japan. 66-B : 46-54.

Hansen J.D., Hara A.H., Tenbrink V.L. (1992). Vapor heat: a potential treatment to disinfest tropical cut flowers and foliage. HortScience 27:139-143.

Hirose T., Abe I., Kohno M., Suzuki T., Oshima K., Okakura T., (1975). The use of microwave heating to control insects in cigarette manufacture. J. Microwave Power. 10: 181-190.

Magan R.L., Ingle S.J., (1992). Forced hot-air quarantine treatment for mangoes infested with West Indian fruit fly (Diptera: Tephritidae). J. Econ. Entomol. 85:1859-1864.

McFarlane J.A., (1989). Preliminary experiments on the use of solar cabinets for thermal disinfestation of maize cobs and some observations on heat tolerance in *Prostephanus truncatus* (Horn) (Coleoptera: Bostrichidae). Trop. Agric. 29:75-89.

Nelson S.O., (1972). Frequency dependence of the dielectric properties of wheat and the rice weevil. Ph.D. dissertation, Iowa State University, publication No.71-19,997, Ann Arbor, Michigan.

Nelson S.O., (1974). Insect-control possibilities using microwaves and lower frequency RF energy. pp. 27-29. In: Gaylord, T.K. (ed). 1974 IEEE S-Mtt International microwave symposium digest of technical papers. Inst. Electrical and electronics Eng., New York.

Nelson S.O., (1976). Microwave dielectric properties of insects and grain kernels. J. Microwave Power. 11:299-303.

Nelson S.O., (1996). A review and assessment of microwave emergy for soil treatment to control pests. Trans. ASAE. 39: 281-289.

Olsen R.G., Hammer W.C., (1982). Thermographic analysis of waveguide-irradiated insect pupae. Radio Science 17:95S-104S.

Philbrick C.T., (1984). Comments on the use of microwave as a method of herbarium insect control: possible drawbacks. Taxon 33: 73-76.

Tilton E.W., Vardell H.H., (1982). Combination of microwaves and partial vacuum for control of four stored-product insects in stored grain. J. Georgia Entomol. Soc. 17: 96-106.

Thuéry J., (1992). Microwaves : Industrial, scientific, and medical applications. Artech House, Boston, Ma. 670 pp.

Wilkin D.R., Nelson (G.), 1987. Control of insects in confectionery walnuts using microwaves. pp. 247-254. In: BCPC Mono No. 37, Stored Products Pest Control.

Optical Filters against Grey Mould of Greenhouse Crops

Philippe C. Nicot, Nicolas Morison and Marie Mermier

1 Introduction

Grey mould is a severe and frequent disease in the greenhouse. The pathogen, *Botrytis cinerea* Pers., is a polyphagous fungus that attacks ornamental and vegetable crops. Control of this disease is difficult because both the microclimate and the cropping methods in greenhouse production are conducive to its development (Jarvis, 1992). No resistant commercial varieties are presently available. Hence, traditional chemical methods continue to be the principal means of control even though it is complicated by resistance to fungicides and the small number of registered products for greenhouse crops (Nicot and Baille, 1996).
An alternative control strategy would consist of slowing down the development of the epidemic by inhibiting or reducing the production of spores by the fungus. Indeed, one of the characteristics of this disease is the rapidity of its propagation in the greenhouse due to abundant sporulation on diseased plants. On tomato plants, for example, the fungus can produce several million spores per gram of fresh plant tissue in less than a week if microclimatic conditions are favourable (Nicot et al. 1996). This inoculum is very easily transported by air currents and propagates the disease to other plants or other organs.
One way of limiting spore production by the pathogen might be to modify the spectrum of light that penetrates into the greenhouse. Studies conducted in the 1970s demonstrated that near-ultraviolet (nUV) light (wavelengths between 300 and 400 nm) stimulates spore production by *B. cinerea*, while blue light inhibits it (Jarvis 1992). Polyethylene films similar to those commonly used to cover greenhouses, but containing additives that absorb near-UV light, have been produced and tested in Japan and in Israël since the 1980s. These films inhibit sporulation of the fungus in vitro and the intensity of their effect is related to their ability to transmit blue light and absorb near-UV light (Reuveni and Raviv 1992; Sasaki et al. 1985).
These first results appeared promising and we undertook a detailed study of the potential of these films for the protection of greenhouse crops against grey mould in commercial conditions. This evaluation was conducted in several steps.

2 Selection of Films Inhibiting Spore Production by *Botrytis cinerea*

2.1 Effect of Experimental Films on the Development of *B. cinerea*

In a preliminary study, we examined the effect of nUV-filtering films on several developmental stages of the fungus. Six samples of polyethylene films were used: a control film, that transmitted most of the radiation with wavelength greater than 230-240 nm, and five films containing various additives to block near-UV radiation (Nicot et al. 1996). These films were evaluated for their effect on spore germination, mycelial growth and sporulation. The tests were conducted *in vitro* and the fungus was grown on potato dextrose agar (PDA), a solid nutrient medium commonly used in mycology.

Among the UV-absorbing films, several had a statistically significant inhibitory effect on spore germination, in comparison with the control film. However, the extent of the inhibition was too limited to be of practical interest for control of grey mould. Germination rates after 24 h of incubation were all greater than 90% regardless of the film (Nicot et al. 1996). No significant difference among films was observed for mycelial growth. However, all UV-absorbing films strongly inhibited spore production on PDA, in comparison to the control film. The inhibition factor (amount of spores produced under the control film divided by the amount produced under a given film) was greater than 1000 for all five nUV-filtering films. For practical reasons, only one film was retained for further work, on the basis of its biological efficiency and its physical properties (Nicot et al. 1996). This film efficiently blocks radiation at wavelengths below 380-390 nm (including the fraction of spectrum that most stimulates sporulation). It has a high transmittance of violet-blue light, which is known to inhibit sporulation in vitro (Tan 1974).

2.2 Durable Inhibition or Delay in Spore Production?

The next step was to verify whether the effect of the nUV-filtering film on *Botrytis* was merely to delay spore production, or to inhibit it for long periods. This point was considered important because the growing season for greenhouse crops, such as tomato, can be very long (10-11 months), and since microclimatic conditions conducive to *B. cinerea* sporulation may occur and persist frequently during that time. In one study, it was shown that spore production on PDA remained approximately 1000 times greater under the control film than under the nUV-filtering film throughout a 5-week period (Fig. 1).

2.3 Inhibition on Plant Tissue?

Results of in vitro experiments cannot always be transposed to situations involving living organisms. To evaluate the usefulness of nUV-filtering films for the protec-

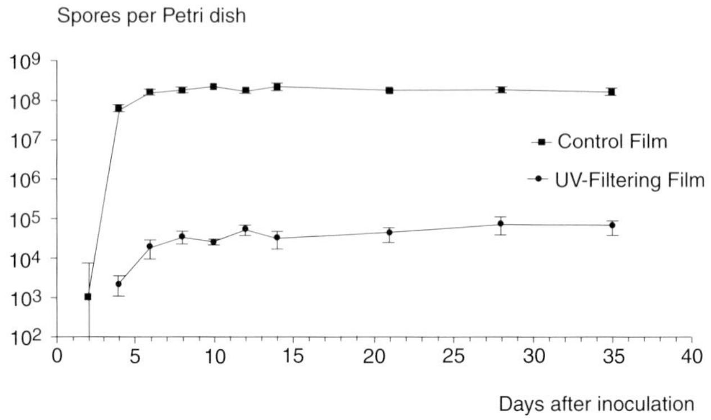

Fig. 1. Effect of a polyethylene film that filters near-UV radiation with wavelength below 380 nm (filtering film) on the kinetics of sporulation of *Botrytis cinerea* grown on potato dextrose agar medium.* Error bars indicate the standard error of the mean.

tion of greenhouse crops, it appeared necessary to determine whether inhibition of *B. cinerea* sporulation was as efficient when the fungus develops on plant tissue as when it is grown on nutrient agar. This was first verified on tomato. On stem segments, the inhibition of sporulation was highly significant, with a difference 1000 fold or greater between numbers of spores produced under the control and the nUV-filtering films. The inhibition factor remained above 1000 for the duration of the experiment (Fig. 2).
The results of all these preliminary studies confirmed the potential of nUV-filtering films to reduce the production of secondary inoculum by *B. cinerea* in greenhouses. A second step consisted of specifying the potential constraints and limits of this method in a commercial production situation.

3 Factors of Efficacy of nUV-Filtering Films

3.1 Effect of the Host Plant

Among biological factors likely to affect the effectiveness of nUV-filtering films in inhibiting sporulation of *B. cinerea* on diseased plants in a greenhouse, the first one that was considered was the nature of the substrate on which the fungus develops.

* reprinted from *Plant Disease*, 1996, 80 (5), with permission from the American Phytopathological Society.

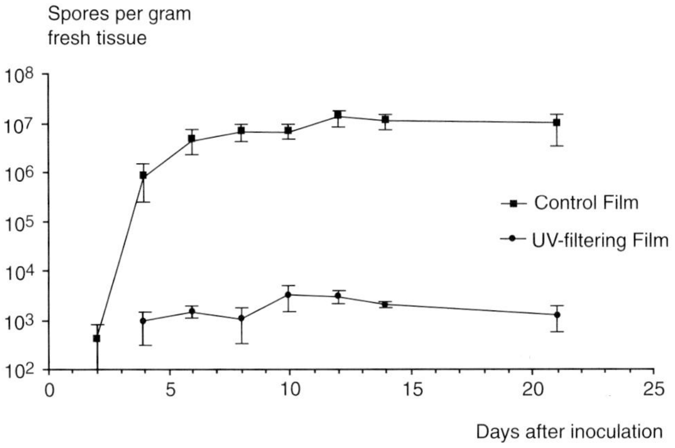

Fig. 2. Effect of a polyethylene film that filters near-UV radiation with wavelength below 380 nm (filtering film) on the kinetics of sporulation of *Botrytis cinerea* grown on 2 cm tomato stem segments.* Error bars indicate the standard error of the mean.

Our results suggested that sporulation of *B. cinerea* may be affected differently, depending on the host species. On tomato, for example, the fungus produced 15.7 million spores per 2-cm stem segment after 7 days under the control film. Under the nUV-filtering film, spore production per stem segment was approximately 6000, giving an inhibition factor at 2600. In comparison, the inhibition factors were only 1326 and 597 on melon and cucumber, respectively (Nicot et al. 1996). The nature and age of the plant organ colonised by *B. cinerea* may also influence the efficiency of the nUV-filtering film. Efficiency appeared to be lower on younger plants, and lower for flowers and cotyledons than for stems, suggesting that the nutritional composition of the substrate may be a significant factor (Nicot et al. 1996). Based on these data, one may speculate that if nutrients can interfere with the efficiency of nUV-filtering films, the composition and activity of the microflora developing on the macerated tissues of grey mould lesions may also play a significant role.

* reprinted from *Plant Disease*, 1996, 80 (5), with permission from the American Phytopathological Society.

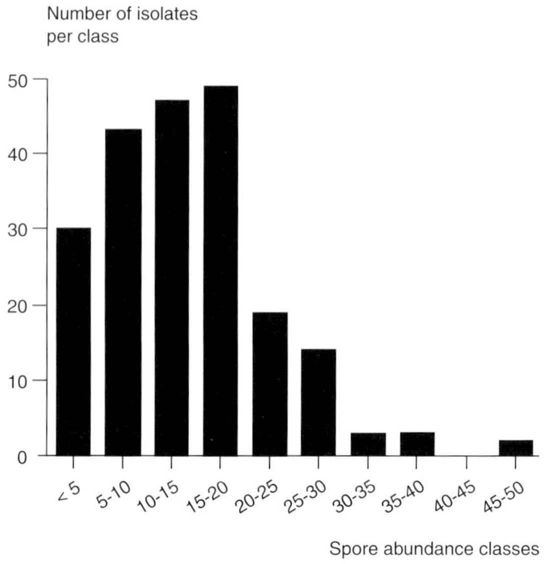

Fig. 3. Diversity in the ability to sporulate among 210 isolates of *Botrytis cinerea* growing on tomato stem segments incubated under a polyethylene film transparent to UV light. Sporulation is quantified as millions of spores g^{-1} of fresh stem tissue.

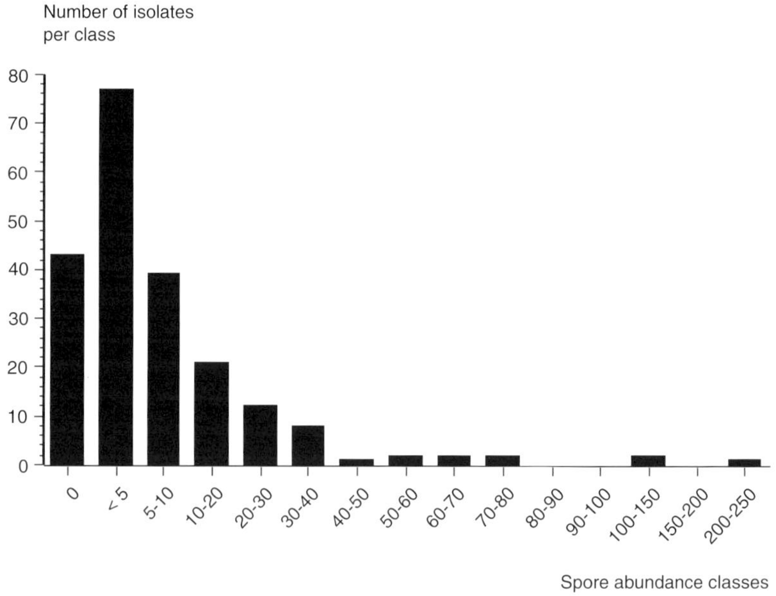

Fig. 4. Diversity in the ability to sporulate among 210 isolates of *Botrytis cinerea* growing on tomato stem segments incubated under a UV-filtering polyethylene film. Sporulation is quantified as thousands of spores g^{-1} of fresh stem tissue.

3.2 Phenotypic Diversity in the Sensitivity of the Pathogen to the Effect of the nUV-Filtering Films

The second biological factor that was studied was the possible variability in sensitivity among populations of *B. cinerea* to the effect of the nUV-filtering films. To that effect, 210 isolates of the fungus were collected in tomato greenhouses in southeastern France. Spore production was quantified on tomato stem segments after 7 days of incubation under a control or an nUV-filtering film. Under the control film, sporulation ranged from less than 5 million to approximately 50 million spores g^{-1} of fresh stem tissue. The majority of tested isolates produced between 5 and 25 million spores g^{-1} of tissue (Fig. 3). No sporulation was detected for about 20% of the isolates tested under the nUV-filtering film. Furthermore, the great majority (ca. 85%) of isolates produced less than 20 000 spores g^{-1} of tissue, i.e. roughly a thousand times less than under the control film (Fig. 4). One can assume that such high levels of reduction in spore production should have an important impact on the development of epidemics in the greenhouse. However, a few isolates were able to produce over 100 000 spores per gram of tissue under the nUV-filtering film, i.e. only 10 to 20 times less than under the control film. For these few strains, one may question whether using nUV-filtering films would have a significant epidemiological effect.

3.3 Effect of Selection Pressure on Phenotypic Stability of *B. cinerea*

B. cinerea is often considered to be a fungus whose phenotypic characteristics can rapidly change following repeated transfers on nutrient agar media, even in the case of monoconidial or monoascospore clones (Salinas and Schot 1967). This phenomenon could be linked to the presence of numerous nuclei in the cells of this fungus (a heterokaryotic state has been suspected by many authors, including Faretra et al. 1988), to complex ploidy levels inside the nuclei (Van der Vlugt-Bergmans et al. 1993), or even to presence of various transposable elements (Levis et al. 1997).
Considering this variability, one might fear a progressive acclimation of *B. cinerea* to the nUV-filtering films. To test this hypothesis, an isolate of *B. cinerea* was grown under nUV-filtering film on PDA medium. The spores produced after 7 days' incubation were collected aseptically and used to reinoculate fresh petri dishes, which, in turn, were incubated under the nUV-filtering film. The same procedure was repeated several times. In parallel, the same procedure was used to obtain successive generations of spores produced under the control film. Spore production was quantified for each generation. While it remained remarkably stable over 20 generations grown under the control film, spore production under the nUV-filtering film increased and plateaued at the fifth to sixth generation. From the beginning to the end of the experiment, spore production under the nUV-filtering film by the isolate of *B. cinerea* increased nearly 100-fold, to stabilise at a level nearly 2% of that under the control film. This test was repeated twice and the acclimation phenomenon was observed again.

In the greenhouse, sporulation on diseased plants only requires a few days when microclimatic conditions are favourable. As these spores can in turn infect new plant organs, numerous generations may occur during a growing season. If other isolates of *B. cinerea* behave similarly to that used in our study, commercial utilisation of nUV-filtering films may lead to a rapid evolution of less sensitive natural populations. A possible hindrance to this evolution could occur if strains able to sporulate abundantly under the nUV-filtering film had lower aggressiveness than the "normal" strains, whose sporulation is inhibited by the nUV-filtering film. Such reduced aggressiveness could result, for example, from a genetic burden (Rapilly 1991) associated with the ability of an isolate to resist the inhibition from the filtering film.

3.4 Relationship between Sensitivity to the nUV-filtering Films and Aggressiveness of *B. cinerea* to Tomato

To evaluate the aggressiveness of *Botrytis* isolates in a situation comparable to conditions of disease development in greenhouse production, we quantified their ability to colonise pruning wounds on tomato plants. A known amount of inoculum was deposited on the cut end of freshly pruned petioles and we measured the size of the lesion that developed from the point of inoculation on a daily basis. In a first step, we compared an isolate whose sporulation was little inhibited by the nUV-filtering film to the reference strain of the laboratory, which is known to be aggressive on tomato and whose sporulation is strongly inhibited by the nUV-filtering film. The disease progress curves showed marked differences between the isolates

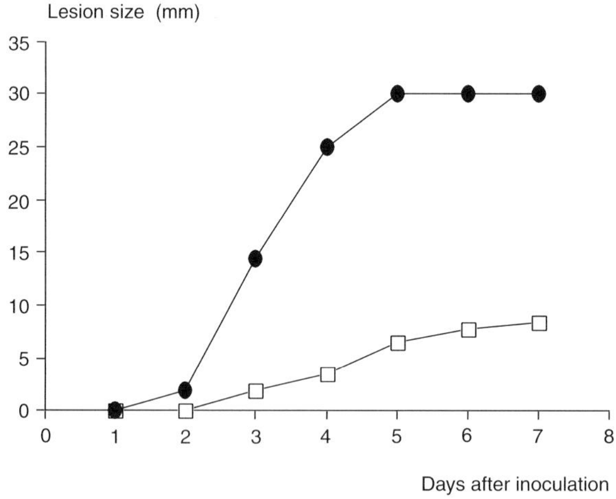

Fig. 5. Evolution of lesions caused by two isolates of *Botrytis cinerea* on pruning wounds on tomato plants. (●) reference isolate, whose sporulation is strongly inhibited by a UV-filtering film; (❑) isolate whose sporulation is little inhibited by the UV-filtering film.

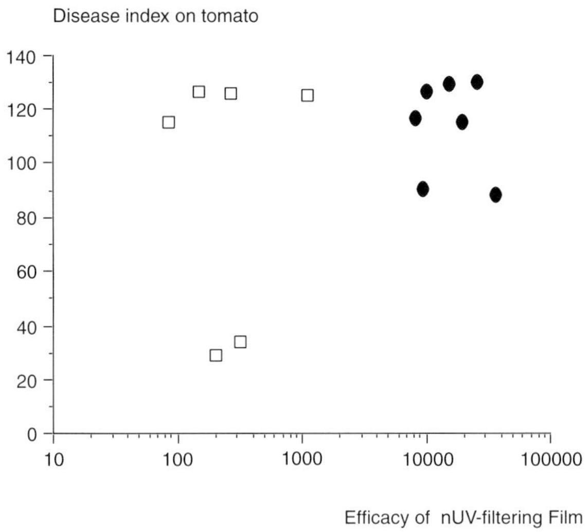

Fig. 6. Relationship between the aggressiveness on tomato (measured as the area under the disease progress curve) of *Botrytis cinerea* isolates and their sensitivity to inhibition of sporulation by a nUV-filtering film (measured by the ratio of the amounts of spores produced under a control film transparent to UV radiation and a nUV-filtering film). (●) isolates whose sporulation is strongly inhibited by a nUV-filtering film; (❑) isolates whose sporulation is little inhibited by the nUV-filtering film.

(Fig. 5). The isolate whose sporulation was little inhibited by the nUV-filtering film was very weakly aggressive compared to the reference isolate.
The same tests were repeated with a greater number of isolates of each class. To easily compare the aggressiveness of these isolates, a disease index was computed for each one based on the area under the disease progress curve. This index varied from 88.4 to 130 for isolates sensitive to the inhibition of the nUV-filtering film, with a group average of 110.7, and from 29.7 to 126, with a group average of 92.7, for isolates whose sporulation was little inhibited by the nUV-filtering film (Fig. 6). No significant difference in aggressiveness was found between the two groups. No genetic burdens appear to restrain the aggressiveness of isolates less sensitive to the nUV-filtering film.

4 Perspectives of Utilisation of nUV-Filtering "anti-*Botrytis*" Films under Conditions of Commercial Greenhouse Production

While confirming the interest of nUV-filtering films for the control of grey mould in greenhouses, the results of the studies described above also suggest that a prudent approach is necessary, and demonstrate the need for additional knowledge on several points.

4.1 Transposition of Laboratory Results to a Greenhouse Situation

The studies conducted in our laboratory were realised in a growth chamber with artificial lighting. Such conditions may differ from natural sunlight in terms of the intensity of nUV radiation received m^{-2} at the level of the plants as well as in terms of the spectral composition of light (Thiel et al. 1996). Both parameters are potentially important factors for the phenomena under study. It is therefore indispensable to verify whether the results obtained in the laboratory could be generalised to the light situation in a greenhouse crop. A possible effect of photoperiod duration also remains to be evaluated.
Temperature is also a factor that might complicate the generalisation of laboratory results to greenhouse conditions. It was shown for several plant pathogenic fungi that the effect of light on sporulation can depend on air temperature (Leach 1967). In unheated greenhouses, a type of shelter used for an important fraction of greenhouse production in various parts of the world, differences between day and night temperatures can be high and can vary considerably during the growing season. In contrast, laboratory work on the efficiency of nUV-filtering films was conducted at nearly constant (18-22 °C) temperatures. The effect of temperature on the *Botrytis*-nUV filtering film interaction remains to be determined on a wider range of temperatures and in fluctuating conditions.

4.2 Epidemiological Impact of nUV-Filtering Films

We observed in the laboratory that isolates of *B. cinerea* are not all affected in the same manner by a nUV-filtering film, and that one may fear progressive adaptation of the fungus to the absence of nUV light. This situation may lead to a decrease in the efficiency of nUV-filtering films for the control of grey mould in the greenhouse. The evaluation of this risk would require a precise quantification of the impact of reduced sporulation on the development of a grey mould epidemic in the greenhouse. A similar study was conducted by Köhl and colleagues in onion plots in field conditions (Köhl et al. 1995). These authors demonstrated that a 50% reduction in the air spores of *B. cinerea* and *B. squamosa* above onion plots obtained by removing diseased tissue resulted in significant delay in the development of disease. Among the *B. cinerea* isolates tested in our study, those least sensitive to the inhibitory effect of the nUV-filtering films had their sporulation reduced approximately tenfold under the filtering film (Fig. 3 and 4). If the development of epidemics for this fungus is comparable in onion fields and in greenhouse crops, it may be expected that using nUV-filtering films would result in substantial disease reduction, even in the presence of less sensitive isolates.
Cultural methods can also influence the efficiency of nUV-filtering films in controlling grey mould. One important issue is aeration in the greenhouses. In cold conditions, when grey mould problems are most severe in the greenhouses, the air is renewed to evacuate water vapour produced by the plants and limit the buildup

of excess humidity, when the outside conditions are favourable. In most cases, air renewal is achieved passively by opening the sides of the shelters during the warmest hours of the day. This action may have several implications for the efficiency of nUV-filtering films. It may allow the intermittent intrusion of unfiltered solar light through the openings. It may also allow the entry of spores from the outside environment, which would contribute to increasing the total inoculum load inside the shelter. Knowledge of the impact of these two phenomena on the amount of inoculum present inside the greenhouses must be improved.

4.3 Potential Effect of "anti-*Botrytis*" nUV-Filtering Films on the Crops

Other questions must be answered to evaluate the prospect of using nUV-filtering films to cover commercial greenhouses. While the detrimental effects of UV radiation on plants have been extensively studied (see for example the reviews by Britt 1996; Manning and Tiedermann 1995; Murthy and Rajagopal 1995), little is known on the possible effect of the absence of nUV light on cultivated plants and the quality of their agricultural products. Beneficial effects of UV light have been reported for several plant species (Hashimoto et al. 1993) and several studies demonstrated that blue light can also influence plant morphogenesis (Kishima et al. 1995). Therefore, the possible effect of nUV-filtering films on greenhouse-grown crops merits particular attention.

4.4 Effects on Useful Insects

Another important issue is the possible impact of nUV-filtering films on pollinating and other useful insects. For several vegetable greenhouse crops, such as tomato, pollination is a key factor for yield and produce quality. Pollination can be achieved mechanically, using hand-held vibrators applied to the flowers, or it can be accomplished by pollinating insects. Over the years, growers of greenhouse vegetables have relied increasingly on bumbles bees to carry out pollination. As the spectrum of light perceived by bees and other pollinating hymenoptera include UV radiation (Kevan 1983), the use of nUV-filtering films might affect their behaviour and disturb their pollination activity. Japanese researchers have compared the behaviour of wild bees inside structures covered with UV-filtering and with control films (Tezuka and Maeta 1993). The changes in behaviour observed under the UV-filtering film were different depending on the species. Under the UV-filtering film, individuals of the species *Ceratina japonica* did not go outside their nest and individuals of *Megachile rotundata* that did go out appeared disoriented and often unable to return to their nest. The third species studied, *Plebeia droryana*, was less affected by the UV-filtering film, but the life of the hive was nevertheless disturbed. Among the key foraging activities of the bees, nectar and resin gathering were particularly disturbed, while pollen collection was hardly affected (Tezuka and Maeta 1993).

The results of this study demonstrate the need for additional knowledge on the effect of nUV-filtering films on insects involved in the pollination of greenhouse crops.

5 Conclusions

Light-filtering films have shown great promise for the control of *B. cinerea*, as they have for several other diseases or pests of greenhouse vegetables (Vakalounakis 1991; Reuveni and Raviv 1997). However, several points remain to be elucidated to evaluate precisely the potential success in using nUV-filtering films for the control of grey mould of greenhouse crops. Studies conducted so far suggest that it is unlikely that such films would completely eliminate air-borne inocula of *B. cinerea* in the greenhouses, or that they could be used as the sole control method against this pathogen. However, it seems reasonable to expect that their efficiency should achieve sufficient reduction in inoculum pressure on greenhouse crops to justify their utilisation in conjunction with other methods, as component of integrated control strategies against grey mould (Nicot and Baille 1996). Reliance on a modified light spectrum could even go beyond the domain of production in plastic-covered greenhouses, and prove useful for glasshouse production. Indeed, certain qualities of glass transmit a fraction of the radiation with wavelengths between 300 and 400 nm (Hite 1973), which appears sufficient to stimulate sporulation of *B. cinerea* in glasshouses. One could imagine the manufacture of glass specifically treated to block nUV light and have high transmittance of blue light, which would have the same qualities as "anti-*Botrytis*" nUV-filtering polyethylene films.

References

Britt A.B., (1996). DNA damage and repair in plant. Ann. Rev. Plant Physiol. Plant Mol. Biol. 47: 75-100.

Faretra F., Antonacci E., Pollastro S., (1988). Sexual behavior and mating system of *Botryotinia fuckeliana*, teleomorph of *Botrytis cinerea*. J. Gen. Microbiol. 134:2543-2550.

Hashimoto T., Kondo N., Tezuka T., (1993). Harmful and beneficial effects of solar UV light on plant growth. Pp. 551-554 in A. Shima et al (eds.) Frontiers of Photobiology, Elsevier Science Publishers, New York.

Hite R.E., (1973). The effect of irradiation on the growth and asexual reproduction of *Botrytis cinerea*. Plant Dis. Rep. 57:131-135.

Jarvis W.R., (1992). Managing Diseases in Greenhouse Crops. American Phytopathological Society, St Paul, MN, 288 p.

Kevan P.G., (1983). Floral colors through the insect eye: what they are and what they mean, pp. 3-30 in C.E. Jones et R.J. Little (eds.) Handbook of Experimental Pollination Biology, Scientific and Academic Editions, New York.

Kishima Y., Shimaya A., Adachi T., (1995). Evidence that blue light induces betalain pigmentation in *Portulaca callus*. Plant Cell Tissue and Organ Culture 43:67-70.

Köhl J., Molhoek W.M.L., van der Plas C.H., Fokkema N.J., (1995). Suppression of sporulation of *Botrytis* spp. as a valid biocontrol strategy. Eur. J. Plant Pathol. 101:251-259.

Leach C.M., (1967). Interaction of near ultraviolet light and temperature on sporulation of the fungi *Alternaria,Cercosporella*, *Fusarium*, *Helminthosporium* and *Stemphyllium*. Can. J. Bot. 45:1999-2016.

Levis C., Fortini D., Brygoo Y., (1997). Flipper, a mobile Fot1-like transposable element in *Botrytis cinerea*. Mol. Gen. Genet. 254:674-680.

Manning W.J., Tiedemann A.V., (1995). Climate change: potential effects of increased atmospheric carbon dioxide (CO_2), ozone (O_3) and ultraviolet-B (UV-B) radiation on plant diseases. Environmental Pollution 88: 219-245.

Murth, S.D.S, Rajagopal S., (1995). UV-B radiation induced alterations in the bioenergetic processes of photosynthesis. Photosynthetica 31: 481-487.

Nicot P.C., Baille A., (1996). Integrated control of *Botrytis cinerea* on greenhouse tomatoes. pp. 169-189 in C.E. Morris, P.C. Nicot and C. NguyenThe (eds.) Aerial Plant Surface Microbiology, Plenum Press, New York.

Nicot P.C., Mermier (M.), Vaissière B.E.), Lagier J., (1996). Differential spore production by *Botrytis cinerea* on agar medium and plant tissue under near-ultraviolet light-absorbing polyethylene film. Plant Dis. 80:555-558.

Rapilly (F.), (1991).L'épidémiologie en Pathologie Végétale: mycoses aériennes, Série "Mieux Comprendre", INRA, Paris, 317 p.

Reuveni R, Raviv M., (1992). The effect of spectrally-modified polyethylene films on the development of *Botrytis cinerea* in greenhouse-grown tomato plants. Biol. Agric. and Hort. 9:77-86.

Reuveni R., Raviv M., (1997). Control of downy mildew in greenhouse grown cucumbers using blue photoselective polyethylene sheets. Plant Dis. 81:999-1004.

Salinas J., Schot C.P., (1987). Morphological and Physiological aspects of *Botrytis cinerea*, Medelingen van de Faculteit Landbouwwetenschappen Rijksuniversiteit Gent 52:771-776.

Sasaki T., Honda Y., Umekawa M., Nemoto M., (1985). Control of certain diseases of greenhouse vegetables with ultraviolet-absorbing vinyl film. Plant Dis. 69:530-533.

Tan K.K., (1974). Blue-light inhibition of sporulation in *Botrytis cinerea*. J. Gen. Microbiol. 82:191-200.

Tezuka T. Maeta Y, (1993). Effect of UVA film on extranidal activities of three species of bees. Jpn. J. Appl. Entomol. Zool. 37:175-180.

Thiel S., Döhring T., Köfferlein M., Kosak A., Martin P., Seidlitz H.K., (1996). A phytotron for plant stress research: how far can artificial lighting compare to natural lighting? J. Plant Physiol. 148:456-463.

Vakalounakis D.J., (1991). Control of early blight of greenhouse tomato, caused by *Alternaria solani*, by inhibiting sporulation with ultraviolet-absorbing vinyl film. Plant Dis. 75:795-797.

Van der Vlugt-Bergmans C.J.B., Brandwagt B.F., Van't Klooster J.M., Wagemakers C.A.M., Van Kan J.A.L., (1993). Genetic variation and segregation of DNA polymorphisms in *Botrytis cinerea*. Mycol. Res. 97:1193-1200.

Photochemical Treatment for Control of Post-Harvest Diseases in Horticultural Crops

Joseph ARUL, Julien MERCIER, Marie-Thérèse CHARLES, Mebarek BAKA and Rohanie MAHARAJ

1 Introduction

Fruits and vegetable add flavor and variety to the human diet and supply essential nutrients such as vitamins and minerals. They are also a source of complex carbohydrates and antioxidants which are important to human health. Increasing awareness of the links between diet and health has resulted in greater consumption of fruits and vegetables. Consumers are also increasingly concerned about the safety of the food they eat and the public demand foods that are free of microbial toxins, pathogens, and chemical residues. Safer and more efficient preservation techniques must be developed to meet these demands.

Fruits and vegetables are highly perishable and maintain an active metabolism in the postharvest phase. The major factors that reduce the storage life of fresh produce are fungal infection, senescence, and transpiration. Storage diseases, especially those caused by fungal pathogens, are responsible for substantial postharvest losses. Factors which accelerate senescence and favor microbial growth, such as physiological and mechanical injuries, as well as exposure to undesirable storage conditions (high ambient temperature and high humidity), can promote postharvest decay. Postharvest deterioration affects the cost and availability of the produce to the consumer and the ability of the producers to service distant markets.

The horticultural industry is undergoing new developments in mechanical harvesting and bulk handling. Unfortunately, these developments sometimes aggravate the incidence of postharvest diseases because of increased wounding and creation of environments favorable for disease proliferation. Mechanical harvesting is responsible for a high incidence of wounding, which facilitates invasion by pathogens. This necessitates prompt refrigeration and treatment with fungicides to reduce losses due to decay. In the context of public pressure to reduce the use of fungicides, there is an urgent need to develop alternative methods to control storage diseases. From the standpoint of producers, there is an urgent need to develop alternatives since they are facing possible deregistration of effective fungicides such as benomyl, and fungicide-tolerant strains of postharvest pathogens are evolving (McDonald et al. 1979; Eckert and Wild 1981; Spotts and Cervantes 1986). These circumstances have stimulated the development of new approaches to combating fungal pathogens. They include: activation of the natural defense mechanisms of the harvested produce, use of natural biocides, the use of biological antagonists and genetic transformation.

2 Natural Defense Mechanisms of the Host and Activation

The intrinsic and variable resistance of fruits and vegetables to postharvest diseases might be associated with one or a combination of properties of the host. Disease resistance of the crop may include both constitutive and inducible mechanisms. The constitutive resistance mechanism involves structural barriers (wax, cuticule, epiderm) and preformed inhibitors. The inducible defenses, that are triggered upon infection, include: (1) biochemical structural barriers such as lignification, suberization, and callose formation; (2) activation of antifungal hydrolases such as chitinases, β-1,3-glucanases, and peroxidases; (3) inhibitors of plant cell wall degrading enzymes and toxins secreted by the pathogen; (4) synthesis of proteinase inhibitors; and (5) hypersensitive cell death and phytoalexins (Bailey and Deverall 1983; Paxton and Groth 1994; Boller 1987; Bowles 1990). Among them, preformed compounds and phytoalexins have received greater attention. There are numerous reviews on preformed inhibitors and phytoalexins and their role in plant immunity to diseases (Keen 1981; Bailey and Mansfield 1982; Stoessl 1983; Darvill and Albersheim 1984).

The expression of resistance in harvested crops appears to be intimately linked to the ripening and senescence process. Fruits and vegetables become vulnerable to diseases as senescence advances due to a decreased ability of the tissues to synthesize active and preformed inhibitors. Preformed inhibitors provide resistance to young tissue but they degrade with age (Goodliffe and Heale 1977; Davies 1977; Dennis 1977; Kozukue et al. 1994). The degradation of preformed inhibitors appears to be responsible for the resumption of latent infections in many crops (Eckert and Ratnayake 1983). Thus, maintaining constitutive and inducible defense responses of the host by slowing down senescence could be of significance in reducing postharvest decay. This can be accomplished by manipulating storage conditions such as temperature, humidity, and gas composition of the atmosphere. The ideal postharvest environment should maintain the resistance of fruits and vegetables by delaying senescence and preventing invasion by microorganisms (Eckert 1978).

Phytoalexins are also antimicrobial compounds but they are synthesized de novo and accumulate in some plant tissues in response to microbial infection or treatment with an elicitor (Paxton 1981). At least 100 plant species representing 21 families have been shown to accumulate phytoalexins (Bailey and Mansfield 1982; Haard and Cody 1978, Vernenghi et al. 1987). In pathogen-plant interactions, there is a delay in the accumulation of phytoalexins which reduces the effectiveness of this defense mechanism. However, if fruits and vegetables are treated with an elicitor or elicitors which can trigger the accumulation of phytoalexins to inhibitory levels, stimulate hydrolytic enzymes and/or induce structural barriers long before the pathogen colonizes the tissue, the tissue would have gained a head start in fighting the infection. This possibility was originally tested in carrot slices, *Capsicum* fruit and mango fruit. Harding and Heale (1980) treated carrot slices with heat-killed conidia and culture fluids of *B. cinerea* and showed that the disease resistance increased with concomitant elicitation of the phytoalexin, 6-methoxymellein. Adikaram et al. (1988) were able to induce the accumulation of capsicannol phytoalexin and suppress *B. cinerea* rot development by treating *Capsicum* fruits with glucan elicitors. Mango fruits were less susceptible to infection when they were treated with *C. gloeosporioides* cell wall hydrolysate (Boulet et al. 1989). The possibility of inducing systemic disease resistance was shown recently in stored carrots preinoculated with *B. cinerea* (Mercier and Arul 1993). Although it is inconceivable to preinoculate postharvest crops with pathogens, it would still be possible to treat crops with nonpathogens, attenuated strains or antagonists. The possibility of controlling postharvest diseases by chitosan-induced defense reactions in the plant tissue was demonstrated by El Ghaouth et al. (1992a,b;1994;1997). They observed production of enzymes that hydrolyze fungal cell walls and the formation of physical barriers. Pre-storage heat treatment also shows potential for protecting crops against pathological disorders, presumably by intensifying natural resistance to infection (Spotts and Chen 1987; Kim et al. 1991; Fallik et al. 1993; Ben-Yehoshua et al. 1988).

Increased resistance to diseases through treatment with UV light was observed in a variety of crops: walla-walla onions (Lu et al. 1987); carrot slices and roots (Mercier et al. 1990; 1993a,b); sweet potatoes (Stevens et al. 1990); pepper and strawberry fruits (Arul et al. 1992); kumquat and citrus fruits (Ben-Yehoshua et al. 1992; Rodov et al. 1992); grapefruits (Droby et al. 1993); peaches (Lu et al. 1993); and tomato fruits (Liu et al. 1993; Charles et al. 1996). These works suggest that control of postharvest diseases through pre-storage treatment with elicitors is feasible and may offer a new strategy for disease control. A variety of biological and physical agents can be used to elicit disease resistance responses in harvested crops. They include: fungal wall fragments, antagonists, attenuated strains, natural compounds, heat treatment, and UV light. In the following, we discuss the physiological basis of using UV light in controlling diseases and the potential of this approach.

3 Activation of Disease Resistance by UV Light

The action spectrum of UV-C ranges from 220-280 nm, In this range, radiation is absorbed primarily by DNA but also by RNA, proteins, and unsaturated fatty acids. Since the latter are relatively abundant in the cell compared to DNA, they are less susceptible at low doses of UV; but at higher doses, inactivation of DNA, RNA and proteins, and membrane damage can occur. At low doses, DNA is more susceptible and can cause a number of photochemical lesions. This can lead to derepression of the genes controlling synthesis of phytoalexins and other secondary metabolites, thereby stimulating beneficial reactions in the tissue (Hadwiger and Schwochau 1971a,b). Thus, UV dose-response relationships can therefore be biphasic, such that high doses are detrimental and lower doses stimulate beneficial reactions. This phenomenon is known as hormesis, and the beneficial doses are termed hormic doses.

Low doses of UV light are known to elicit phytoalexin compounds in many plant bodies: pisatin in pea tissue (Hadwiger and Schwochau 1971b); hydroxyphaseollin in soybean hypocotyls (Bridge and Klarman 1973); kievitone in cowpea hypocotyls (Munn and Drysdale 1975); resveratol in grape vines (Langeake and Price 1977); phaseollin and phaseollin isoflavan in French bean hypocotyls (Andebrhan and Wood 1980); stilbenes in callus of peanut (Fritzemeier et al. 1993); and oryzalexins and momilactones in rice leaves (Kodoma et al. 1988). UV light activates a number of biosynthetic pathways and several key enzymes. Chief among them are phenylalanine ammonia lyase (PAL), 4-coumarate CoA ligase (CL) and chalcone synthase (CHS), catalysts for the synthesis of a number of chemical classes which include phenylpropanoids, coumarins and flavonoids (Chappel and Hahlbrock 1984). UV light also triggers the synthesis of pathogenesis-related proteins in tobacco and tomato leaves (Brederode et al. 1991; Christ and Monsinger 1989). UV light induces the full spectrum of known acidic and basic defense proteins and induced resistance to infection in tobacco (Yalpani et al. 1994). Although there is enough evidence to suggest that UV light is an elicitor of phytoalexins and antifungal hydrolases, the elicitation of defense reactions in postharvest crops by UV light is somewhat sketchy.

4 Disease Resistance and Elicitation of Defense Reactions in Post-Harvest Crops by UV Light

UV radiation exhibits a hormic effect in disease resistance. Figure 1 illustrates the effect of various UV doses on the disease resistance of whole carrot roots to *Botrytis cinerea*. The hormic or optimum dose for maximum disease resistance for carrots roots was 0.88 Mergs cm^{-2} (Mercier et al. 1993b). The carrot roots treated with a hormic dose of UV were able to resist invasion by *B. cinerea* when they were challenged with the fungus (Fig. 2). The optimum doses of UV light for the control of postharvest diseases are reported to be in the range of 0.2-7.5 Mergs cm^{-2}, and appear to depend on the type of produce, maturity, and picking season. Beyond the optimum dose, UV radiation results in tissue damage and increased susceptibility to diseases (Stevens et al. 1990; Lu et al. 1991; Rodov et al. 1992; Arul et al. 1992; Droby et al. 1993; Liu et al. 1993; Mercier et al. 1993a; Baka et al. 1999; Mercier et al. 2000a). The disease resistance induced by UV light appears to involve accumulation of antimicrobial secondary metabolites (phytoalexins and phenolic compounds), stimulation of antifungal hydrolases (chitinases and glucanases), and induction of biochemical (lignification and suberization) and physical barriers.

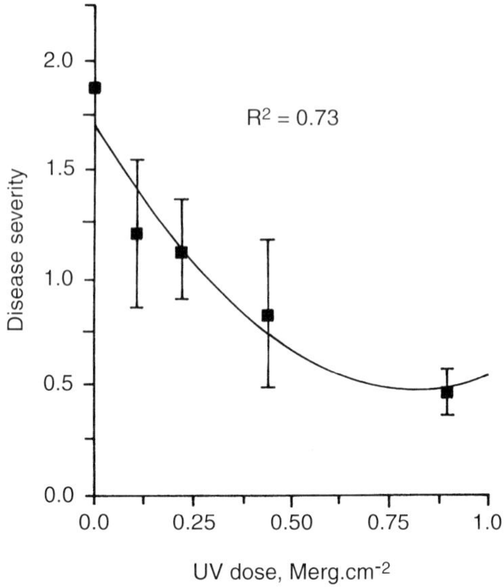

Fig. 1. Effect of UV on the susceptibility of carrots to *B. cinerea.* Carrots were inoculated 25 days after treatment and the infection was assessed 48 days after infection.

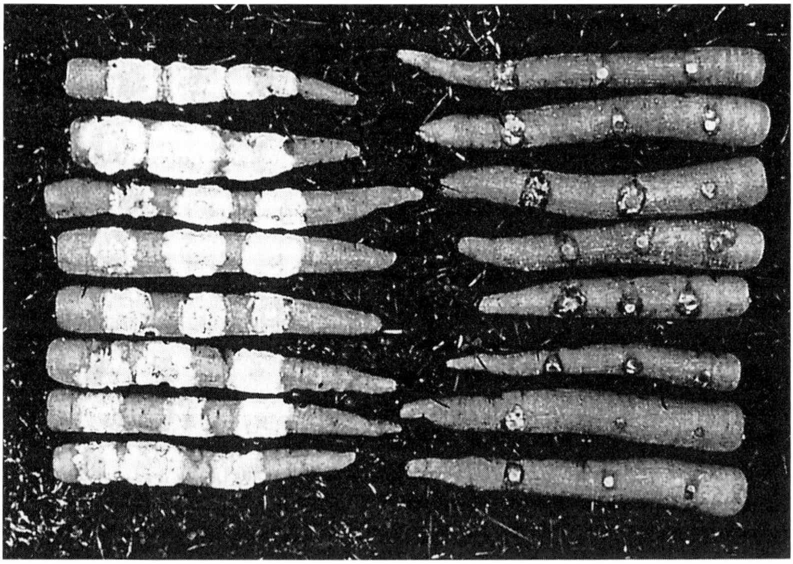

Fig. 2. Induction of resistance to *B. cinerea* in carrot by UV treatment. The infection was assessed 25 days after inoculation; control (*left*), UV (*right*) (Mercier et al. 1993b).

Accumulation of phytoalexins appears to be an early defense reaction of the plant tissue in response to UV. Mercier et al. (1993a,b) showed that UV-treated carrot slices were resistant to *Botrytis cinerea* and *Sclerotinia sclerotiorum*, and UV-treated whole carrot roots to *B. cinerea* with concomitant accumulation of 6-methoxymellein, a principal phytoalexin in carrots (Table 1, Fig. 3). Ben-Yehoshua et al. (1992) found an inverse relation between scoparone (phytoalexin) concentration and fruit decay in citrus fruits. The accumulation of 6-methoxymellein in carrot roots stored at 1 °C, 25 days after treatments, was 298.9 μg g^{-1} tissue (Mercier et al. 2000b). The maximum level of accumulation of rishitin (phytoalexin) in tomato fruits stored at 13 °C, was only modest (47.5 μg g^{-1} tissue), and the maximum was achieved 15 days after UV treatment (Fig. 4; Charles et al. 2000a). However, phytoalexin accumulation in response to infection was quite the opposite in these two systems. It is well established that phytoalexins are accumulated in response to infection but the accumulation is often delayed in plant-pathogen interactions. UV-treated carrots and tomato fruits responded to challenge inoculation with *B. cinerea*, as did the untreated produce. The accumulation of rishitin in UV-treated tomato fruit in response to infection was swift. The increase was about three times as that in the control fruits (108.1 μg g^{-1} freshwt), 10 days after inoculation. Beyond this period, the level of rishitin declined rapidly (Charles et al. 2000a). On the other hand, the increase in 6-methoxymellein in UV-treated carrots was similar to that in the control (Arul et al. 2000). It appears that phytoalexin response to UV treatment may be different from one type of produce to another.

The maximum resistance induced by UV treatment expressed after a certain lag period suggests that biochemical processes are involved. The lag depends on the type of produce and storage temperature. Carrot slices inoculated immediately after UV treatment exhibited resistance to *B. cinerea* 7 days later at 4 °C. The resistance was lower at 1 °C (Mercier et al. 1993a). Mercier et al. (1993b) observed significant resistance to *B. cinerea* by carrot roots stored at 1 °C, 25 days after UV treatment. UV-treated tomato fruits stored at 13 °C were more susceptible to *B. cinerea* in the first 48 h than were control fruits. The resistance increased 3 days after treatment and the maximum resistance was observed 15 days after UV treatment, corresponding to the maximum accumulation of rishitin in the tissue (Fig. 4;

Table 1. Effects of UV treatment and temperature on the accumulation of 6-methoxymellein (6-MM) after 2 weeks of storage and inhibition of fungal growth in carrot slices (Mercier at al. 1993a).

Pathogen	Storage temperature (°C)	6-MM level μg g^{-1} fresh weight	Inhibition (%)
B. cinerea	1	36.4	62.8
	4	78.5	80.8
S. sclerotiorum	1	36.4	64.5
	4	78.5	72.0

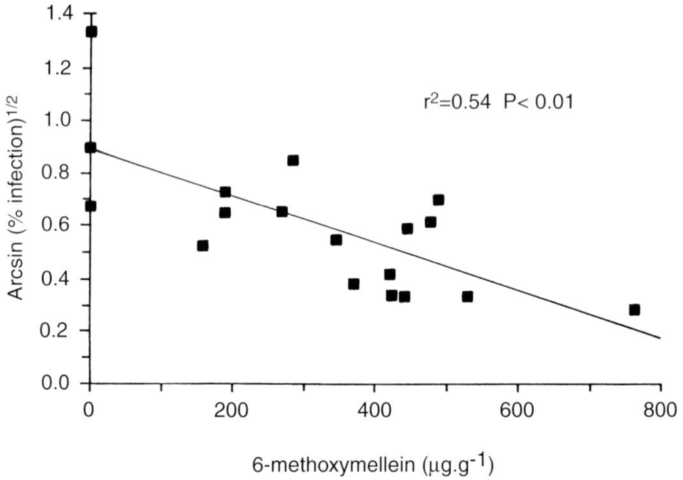

Fig. 3. Relationship between the levels of 6-methoxymellein induced by UV and the susceptibility of freshly harvested carrots to *B. cinerea*. 6-methoxymellein assay and inoculation were performed 25 days after treatment, and the infection was assessed 48 days after inoculation.

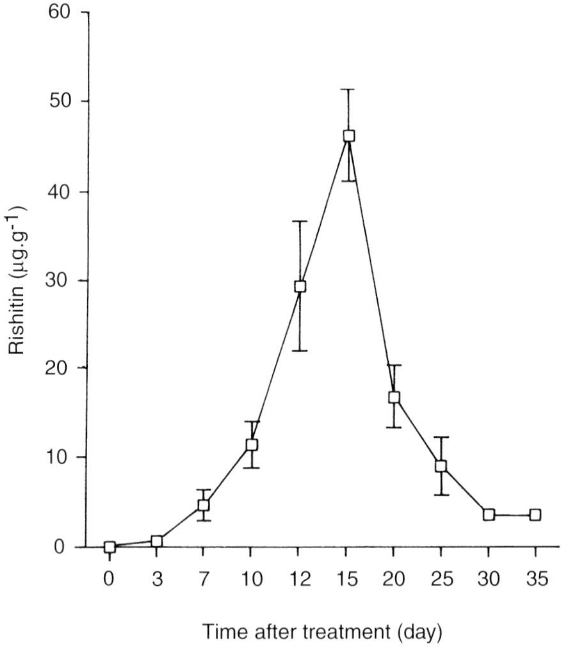

Fig. 4. Accumulation of rishitin in response to UV treatment in tomato fruits during storage (Charles et al. 2000a).

Charles et al. 2000a). Maximum resistance to *Penicillium digitatum* in citrus fruits was observed between 2 and 8 days after UV treatment. This also corresponded to the accumulation of scoparone (Ben-Yehoshua et al. 1992). A delay of 48-72 h in the resistance response was also observed in grapefruits (Droby et al. 1993) and in peaches (Lu et al. 1993). Maximum protection against black rot occurred 1 to 7 days after treatment in sweet potatoes (Stevens et al. 1990). The delay in the expression of disease resistance appears to be influenced by the storage temperature after treatment. This is to be expected, since temperature influences biochemical processes, including phytoalexin synthesis. The accumulation of 6-methoxymellein and disease resistance of carrot slices were higher at 4 °C than at 1 °C (Table 1). Droby et al. (1993) observed that UV-treated grapefruits stored at 6 °C did not exhibit any resistance compared to the fruits stored at temperatures from 11 to 20 °C.
The responsiveness of postharvest tissues to UV light decreases with physiological age (Mercier et al. 1993b; Liu et al. 1993). The phytoalexin response and resistance to *B. cinerea* of aged carrots (stored for 4 months) to UV dose was lower than that of fresh carrots (Figs. 5, 6). Creasy and Coffee (1988) have also shown that the phytoalexin production potential of grape berries decreased with maturity. While the coding regions of the nuclear DNA are known to remain intact during senescence (Wittenback 1977; Venkatarayappa et al. 1984), Markrides and Goldthwaith (1981) showed that the content of RNA and proteins of bean leaves decreased with age. Thus, even if DNA is derepressed by UV, the amounts of mRNA and proteins may not be adequate for the expression of the response.
It appears that the induced resistance and the accumulation of phytoalexins by UV light is a localized rather than systemic reaction. Based on the limited number of types of produce studied so far, it is very likely that UV induced reactions are loca-

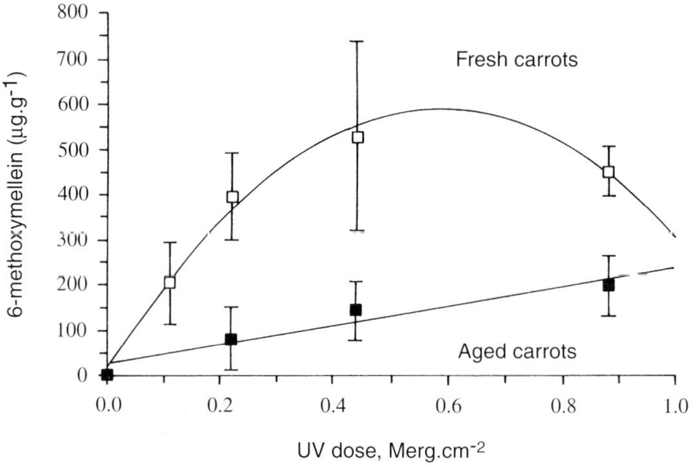

Fig. 5. Effect of UV irradiation on the accumulation of 6-methoxymellein in the peel of fresh and aged (stored for 4 months) carrots in 25 days of storage at 1 °C (Mercier et al. 1993b).

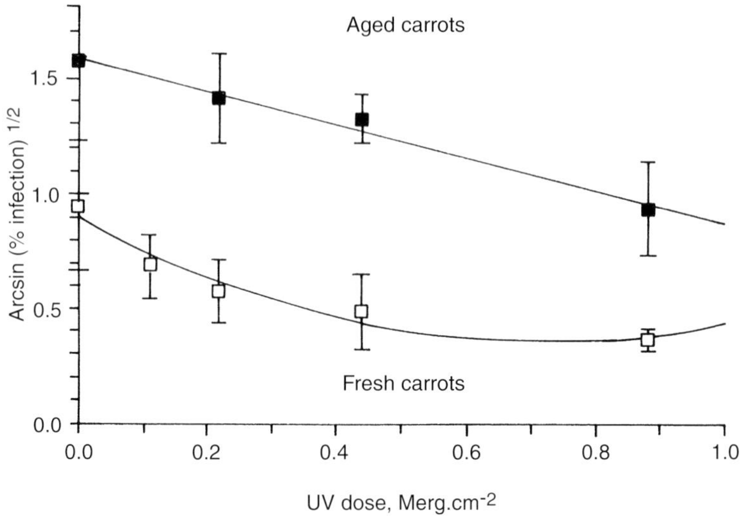

Fig. 6. Effect of UV on the susceptibility of fresh and aged carrots to *B. cinerea*. The carrots were treated with UV and stored at 1 °C. After 25 days the roots were inoculated with mycelial plugs. The infection was assessed after 48 and 25 days for freshly harvested and aged (stored for 4 months) carrots, respectively (Mercier et al. 1993b).

lized. Table 2 shows the accumulation of 6-methoxymellein and resistance to *B. cinerea* of carrot roots which were either totally or partially exposed. No induced resistance was observed when the roots were partially exposed. There could be differences in the ability of various cultivars to accumulate phytoalexins in response to UV treatment. Mercier et al. (1993c) reported that there were differences in the phytoalexin (6-methoxymellein) production potential among carrot cultivars. However, the varietal differences were rather small. The carbohydrate profiles of the cultivars did not affect their production of 6-methoxymellein (Table 3). Furthermore, the 6-methoxymellein response was similar for all the cultivars. Stevens et al. (1990) also observed differences in the induced resistance among sweet potato cultivars.

Table 2. Effect of total and partial exposure of carrot roots to UV light (Mercier et al. 2000b).

Treatment	6-methoxymellein, 25 days after treatment ($\mu g\ g^{-1}$ fresh wt)	Lesion diameter, 50 days after inoculation (mm)
Control	5.7	51.1
UV (0.88 Merg cm^{-2})		
Total surface exposed	298.9	12.1
Crown only	4.3	44.1
Lower half only	12.6	52.4

Table 3. Accumulation of 6-methoxymellein in response to UV and sugar profiles in carrot cultivars (Mercier et al. 1993a).

Cultivar	6-methoxymellein ($\mu g\ g^{-1}$fresh wt)	Sugars ($\mu g\ g^{-1}$ fresh wt) G+F+S[a]	Total non structural carbohydrates
Caropak	78.1	30.7	56.0
DessDan	57.5	26.8	53.6
SixPak II	63.4	38.8	63.1
Spartan Bonus	84.7	38.3	62.7
XPH 875	99.7	34.6	63.4

[a] Total glucose, fructose and sucrose.

It is known that UV radiation triggers the phenylpropanoid pathway in which the key enzyme is phenylalanine ammonia lyase (PAL) (Chappel and Hahlbrock 1984). Droby et al. (1993) observed a transient increase in PAL activity in the peel of UV-treated grapefruit, peaking 24 h after the treatment. Stimulation of this pathway can lead to the synthesis of phenolics, coumarins and flavonoids. Flavonoids strongly absorb UV light and are also antioxidants. Phenolics are antioxidants as well as antimicrobial agents, whereas furanocoumarins possess fungistatic properties. UV-treated tomatoes exhibited higher levels of total phenols at various stages of ripening (Table 4). It is possible that the accumulation of phenolics could also contribute to the resistance expressed by UV-treated produce. While phenolics are accumulated from the phenylpropanoid pathway, quite a number of phytoalexins such as 6-methoxymellein, rishitin, etc. derive from the acetate pathway (Bailey and Mansfield 1982).

In the few products studied in detail, UV treatment induces a gradual resistance to diseases; however, the persistance of this resistance is quite variable between products. It appears that the resistance does not persist for significant periods in citrus fruits and the tissues become more susceptible as the phytoalexins degrade (Ben-Yehoshua et al. 1992: Rodov et al. 1992; Droby et al. 1993). However, the resistance persists longer in produce such as tomato, even when the phytoalexins drop below inhibitory levels (Charles et al. 2000a). This suggests that other defense

Table 4. Effect of UV treatment on total phenols in tomato at different stages of maturity (Maharaj et al. 2000).

Treatment	Total phenols ($\mu g\ g^{-1}$ fresh wt)			
	Preclimacteric	Climacteric	Postclimacteric	Ripe
Control	9.9	5.3	17.3	23.0
UV	8.8	14.3	24.1	27.4

mechanisms are involved in the persistence of disease resistance in this crop. The resistance of UV treated tomato fruits may be partly a consequence of a limited delay in the ripening process (Maharaj et al. 1999; Liu et al. 1993).
Defense enzymes such as chitinases and β-1,3-glucanases are capable of hydrolyzing the fungal cell wall and can, therefore, play a role in disease resistance. This defense is generally considered to be induced in response to pathogen attack. Results indicate that these defense enzymes are constitutive in tomato fruits, bell peppers, and carrot roots (Baka et al. 2000; Charles et al. 2000b; Mercier et al. 2000b). Both freshly harvested tomatoes and carrots exhibited activities of chitinases and glucanases, the activities being stronger in carrots. However, these activities decreased with storage time. The effect of UV treatment on defense enzymes varied among produce types. UV treatment intensified the activities of basic chitinases and stimulated de novo basic glucanase activities in tomato fruits, but did not affect the activities of acidic proteins. On the other hand, UV treatment did not intensify or stimulate the activities of any of the enzymes in carrots and bell peppers. Interestingly, the activities of the defense enzymes were maintained for longer periods in the UV-treated produce than in the controls.
The prolonged resistance in UV-treated tomato was attributed to the formation of biochemical barriers (lignification and suberization) and ultrastructural changes in the epidermal tissue (Charles et al. 2000c, d). The biochemical barriers developed slowly and were evident 14 days after UV treatment when the phytoalexin, rishitin, started to degrade. In addition, these defenses were strongly activated when the fruits were infected (Fig. 7). The ultrastructure of the epidermal tissue to tomato was also modified by UV treatment (Charles et al. 2000d). The principal characteristic of this modification was the formation of a cell wall stacking zone (CWSZ) (Fig. 8). The CWSZ involved epidermal and subepidermal cells, and the cell walls appeared to be impregnated with lignin. These biochemical and physical barriers play an important role in blocking the progress of infection. On the other hand, the fungus rapidly overcame the natural defenses of untreated tomato fruits and colonized the tissue extensively (Fig. 8).

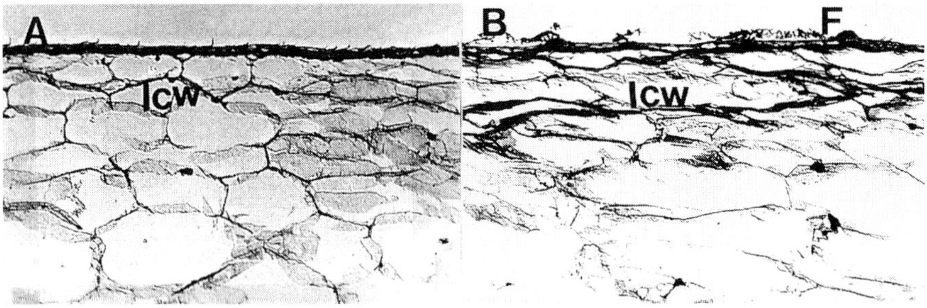

Fig. 7. A, B Thickening of the cell wall by lignification in UV-treated tomato. **A** : 14 days after treatment. **B** : 4 days after inoculation with *B. cinerea* (inoculation was performed 10 days after UV treatment). Maule reaction is more intense in the epidermal and subdermal cells. lcw Lignified cell wall; F fungus (Charles et al. 2000c).

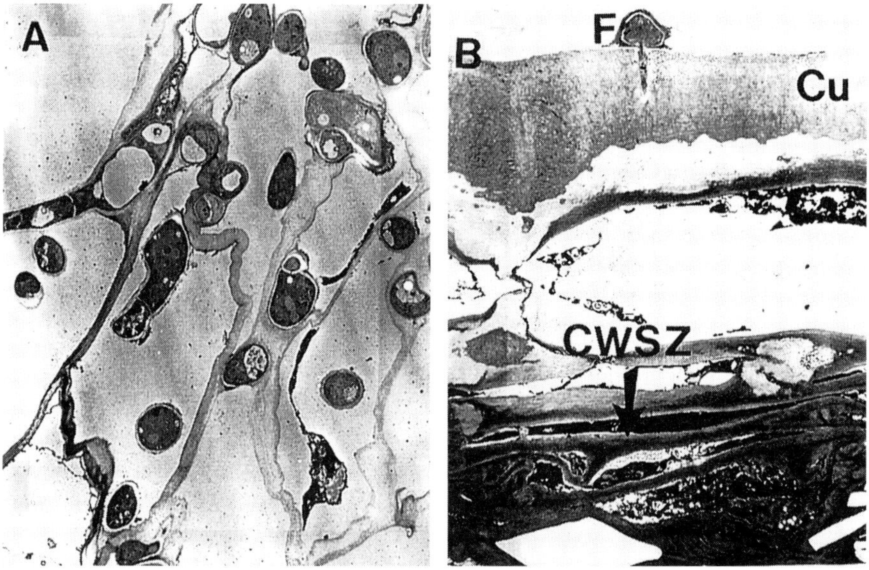

Fig. 8. A, B Transmission electron micrographs showing the colonization of tomato tissue. **A** Control. **B** A site of penetration on the cuticule of UV-treated tomato fruit by *B. cinerea* and the cell walls involved in CWSZ. Cu Cuticule; F fungus; CWSZ cell wall stacking zone (Charles et al. 2000d).

5 Potential of UV Treatment in the Control of Post-Harvest Diseases — A Critical Evaluation

It is evident that UV treatment induces disease resistance in several crops. UV stimulates several defense mechanisms and sensitizes the tissue, thereby giving it a head start in fighting the infection. However, its application in practice would depend eventually on whether UV treatment can provide disease control for as long as fungicides do. At this point, the duration of UV-induced resistance appears to be variable among crops. This is understandable because the intrinsic resistance of crops is also variable. When multiple defense mechanisms are expressed in response to UV, as in the case of tomato fruits, the protective effect persists for longer periods. On the other hand, when a single defense reaction is involved, as appears to be the case for citrus fruits which respond by accumulating phytoalexins, the protective effect may last only for short periods. This is because the phytoalexins degrade during ripening.

There are other concerns which could hamper the application of this technique. For example, red-brown discolorations in tomato fruits and orange-brown discolorations in carrots have been observed. Ben-Yehoshua et al. (1992) have also observed brown or bronze discolorations in UV-treated citrus fruits. These discolorations usually develop during storage after the treatment. This suggests a similarity between the symptoms of UV treatment and the hypersensitive reaction to infection which also provokes the synthesis of antifungal phytoalexins. Another aspect of discoloration is that it is quite variable. We observed that the intensity of discoloration also depends on the source of carrots, suggesting variability in the concentrations of photoreceptors and the compounds masking the photoreceptors. Although the chemical compounds involved in discoloration have not been identified, it is likely that they are flavonoids, particularly catechins, which take part in enzymatic browning. Furthermore, ultrastructural changes in the epidermis lead to changes in the light reflectance characteristics of the fruit surface, which can affect the appearance of the fruits.

It is also important to determine whether UV treatment results in the accumulation of undesirable secondary metabolites such as terpenoids and glycoalkaloids (e.g., α-solanine in potato tubers). However, this may not be of great concern because secondary metabolites degrade with ripening process (e.g., tomatine and rhishitin in tomato fruits). Secondary metabolites are generally accumulated in the peel and, in many cases, fruits and vegetables are peeled before consumption. The peeling operation alone could remove most of the secondary metabolites (Mercier et al. 1994). Furthermore, the fruits and vegetables can be conditioned to degrade the metabolites before releasing the produce to the market. Finally, the induction of disease resistance by UV light appears to be localized. This means that the entire surface of the fruit or vegetable has to be exposed to UV uniformly. Because of the irregular shape and size of fruits and vegetables, this may pose a practical problem.

In conclusion, UV treatment can be a means of controlling storage diseases for certain crops possessing several natural defenses. UV treatment can trigger these defense mechanisms and provide protection against disease for longer periods. In our opinion, UV treatment may not be a universal option because of the short periods of protection it offers to certain crops. UV treatment may appear to be an easy technology but it requires a significant understanding of the physiology of the crop, so that the resistance induced by UV can be enhanced, and the undesirable side effects such as discoloration can be diminished. More research is needed to establish the practical protective potential of UV.

References

Adikaram N.K.B., Brown A.E., Swinburne T.R., (1988). Phytoalexin induction as a factor in the protection of *Capsicum annuum* L. fruits against infection by *Botrytis cinerea* Pers. J. Phytopathology 122:267-273.

Andebrhan T., Wood R.K.S., (1980). The effects of ultraviolet radiation on the reaction of *Phaseolus vulgaris* to species of *Colletotrichum*. Physiol. Plant Pathol. 17:105-110.

Arul J., Mercier J., Baka M., Maharaj R., (1992). Photochemical therapy in the preservation of fresh fruits and vegetables: Disease resistance and delayed senescence. Proc. Int. Symp. Physiological Basis of Postharvest Technologies. Abstr., University of California, Davis, CA, p. 42.

Arul, J., Corcuff R., Roussel D., (2000). Unpublished data.

Baka M., Mercier J., Corcuff R., Castaigne F., Arul J., (1999). Photochemical treatment to improve storability of fresch strawberries. J. Food Sci. 64: 1068-1072.

Bailey J.A., Mansfield J.W., (1982). Phytoalexins. John Wiley and Sons, New York, 334 p.

Bailey J.A., Deverall B.J., (1983). The dynamics of host defense. Academic Press, North Ryde, Australia, 233 p.

Baka M., El Ghaouth A., Arul J., (2000). Unpublished data.

Ben-Yehoshua S., Shapiro B., Kim J.J., Sharoni J., Carmeli S., Kashman Y., (1988). Resistance of citrus fruit to pathogens and its enhancement by curing. Proc. 6th Int. Citrus Congr., R. Goren and K. Mendel (Ed.). Baladan Publ. Philadelphia, pp. 1371-1379.

Ben-Yehoshua S., Rodov V., Kim J.J., Carmeli S., (1992). Preformed and induced antifungal materials of citrus fruits in relation to the enhancement of decay resistance by heat and ultraviolet treatments. J. Agric. Food Chem. 40:1217-1221.

Boller T., (1987). Hydrolytic enzymes in plant disease resistance. In: Plant-Microbe Interactions, Vol. 2, T. Kosunge et E.W. Nester (Eds.), MacMillan, New York, 385 p.

Boulet M., Arul J., Verret P., Kane O., (1989). Induced resistance of stored mango (*Mangifera indica* L.) fruits to mold infection by treatment with *Colletotrichum gloeosporioides* L. cell-wall hydrolysate. Can. Inst. Food Sci. Technol. J. 22:161-164.

Bowles D.J., (1990). Defense-related proteins in higher plants. Annu. Rev. Biochem. 59:873-907.

Brederode F.T., Linthorst H.J.M., Bol J.F., (1991). Differential induction of acquired resistance and PR gene expression in tobaco by virus infection, ethephon treatment, UV light and wounding. Plant Mol. Biol. 17:1117-1125.

Bridge M.A., Klarman W.L., (1972). Soybean phytoalexin, hydroxyphaseollin, induced by ultraviolet irradiation. Phytopathology. 63:606-609.

Chappel J., Hahlbrock K., (1984). Transcription of plant defense genes in response to UV light or fungal elicitor. Nature 311:76-78.

Charles M.T., Benhamou N., Arul J., (2000a). Induction of resistance to gray mold and accumulation of the phytoalexin rishitin in postharvest tomato fruit by UV treatment. Phytopathology (in press).

Charles M.T., Asselin A., Ait-Barka E., Arul J., (2000b). Changes in protein profile of tomato fruits in response to UV light and *Botrytis cinerea* infection : Pathogenesis-related proteins. Can. J. Plant Pathol. (submitted).

Charles M.T., Goulet A., Arul J., (2000c). Biochemical barriers induced by UV light and *Botritys cinerea* in postharvest tomato fruit. Phytopathology (submitted).

Charles M.T., Benhamou N., Arul J., (2000d). Ultrastructural changes of tomato fruit tissues treated with UV light and inoculated with *Botrytis cinerea*. Phytopathology (submitted).

Cheema A.S., Haard N.F., (1978). Induction of rishitin and lubimin in potato tuber discs by non-specific elicitors and the influence of storage conditions. Physiol. Plant Pathol. 13 :233-240.

Christ U., Monsinger E., (1989). Pathogenesis-related proteins of tomato: I. induction by *Phytophthora infestans* and other biotic and abiotic inducers and correlations with resistance. Phys. Mol. Plant Pathol. 35 :53-65.

Creasy L.L., Coffee M., (1988). Phytoalexin production potential of grape berries. J. Amer. Soc. Hort. Sci. 113-230-234.

Darvill A.G., Albersheim P., (1984). Phytoalexins and their elicitors. A defense against microbial infection in plants. Ann. Rev. Plant. Physiol. 35:243-275.

Davis W.P, (1977). Infection of carrot roots in cold storage by *Centrospora acerini*. Ann. Appl. Biol. 85:163-168

Dennis C., (1977). Susceptibility of stored crops to microbial infection. Ann. Appl. Biol. 85:430-432.

Droby S., Chalutz E., Horev B., Cohen L., Gaba V., Wilson C.L., Wisniewski M., (1993). Factors affecting UV-induced resistance in grapefruit against the green mould decay caused by *Penicillium digitatum*. Plant Pathology 42:418-424.

Eckert J.W, (1978). Pathological diseases of fresh fruits and vegetables. In Postharvest Biology and Biotechnology, H.O. Hultin and M. Milner (Eds.). Food and Nutrition Press, Inc., Westport, Connecticut, pp. 161-209.

Eckert J.W., Wild B.L., (1981). Problems of fongicide resistance in *Penicillium* rot of citrus fruits. In: Pest Resistance to Pesticides: Challanges and Prospects, G.P.Georghiou and T. Saito (Eds.). Plenum Press, New York.

Eckert J.W., Ratnayake M., (1983). Host-pathogen interaction in postharvest diseases. In: Postharvest Physiology and Crop Protection, M. Lieberman (Ed.). Plenum Press, New York, pp. 247-264.

El Ghaouth A., Arul J., Grenier J., Asselin A., (1992a). Antifungal activity of chitosan on two postharvest pathogens of strawberry fruits. Phytopathology 82:398-492.

El Ghaouth A., Arul J., Asselin A., Benhamou N., (1992b). Antifungal activity of chitosan on post-harvest pathogens: Induction of morphological and cytological alteractions in *Rhizopus stolonifer*. Mycol. Res. 96:769-779.

El Ghaouth A., Arul J., Wilson C., Benhamou N., (1994). Ultrastructural and cytological aspects of the effect of chitosan on decay of bell pepper fruit. Physiol. Mol. Plant. Pathol. 44:417-432.

El Ghaouth A., Arul J., Wilson C., Benhamou N., (1997). Biochemical and cytochemical aspects of the interactions of chitosan and *Botrytis cinerea* in bell pepper fruit. Postharvest Biol. Technol. 12:183-194.

Fallik E., Klein J., Grinberg S., Lomaniec E., Lurie S., Lalazar A., (1993). Effect of postharvest heat treatment of tomato on fruit ripening and decay caused by *Botrytis cinerea*. Plant Dis. 77:985-988.

Fritzemeier, K.H., C.H. Rolfs, J. Pfan and H. Kindl 1983. Action of ultraviolet-C on stiebene formation in callus of *Arachis hypogaea*. Planta 159:25-29.

Goodliffe J.P., Heale J.B., (1977). Factors affecting the resistance of cold-stored carrots to *Botrytis cinerea*. Ann. Appl. Biol. 87:17-28.

Haard N.F., Cody M., (1978). Stress metabolites in postharvest fruits and vegetables. In: Postharvest Biology and Biochenology, H.O. Hultin and M. Milner (Eds.), and Nutrition Press, Westport, CT, pp. 111-135.

Hadwiger L.A., Schwochau M.E., (1971a). Specificity of deoxyribonucleic acid intercalating compounds in the control of phenylalanine ammonia-lyase and pisatin levels. Plant Physiol. 47:346-351.

Hadwiger L.A., Schwochau M.E., (1971b). Ultra-violet light-induced formation of pisatin and phenylalanine ammonia-lyase. Plant Physiol. 47:588-590.

Harding V.K., Heale J.B., (1980). Isolation and identification of the antifungal compounds in the induced resistance response of carrot root slices to *Botrytis cinerea*. Physiol. Plant Pathol. 17:277-289.

Kim J.J., Ben-Yehoshua S., Shapiro B., Henis Y., Carmeli S., (1991). Accumulation of scoparone in heat treated lemon fruit inoculated with *Penicillium digitatum* Sacc. Plant Physiol. 97:880-885.

Kodama O., Suzuki T., Miyakkawa J., Akatsuka T., (1988). Ultraviolet-induced accumulation of phytoalexins in rice leaves. Agric. Biol. Chem. 52:2469-2473.

Kozukue N., Kozukue E., Yamashita H., Fujii S., (1994). Alpha-tomatine purification and quantification in tomatoes by HPLC. J. Food Sci. 59:1211-1212.

Langcake P., Pryce R.J., (1977). The production of resveratol and viniferins by grapevines in response to ultra-violet irradiation. Phytochemistry 16:1193-1196.

Liu J., Stevens C., Khan V.A., Lu J.Y., Wilson C.L., Adeyeye O., Kabwe M.K., Pusey P.L., Chalutz E., Sultana T., Droby S., (1993). Application of ultraviolet-C light on storage ros and ripening of tomatoes. J. Food Prot. 56:868-873.

Lu J.Y., Stevens C., Yakubu P., Loretan P.A., (1987). Effects of gamma, electron beam and ultraviolet radiation on control of storage rots and quality of Walla Walla onions. J. Food Proc. Preserv. 12:53-62.

Lu J.Y., Stevens C., Khan V.A., Kabwe M., Wilson C.L., (1991). The effect of ultraviolet irradiation on shelf-life and ripening of peaches and apples. J. Food Quality 14:299-305.

Lu J.Y., Lukombo S.M., Stevens C., Khan V.A., Wilson C.L., Pusey P.L., Chalutz E., (1993). Low dose UV and gamma radiation on storage rot and physicochemical changes in peaches. J. Food Qual. 16:301-309.

Maharaj R., Arul J., Nadeau P., (1999). Effect of photochemical treatment in the preservation of fresh tomato *Lycopersicon esculentum* cv. Capello by delaying senescence. Postharvest Biol. Technol. 15: 13-23.

Maharaj R., Nadeau P., Arul J., (2000). Unpublished data.

Markrides S.C., Goldthwaith J., (1981). Biochemical changes during bean leaf growth, maturity and senescence. J. Exp. Bot. 32:725-735.

McDonald R.E., Risse L.A., Hillebrand B.M., (1979). Resistance to thiabendazole and benomyl of *Penicillium digitatum* and *P. italicum* isolated from citrus fruit from several countries. J. Amer. Soc. Hort. Sci. 104:333-337

Mercier J., Ponnampalam R., Arul J., (1990). Induction of phytoalexin production and disease resistance in carrot slices by UV light. Proc. Am. Can. Phytopathol. Soc., Abstr. East Lansing, MI, C18.

Mercier J., Arul J., (1993). Induction of systemic disease resistance in carrot roots by pre-incubation with storage pathogens. Can. J. Plant. Pathol. 15:281-283.

Mercier J., Arul J., Ponnampalam R., Boulet M., (1993a). Induction of 6-methoxymellein and resistance to storage pathogens in carrot slices by UV-C. J. Phytopathology 137;44-54.
Mercier J., Arul J., Julien C., (1993b). Effect of UV-C on phytoalexin accumulation and resistance to *Botrytis cinerea* in stored carrots. J. Phytopathol. 139:17-25.
Mercier J., Ponnampalam R., Bérard L.S., Arul J., (1993c). Polyacetylene content and UV-induced 6-methoxymellein accumulation in carrot cultivars. J. Sci. Food Agric. 63-313-317.
Mercier J., Arul J., Julien C., (1994). Effect of food preparation on the isocoumarin, 6-methoxymellein, content of UV-treated carrot. Food Res. Int. 27:401-404.
Mercier J., Baka M., Reddy B., Corcuff R., Arul J., (2000a). Short-wave ultraviolet irradiation for control of decay caused by *Botrytis cinerea* in bel pepper: induced resistance and germicidal effects. J. Amer. Soc. Hort. Sci. (in press).
Mercier J., Roussel D., Charles M.T., Arul J. (2000b). Systemic and local responses associated with UV -and pathogen- induced resistance to *Botrytis cinerea* in stored carrot. Phytopathology 90: 981-986.
Munn C.B., Drysdale R.B., (1975). Keivitone production and phenylalanine ammonia-lyase activity in Cowpea. Phytochemistry 14:1303-1307.
Paxton J.D, (1981). Phytoalexins - a working redefinition. Phytopathol. Z. 101-106-108.
Paxton J.D., Groth J., (1994). Constraints on pathogens attacking plant. Crit. Rev. Plant Sci. 13:77-95.
Rodov V., Ben-Yehoshua S., Kim J.J., Shapiro B., Ittah Y., (1992). Ultraviolet illumination induces scoparone production in kumquat and orange fruit and improves decay resistance. J. Amer. Soc. Hort. Sci. 117:788-792.
Spotts R.A., Cervantes L.L., (1986). Populations, pathogenecity, and benomyl resistance of *Botrytis* spp., *Penicillium* spp. and *Mucor piriformis* in packaging houses. Plant Dis. 70:106-108.
Spotts R.A., Chen P.M., (1987). Prestorage heat treatments for control of decay of pear fruit. Phytopathology 77:1578-1582.
Stoessl A., (1983). Secondary plant metabolites in preinfectional and postinfectional resistance. In: The Dynamics of Host Defence, J.A. Bailey and B.J. Deverall (Eds.), Academic Press, New York, pp. 71-122.
Stevens C., Khan V.A., Tang A.Y., Lu J.Y, (1990). The effect of ultraviolet radiation on mold rots and nutrients of stored sweet potatoes. J. Food Protect. 53-223-226.
Venkatarayappa T., Fletcher R.A., Thompson J.E., (1984). Retardation and reversal of senescence in bean leaves by benzyladenine and decapitation. Plant Cell Physiol. 25-407-418.
Vernenghi A., Ramiandrasoa F., Chuilon S., Ravise A., (1987). Phytoalexines des citrus. Fructas 42:103-108.
Wittenback V.A., (1977). Induced senescence of intact wheat seedlings and its reversibility. Plant Physiol. 59:1039-1042.
Yalpani N., Eneyedi A.J., Leon J., Raskin I., (1994). Ultraviolet light and ozone stimulate accumulation of salicylic acid, pathogenesis related proteins and virus resistance in tobacco. Planta 193:372-376.

Control of Insects in Post-Harvest: Radio Frequency and Microwave Heating

Francis FLEURAT-LESSARD

1 Introduction

The sanitary and health problems caused by the insects that regularly infest raw and processed plant products (e.g. cereals, pulses, dried fruits and vegetables, tobacco, spices, etc.), and that invade the processing plants are difficult to solve. Except for fumigants applied under certain conditions and with strict safety precautions, the use of chemical insecticides on processed products is forbidden. Preventive treatments in empty stores are improperly applied and often inefficient, at least at certain times of year. Only physical control methods can be used without restriction throughout the entire process from storage to marketing.

The search for applications of electrical energy in the physical control of insects dates back to the 19th century (d'Arsonval 1893). Among electromagnetic radiation (Fig. 1) microwaves and radio frequencies (RF) rapidly transfer energy in the form of heat in dielectric materials. Food materials with a low moisture content have dielectric properties that are favourable to rapid transfer of microwave or RF

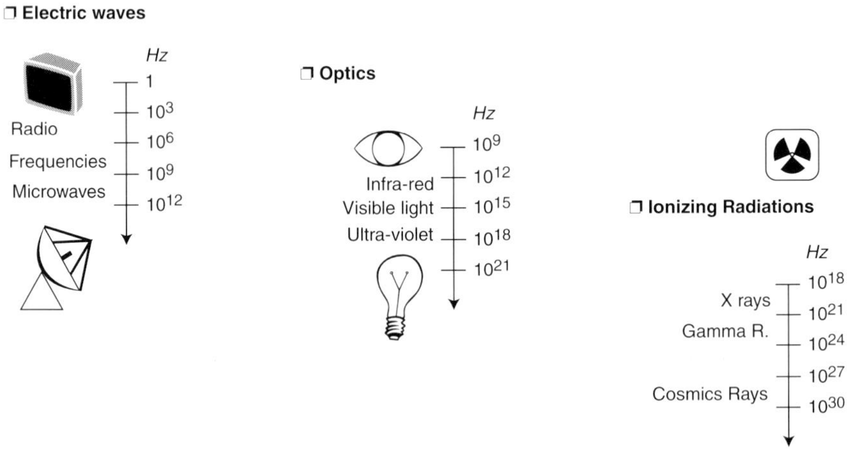

Fig. 1. Electromagnetic spectra in the radio and microwave range.

energy. The heat generated can thus be used to control the insects that infest them. Since the first US patent (Davis 1934, cited by Nelson and Whitney 1960) for a process to control insects in seed grain by high frequency waves, there have been numerous studies on the effects of electromagnetic energy on insects (Baker et al. 1956; Fleming 1944; Frings 1952; Grison and Martouret 1951; Nelson and Whitney 1960; Van den Bruel et al. 1960; Watters 1962). Several reviews have since been published on the subject (Kilgore and Doutt 1967; Watters 1962; Nelson 1996).

2 Biophysical Interactions between Electromagnetic Waves and Biological Matter

RF waves or microwaves are non-ionising, i.e. the energy carried by these types of waves do not disrupt atomic bonds to cause mutations (see Lewandovski, Chap. 7). When these waves impinge on a material, part of their energy is absorbed by dielectric components and dissipated as heat within the material, resulting in internal heat generation. Because the transfer of energy to the material does not require an intermediate heat carrier, the effect is much more rapid. Water is the "dipole" that is the

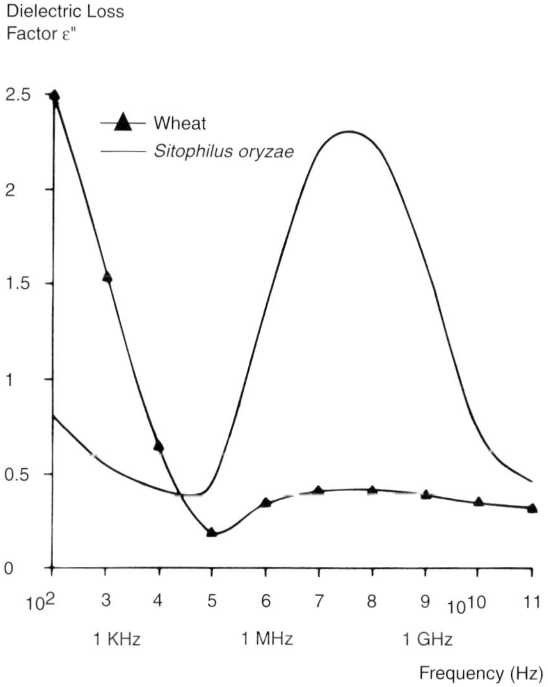

Fig. 2. Variation in dielectric properties of grain and the rice weevil, *S. oryzae*, with respect to wavelength in the radio and microwave band of the electro-magnetic waves spectrum (ε"= dielectric loss factor) (After Nelson and Charity 1972).

most intensively vibrated by the electromagnetic field. Consequently, the energy transfer and heat generation are higher in materials containing high moisture levels. Thus, it is possible to selectively heat organisms of high moisture content in a much drier food matrix, such as would be the case of a live beetle in a kilogram of wheat. The energy conducted by RF or microwaves is dissipated in biological dielectric biological material by two processes or "losses". Losses by conduction drop rapidly as frequency increases. In the RF and microwave bands, such losses are much lower than dielectric losses (Fig. 2), these latter being due to intermolecular friction resulting from the response of asymmetrical molecules (dipoles) to the alternating electric field. The frequency at which the amplitude of molecular oscillations is greatest is called the dielectric relaxation frequency and is that at which dielectric losses are energy transferred are greatest. The relaxation frequency is a function of the material's temperature. This is particularly relevant because the material temperature rises during processing. Moreover, the different types of water (free water, bound water, constituent water) have different dielectric loss response curves (see Pelletier and Collpitts, Chap. 8).
The dielectric permittivity represents the extent to which a material is polarised under the influence of an electric field. It consists of two components: (1) the dielectric constant ε', corresponding to the capacity of the material to absorb energy, and (2) the dielectric loss factor ε'', which is directly related to the dissipation of RF and microwave energy as heat. The RF or microwave power dissipated as heat by a material is described by the following equation:

$$P = 55.63 \times 10^{-12} \,.\, f \,.\, E^2 \,.\, \varepsilon_r. \quad (1)$$

The variables are:
- power per unit volume P (Wm^{-3}) ;
- frequency f (Hz) ;
- electric field intensity inside the material E (Vm^{-1});
- loss factor related to loss in air ε_r.

The heat released by the Joule effect in a material depends upon its specific heat and on the proportion of the electric field that penetrates the surface and is absorbed (some is reflected, which prevents a 100% conversion to heat). The heat budget for the theoretical case of complete energy absorption (i.e. no losses due to reflection) is described by the following equation:

$$Q = d\theta \,.\, M \,.\, c \,.\, 1.1627 \times 10^{-3}. \quad (2)$$

The variables are:
- energy consumed by hour Q (kWh^{-1});
- rise in temperature during exposure $d\theta$ (°C) ;
- mass of product treated M (kg) ;
- specific heat of the product c (Cal g^{-1} °C^{-1}).

The relation between the two equations when the theoretical production is 100% is the following:

$$dt = Q \,.\, P_{th}^{-1}, \quad (3)$$

where dt corresponds to a time of exposure expressed in hours and the theoretical total power delivered to the material P_{th} being expressed in kW (if dt = 1 h, then $P_{th} = Q$).
The theoretical heat transfer to a material is given by the classical heat equation:

$$P = 4.186\ 10^6 \ .\ c\ .\ d\theta.dt^{-1} \quad (4)$$

with $d\theta.dt^{-1}$ expressed in $°Cs^{-1}$ and P in Wm^{-3} .
The theoretical power transported by the electromagnetic wave delivered to the material P_{th} differs more from the observed heat delivered to the material when the energy transfer is weak.
Studies on the use of RF and microwaves to control insects in stored products have been oriented towards selective heating, the idea being to heat the insects without overheating the foodstuffs or reducing their end-use quality. Several studies have shown that the dielectric heating of insects is much greater than that of the cereal products in which they are found. The first studies on the dielectric constants of insects were conducted at 9.4 GHz (Nelson and Charity 1972). The dielectric loss factor of wheat ranged from 0.9 to 1.8 (depending on moisture content), while that of the rice weevil *Sitophilus oryzae* (L.) was about 13. Although these loss factors are extrapolations, they show that insects absorb more microwave energy and heat faster than cereal grains or finished product. Moreover, since the cereal product has a lower loss factor, the RF and microwaves will penetrate more deeply.
The difference in heating rates of the insect and the food material is mainly due to the differences in water content. Stored-product insects have a content of free water between 50 and 65%, while the cereal seed is stored at between 11 and 16%. However, few wavelengths are appropriate for "differential heating". The most effective differential heating occurs at wavelengths around 30 MHz (Nelson and Charity 1972), which is quite close to the 27.12 MHz band that is available for industrial, scientific and medical (ISM) applications. Although it is not possible to develop applications at wavelengths other than those authorised for the ISM applications, it is known that other wavelengths can affect the development and metabolism of insects (Pay et al. 1972; Fleurat-Lessard et al. 1979; Halverson et al. 1996).

3 Properties of Radio Waves and Microwaves for the Control of Insects

3.1 Introduction to Insect Infestation in Stored Products

A multitude of interrelated physical and biological factors must be taken into account in the development of electromagnetic tools to control stored-product insect pests. In summary, there are two types of plant-based foodstuffs: (1) processed products that are ready to eat or ready to be used as an ingredient, and (2) raw products such as seeds. The quality of stored foods is degraded in time by biological agents such as insects, mites and microorganisms, as well as by physical factors such as heat, cold, humidity and light. The final quality of the product depends mainly on the original moisture content and the storage temperature since their levels determine whether or not conditions will be favourable to the development of the biological organisms that cause damages. There are two main types of insect pests. The most numerous are free living at all their developmental stages and can only be considered as impurities of animal origin (e.g. rusty grain beetle, red and confused flour beetles, saw-toothed grain beetle). The other type consists of insects that spend part of their life cycle inside the seed, from which they cannot be separated (weevils, bruchids, grain borers, Angoumois grain moth, drugstore and cigarette beetle, etc.). As mentioned above, insects are dielectric materials with a high dielectric loss factor. Thus, they heat rapidly in an electromagnetic field and die once they reach the lethal temperature threshold close to 60 °C. This threshold corresponds to the temperature at which some proteins begin to coagulate (Anglade et al. 1979; Fleurat-Lessard and Le Torc'h, Chap. 5).

3.2 Choosing between RF and Microwave

Numerous industrial applications of dielectric heating by RF or microwaves have been developed (Nelson 1985), including some applications of insect control (Table 1). The choice of wavelength best adapted to solve a particular insect pest problem depends upon the packaging and characteristics of the product. In practice, this choice is limited by national and international rules on the use and transmission of microwaves and radio waves (Table 2).
The control of insects in food products by RF heating was tried quite a while ago. One can heat low moisture products in a closed cavity designed to prevent leakage

Table 1. Principal applications of dielectric heating by radio waves (RF) or microwaves in the agri-food industries.

Industrial Sector	Application
Seed	Breaking dormancy, germination
Cereals and beans (pulses)	Disinfestation, preventive sanitation
Cookies and cereal products	Drying, heating, preventive disinfestation
Flour, plant powders and spices	Pasteurisation, preventive disinfestation
Dehydrated aromatic plants	Drying - disinfestation
Liquid foods	Pasteurisation, sterilisation, cooking
Cooked meals, vegetables	Bleaching, pasteurisation, cooking
Pharmaceutical powders	Drying, pasteurisation

Table 2. Electromagnetic wavelengths authorised for industrial, scientific and medical uses (ISM) in the RF and microwave frequency bands, and a comparison of the respective advantages of RF and microwaves for energy transfer applications (mainly thermal).

	Radio waves	Microwaves
Authorised wavelengths	13.56 MHz 27.12 MHz 40.68 MHz	433 MHz 915 MHz 2,450 MHz 5,800 MHz 24,125 MHz
Principal cause of heating	Ionic conduction	Vibration of dipoles in an electric field
Depth of penetration	Deep (0.2 to 0.6 m) (potentially variable)	Shallow (the order of 1 dm) (constant)
Shape	Homogeneous	Heterogeneous
Cost	800 to 1000 US$/kW 60% for generator; 40% for application system	1300 to 2000 US$/ kW 40% for generator; 60% for application system
Power of available applicators	600 kW	50 kW
Expected life of tubes	5000 to 10 000 h	2000 to 5000 h
Maintenance	Replacing tubes	Replacing tubes

of electromagnetic waves that could otherwise travel long distances and affect telecommunications. This type of cavity is called an applicator. The first pilot scale tests were aimed at destroying weevils (*Sitophilus* spp.) and the lesser grain borer [*Rhyzopertha dominica* (F.)] in wheat or maize (Nelson and Kantack 1966; Fleurat-Lessard 1980), at stages during which they are lodged inside the kernels. These studies were followed by the construction of industrial prototypes for disinfestation of cereal seeds (Boulanger et al. 1969), and cereal products and beans (pulses) (Fleurat-Lessard 1988; Fleurat-Lessard and Fuzeau 1991).

RF waves in the band 1-30 m are preferred for this type of application, i.e. the disinfestation of low moisture content foodstuffs, since they penetrate more deeply into the exposed product than do microwaves (wavelengths between 1 and 100 cm).

For a given product, the efficiency of heating depends on the degree of coupling of the generated electric field and the water molecules which can depend on the degrees of free and bound water present, since they react differently. One must thoroughly understand the dielectric characteristics of the product to obtain a good coupling of the electromagnetic energy and the product. Each product is different and the choice of wavelengths is limited to the frequency bands allocated to ISM applications (Table 2). It is necessary to optimise the wavelength-product interaction in order to maximise the energy transferred from the generated RF or microwave electric field to the material to be heated. For the case of insects in dry foodstuffs or for treating products in non-metallic packaging and laid out in a sheet or in a regular shape, RF is the wavelength of choice (Hafner 1975). Microwaves are better for combined treatments that are used for preventive sanitation on foodstuffs with slightly more moisture, such as combined drying and disinfestation, or the combined reduction of moisture variation inside the product and disinfestation. However, this method is only possible if the product is less than 10 cm thick, regardless of its shape.

4 Microwave Applications for Insect Control (Single or Combined)

The use of microwaves to control insects is prohibitively expensive if the electrical energy is used only to produce a lethal thermal shock in the insects. That is why the first applications of insect control in stored grain, developed in North America, combined insect control and grain drying. When a slightly moist seed is placed into a microwave field, the moisture migrates rapidly to the surface, from where it can evaporate more easily. This effect is particularly useful near the end of the drying period when water migration is particularly slow when traditional hot air drying is used. Microwaves accelerate moisture migration to a degree that can result in large savings in time and energy.

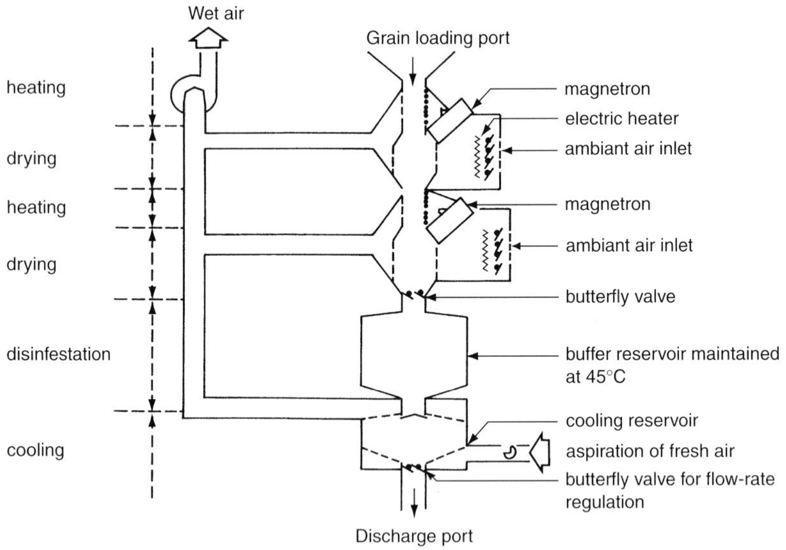

Fig. 3. Pilot-scale apparatus for the drying and disinfestation of maize using microwaves (After Boulanger et al. 1969).

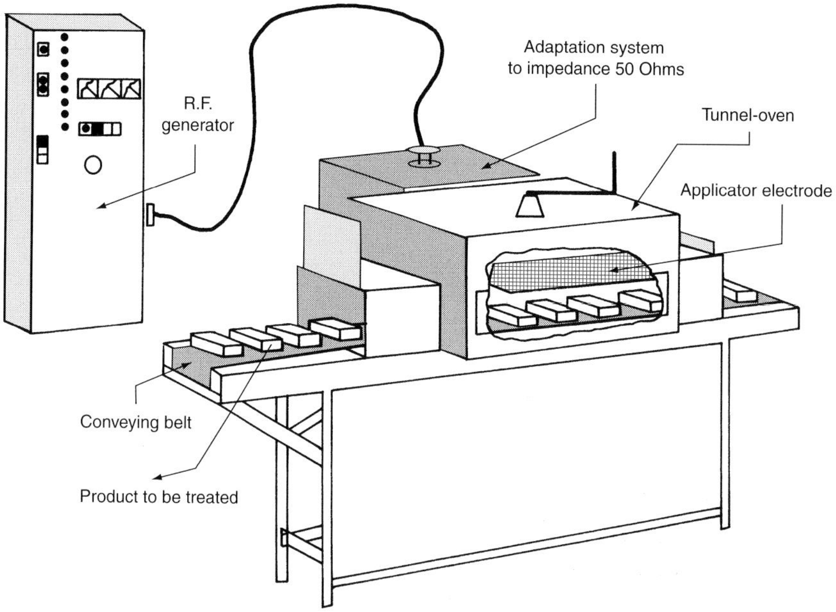

Fig. 4. Pilot-scale apparatus for feasibility studies on the disinfestation of finished or semi-finished processed foods using radio waves at 27.12 MHz.

Simple applicators were built for insect control to demonstrate the feasibility of rapid heating by microwaves (Boulanger et al. 1969). American manufacturers developed several applicators of this type. One was designed to dry as well as to control insects during a single pass of grain through the applicator (Fig. 3). Another was designed to completely dry the product, by heating the product in a vacuum chamber and removing the excess water (Elias 1979). Insect control in dried fruit and nuts (dates, figs, apricots, hazelnuts, walnuts, etc.) using a tunnel-type oven with conveyor is a promising concept for immediate development and application in the food processing industry (Wilkin and Nelson 1987; Reynes and Tabuna 1994). Microwaves have specific applications that cannot be fulfilled by other means. For example, they can be used in museums to control insects in woollen garments of historical or archaeological importance. A careful treatment by microwaves of pieces of fabric or tissue, which could be kept under plastic to prevent reinfestation, has permitted the disinfestation of collections of certain museums that have problems of wool mite and other keratophagous species such as *Dermestes* spp. (Chauvin and de Reyer 1991). An unanticipated use for microwave heating is the pre-sowing or pre-transplant control of weeds and destructive micro-organisms. A case in point is the destruction of fungi that attack seedlings in infected forest plantations and reduce their viability. A tractor-mounted microwave apparatus for soil sterilisation was developed several years ago in France (Thourel 1980; Patay and Lonne 1983). This technique was adapted to the sterilisation of composts used in mushroom production, and to the disinfestation of ascarid eggs in sand boxes found in public areas. The eggs of *Toxocara canis* are deposited in dog faeces and can cause the parasitic disease *Larva migrans viscerale* syndrome in humans, cats and dogs (Bouchet and Leger 1984). These methods are needed as alternatives to methyl bromide, a soil fumigant that will be phased out in the near future because it is an ozone depleting substance.

We had made a pilot scale RF application apparatus equipped with a conveyer belt to study the possibility of using RF at 27.12 MHz to control insects in dry processed food as part of a preventive sanitation program (Fig. 4) concerning:

- Disinfestation and drying of wheat flour (1- or 25-kg packs)
- Insect control in bulk rice after natural drying (for organic rice).
- Insurance treatment for insects in packaged (i.e. ready-for-sale) aromatic plants.
- Disinfestation of spices after packaging.
- Disinfestation and lowering microbial contamination in plant powders before processing.

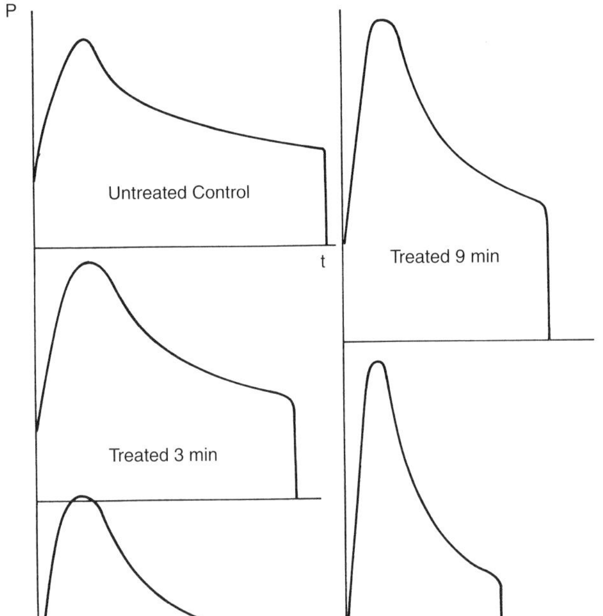

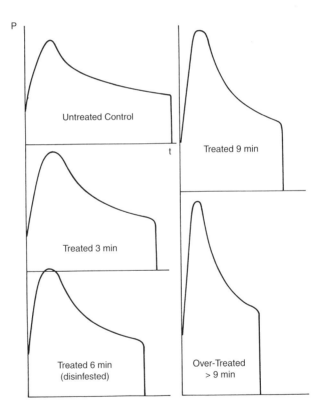

Figure 5. Change in rheological properties of wheat flour following RF dielectric heating (Chopin Alveograph test).

The main risk of this type of treatment is the overheating of the product. Thus, it is not practical for heat-sensitive products such as wheat flour used to make bread, since its rheological properties decline as exposure to microwaves increases. Bread-making qualities may be totally changed by long exposures (Fig. 5). It is possible to disinfest seeds using RF energy without reducing germination only if grain moisture content is below about 14% (Nelson 1996). Adverse effects of dielectric heating depend of the water content of the product. We have also shown that aromatic plants, notably dried plants with light essential oils, or herbs can lose part of their aroma (Fleurat-Lessard 1996). However, for all the products that tolerate heat, pasteurisation and even sterilisation are possible, especially with microwave treatments. Some products even improve with heat treatment. For example, the yellow of durum wheat semolina, a desired quality, is increased after heat treatment (Fleurat-Lessard and Fuzeau 1991). The synergistic action of RF or microwave in combination with modified atmospheres is currently being investigated in our laboratory on packaged foods. For instance, the time needed to reach 99% kill of Mediterranean flour moth eggs was reduced from 2 days to 4-6 h by packaging a food product in an atmosphere of 60% carbon dioxide. The synergistic conditions afforded by the combination of modified atmosphere packaging and RF heating were such that the temperature level reached by the packaged product corresponding to the lethal temperature for insect eggs was reduced at about 50 °C.
One can even imagine the possibility of using RF in such a way that properties of the treated products are different from those of the initial product. This would require sophisticated means of temperature control and real time monitoring of temperature changes. This is possible with an optic thermometer, or with a sensor that is unaffected by RF or microwave electric field such as the fluoroptic thermometers. This would allow the real time adjustment of RF or microwave energy through a feedback loop. It would also be necessary to characterise the dielectric properties of the food matrix.

5 Conclusion

Applications of RF and microwave technologies to insect control and sanitation of food products destined for long term storage are envisioned particularly in the context of strategies for cleanliness and food protection that are presently being implemented in the dry food sector (ISO certification and HACCP). The development of this approach is welcome for all situations in which chemical pesticides are prohibited:
- Cereals and pulses that receive no post-harvest treatment.
- Cereals used as feed ingredients (combined disinfestation and additional drying).
- Processed foods with a high cereal content (breakfast cereals, baked goods, pet foods).
- Dried fruits and related foods.
- Aromatic plants, dried vegetables, spices.

The synergistic action of high-frequency electromagnetic waves and other physical control methods could also be exploited in the preventive treatment of food materials (e.g. high CO_2 atmosphere in air-tight packaging combined with heating of the package by RF or microwaves). On-going studies are presently aimed at the determining exact conditions for post-harvest treatments of foods.

References

Anglade P., Cangardel H., Fleurat-Lessard F., (1979). Applications des O.E.M. de haute fréquence et des micro-ondes à la désinsectisation des denrées stockées. Proc. 14e Symp. Int. Appl. Energétiques Micro-ondes (IMPI), Monaco, juin 1979, pp. 67-69.

Arsonval A. d', (1893). C. R. Acad. Sci., Paris, 116: 630-632; (In Grison and Martouret 1951).

Baker V.H., Wiant D., Taboada O., (1956). Some effects of microwaves on certain insects which infest wheat and flour. J. Econ. Entomol. 49: 33-37.

Bouchet F., Léger N., (1984). Utilisation des micro-ondes comme moyen de lutte contre la toxocarose dans les bacs à sable urbains. Economie et Progrès par l'Electricité, Aderme (éd.), Lyon-Villeurbanne, 12: 17-20.

Boulanger R.J., Boerner W.M., Hamid M.A.K., (1969). Comparison of microwave and dielectric heating systems for the control of moisture content and insect infestations of grain. J. Microwave Power 4: 194-208.

Chauvin G., de Reyer D., (1991). Application de l'énergie micro-onde pour la désinsectisation. Essais sur un lépidoptère, *Tineola bisseliella*, pp.1167-1171. In F. Fleurat-Lessard and P. Ducom (eds.) Proc. 5th Int. Working Conf. Stored Product Protection, INRA/SPV, Bordeaux, 2066 p.

Davis J.M., (1934). U.S. Patent, 1, 972, 050, 28 August In S.O. Nelson and W.K. Whitney 1960.

Elias S., (1979). Microwave vacuum drying: quick, quiet, clean, efficient. Food Eng. Int'l 1: 32-33.

Fleming H., (1944). Effect of high-frequency fields on microorganisms. Elec. Eng. 63: 18-21.

Fleurat-Lessard F., Lesbats M., Lavenseau L., Cangardel H., Moreau R., Lamy M., Anglade P., (1979). Effets biologiques des micro-ondes sur deux insectes, *Tenebrio molitor* L. (Col. Tenebrionidae) et *Pieris brassicae* L. (Lep. Pieridae). Ann. Zool. Ecol. Anim. 11: 457-478.

Fleurat-Lessard F., (1980). Lutte physique par l'air chaud ou les hautes fréquences contre les insectes des grains et des produits céréaliers. Bull. Techn. Info. 349: 345-352.

Fleurat-Lessard F., (1988). Désinsectisation préventive des produits agro-alimentaires par les ondes de haute fréquence (27 MHz), pp. 82-89. In EDF-DER (éd.) C.R. Réunion du club "rayonnements" EDF. December 1988, EDF, Moret-sur Loing, France, 134 p.

Fleurat-Lessard F., (1996). La destruction préventive des insectes par les micro-ondes et les ondes à haute fréquence. Applications dans le domaine agro-alimentaire. CR Conférence CILDA, Paris, Octobre 1996. Bull. CILDA 27: 19-48.

Fleurat-Lessard F., Fuzeau B., (1991). Une nouvelle technique disponible pour la stabilisation sanitaire des produits agro-alimentaires conditionnés: l'étuvage en four-tunnel à hautes fréquences (H.F.). C.R. réunion club EDF " rayonnements ", Gradignan, 23 mai 1991. EDF-DER (éd.), Moret-sur Loing, France, 14 p.

Frings H., (1952). Factors determining the effects of radio-frequency Electro-magnetic fields on insects and materials they infest. J. Econ. Entomol. 45: 396-408.

Grison P, Martouret D., (1951). Essai de destruction des insectes des grains par les courants à haute fréquence. C. R. Acad. Agric. Fr., 20 oct. 1951, 3 p.

Hafner T., (1975). Traitement sans dommage par chauffage HF des denrées alimentaires à stocker. Rev. Brown Boveri 1/2: 52-55.

Halverson S.L., Plarre R., Burkholder W.E. , Bigelow T.S., Misenheimer M.E., Nordheim E.V., (1996). Effects of SHF and EHF microwave energy on the mortality of *Sitophilus zeamais* in soft white wheat. Paper No. 963013, ASAE Annual International Meeting, Phoenix, Arizona, July 14-18, 1996, 13 p.

Kilgore W.W., Doutt R.L., (1967). Pest control: biological, physical, and selected chemical methods. Academic Press, New York, 477 p.

Nelson S.O., (1985). RF and microwave energy for potential agricultural applications. J. Microwave Power 20: 65-70.

Nelson S.O., (1996). Review and assessment of radio-frequency and microwave energy for stored-grain insect control. Trans. ASAE 39:1475-1484.

Nelson S.O., Whitney W., (1960). Radio-frequency electric fields for stored grain insect control. Trans. ASAE 3: 133-137, 144.

Nelson S.O., Kantack B.H., (1966). Stored-grain insect control studies with radio-frequency energy. J. Econ. Entomol. 59: 588-594.

Nelson S.O., Charity L.F., (1972). Frequency dependence of energy absorption by insect and grain in electric fields. Trans. ASAE 15: 1099-1102.

Patay L., Lonne J., (1983). Désherbage et désinfection des sols par localisation d'énergie micro-ondes. Economie et Progrès par l'Electricité, Aderme (ed.), Lyon-Villeurbanne, 8/9: 7-8.

Pay T., Beyer E., Reicherdelfer C., (1972). Microwave effects on reproductive capacity and genetic transmission in *Drosophila melanogaster*. J. Microwave Power 7: 75-82.

Reynes M., Tabuna H.. Mise au point d'une technique de désinfestation des dattes basée sur l'emploi des micro-ondes. COMAGEP Marrakech Proceedings, 1: 621-624.

Thourel L., (1980). Utilisations des ondes électromagnétiques dans l'industrie agro-alimentaire. Ind. Céréales 4: 23-29.

Van den Bruel W., Bollaerts D., Pietermaat F., Van Dijck W., (1960). Etude des facteurs déterminant les possibilités d'utilisation du chauffage diélectrique à haute fréquence pour la destruction des insectes et des acariens dissimulés en profondeur dans les denrées alimentaires empaquetées. Parasitica 16: 29-61.

Watters F.L., (1962). Control of insects in foodstuffs by high-frequency electric fields. Proc. Ent. Soc. Ontario 92: 26-32.

Wilkin D.R., Nelson G., (1987). Control of insects in confectionery walnuts using microwaves pp. 247-254. In T.J. Lawson (ed.) Stored products pest control. BCPC monograph. No. 37, Thornton Heath, U.K., 277 p.

Electrical Weed Control: Theory and Applications

Clément VIGNEAULT and Diane L. BENOÎT

1 Introduction

The concept of using electrical energy to kill weeds was developed in the late 1800s and several patents have been registered in the United States since 1890. The most recent electrical weed control system consists of a tractor-driven device (Lasco Inc.) designed to destroy persistent weeds in row crops following conventional chemical treatment. The machine, called the Lasco LW5 Lightning Weeder, was manufactured in Vicksburg, Mississippi, and first appeared on the market in 1980 (Fig. 1).

Some environmentalists believe that using electricity is the ideal way to control early-emerging weeds. This approach leaves no chemical residues into the environment or food chain, and it does not disturb the soil surface or promote erosion.

An analysis of the annual operating costs of destroying weeds that are taller than the crop on a basis of 100 ha, showed that electrical weed control is on a par with a roller applicator, but is more expensive than a recirculating sprayer (Kaufman and Schaffner 1980). Three passes are recommended to ensure optimum weed control using the electrical method.

Fig. 1. Tractor-mounted electrical weed control machine, in operation.

The use of electricity to control small weeds in row crops was studied by Hackam (1985), and a machine similar to that manufactured by Lasco was developed for this purpose. The single electrode of the Lasco system was replaced with a series of small electrodes that could be placed closer to the soil surface between the crop rows. Deflectors and shields were used to prevent contact between the crop and the electrode. However, testing of this prototype revealed excess power consumption and mechanical failure of the generator drive train when the machine was operated in fields with high weed densities.
An analysis of the theoretical concepts of electrical weed suppression is essential in order to identify key parameters for machine design and use. This chapter reviews the theory of electrical weed control and present knowledge of the factors influencing the efficiency of the electrical method. It also compares the energy requirements of the electrical method with other weed control techniques.

2 Theory of Electrical Weed Control

When a high-voltage electrode touches a weed, electric current passes through the plant and is returned to the transformer via the soil by a ground contact device. Due to the electrical resistance of the plant, the electrical energy is converted to heat in accordance with Eq. (1).

$$E=\frac{V^2 T_c}{R_p} \tag{1}$$

The energy transmitted to a single plant (E) is directly related to the electrical resistance of the plant (Rp), the contact time (Tc) and the electrode voltage (V).
Plant death is caused primarily by the increase in temperature and vaporization of the water and other volatile liquids it contains. This results in a buildup of pressure within the cells, and subsequent rupture of the cell membranes (Diprose et al. 1980; Dykes 1980). In extreme cases, when an excess of energy is absorbed by the plant, localized hot spots occur and eventually the plant stem burns and ruptures. The severity of damage depends on the plant species, morphology, age and population density. The contact time and voltage can affect the extent of damage (Dykes 1980; Sanwald and Koch 1978).

2.1 Impact of the Morphology of the Target Species

In plants with an extensive root system, the electric current travels deep into the root system before being dissipated into the soil (Diprose et al. 1980; Drolet and Rioux 1983; Dykes 1980). Plants with large or specialized underground organs will suffer little root damage (Diprose et al. 1980; Drolet and Rioux 1983). Root damage is more severe under dry than under moist soil conditions (Diprose and Benson 1984; Dykes 1980; Vigoureux 1981). The rhizomes of quackgrass (*Agropyron repens* L. Beauv.) survive several electrical treatments (Diprose and Benson 1984). The use of electricity is ineffective in controlling common burdock (*Arctium minus* Hill Bernh.) in field-grown crops, because the lower branches are not touched by

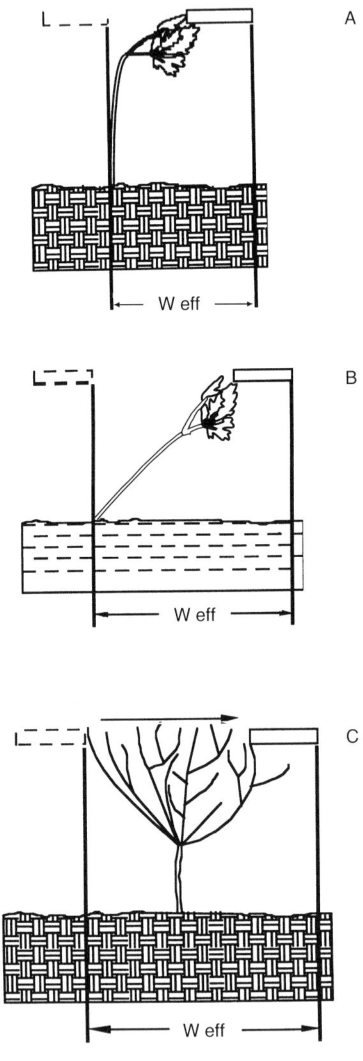

Fig. 2 A-C. Three possible scenarios to describe the maximum and minimum contact times. **A** Erect plant where only the foliage bends. **B** Erect plant where the entire plant bends at its base. **C** Bushy plants with multiple contacts. W_{eff} Effective electrode width.

the electrode (Drolet and Rioux 1983). Bull thistle [*Cirsium vulgare* (Savi) Tenore] and Canada thistle [*Cirsium arvense* (L.) Scop.] escaped destruction, with regrowth of shoots occurring from the undamaged root system.

2.2 Impact of Electrode Contact Time

When an electrode passes over the weed population, individual weeds remain in contact with the electrode for a distance determined by the forward speed of the tractor. Diprose and Benson (1984) derived theoretical limits of contact for an upright plant by considering two extreme cases: (1) the plant remains upright and only bends at the point of contact; 2) the plant stays rigid but bends at the base (Fig. 2a,b). For very bushy plants, contact may begin before the leading edge of the electrode is positioned over the plant base and may cease after the trailing edge of the electrode has passed the base (Fig. 2c). The distance travelled by the electrode while maintaining contact with the plant is called the effective electrode width (W_{eff}). It is a function of electrode width, height of horizontal movement and plant geometry and rigidity.
The number of plants in contact with the electrode at any given time (n_c) is equal to the product of electrode length (L), effective width (W_{eff}) and weed density (D).

$$n_c = LW_{eff}D \tag{2}$$

2.3 Effect of Weed Density

Load resistance across the transformer (R_l) is equal to the sum of the parallel resistances (R_{pi}) created by n_c plants touching the electrode simultaneously, plus the resistance of the soil (R_s) [Eq. (3)]. Thus, the load resistance across the transformer is largely reduced by multiple contacts of weeds with the electrode.

$$R_l = \left[\sum_{i=1}^{n_c} \frac{1}{R_{pi}}\right]^{-1} + R_s \tag{3}$$

Plant resistance varies widely with species, morphology, age and lignin or cellulose content (Diprose and Benson 1984; Dykes 1980; Vigneault 1985). However, in a simplified case where all plants have the same resistance, Eq. (3) reduces to Eq. (4).

$$R_l = \frac{R_p}{n_c} + R_s \tag{4}$$

2.4 Importance of Generator Power

In the ideal case, where all plants have the same electrical resistance, and disregarding the energy losses due to the resistance of the transformer, the power produced by a generator is given by Eq. (5).

$$P = \frac{V^2}{R_l} = \frac{n_c V^2}{R_p + n_c R_s} \tag{5}$$

The following relationships can be deduced from Eq. (2) to (5).
The required generator power (P) increases as the number of weeds touching the electrode (n_c) increases, which is itself proportional to weed density (D).
As weed density increases, the electrical resistance of the plants becomes negligible compared to the soil resistance. As a result, generator power becomes a function of the soil resistance and the voltage applied. Eq. (5) can thus be replaced by Eq. (6). At high weed densities, a significant proportion of generator power is absorbed by the ground, leaving little energy for killing the weeds.

$$P = \frac{V^2}{R_s} \tag{6}$$

Generator power is independent of the forward speed of the tractor. Thus, unlike tillage operations, where power requirements are proportional to ground speed, power demands on the generator cannot be reduced simply by decreasing the forward speed of the tractor.

2.5 Operating Capacity of the Machine

A minimum threshold contact time (T_c) of about 0.2s is required for a lethal effect on young weeds under field conditions (Dykes 1980). This places an upper limit on the machine's forward speed (U), as a function of the effective electrode width (W_{eff}).

$$T_c = \frac{W_{eff}}{U} \leq 0.2s \tag{7}$$
$$U \leq 5.0 W_{eff}$$

The electrode length (L) is, however, limited by the available generator power (P).

$$L = \frac{P T_c}{E D W_{eff}} \tag{8}$$

The operating capacity of the machine in terms of area treated per unit of time (Q) is a function of available generator power (P), energy transmitted to the plant (E) and density of the weed population (D).

$$Q = \frac{P}{ED} \qquad (9)$$

Most of the studies done so far have examined the effectiveness of electrical control of escaped weeds. One field trial investigated the use of electric currents to control weed beet and bolting beet growing in the rows of a sugar beet crop (*Beta vulgaris* L.). Before the electrical method can be applied, however, the weeds must be allowed to grow taller than the crop. In theory, a 5- to 10-cm difference in height between the crop plants and the weeds is necessary for adequate weed control and minimal crop damage. In practice, however, a minimum height differential of 10 to 20 cm is required (Lutman 1980).

Tests revealed that a single pass is sufficient to treat infestation densities of up to 2000 stems ha^{-1}. For weed densities of 2000 to 6000 stems ha^{-1}, two passes are generally required (Guiraud and Givelet 1981; Vigoureux 1981, 1982). Vigoureux (1981, 1982) obtained 98 to 99.9% control of bolting beet at infestation densities of between 100 and 5000 stems ha^{-1}. In contrast, only 24% weed control was achieved at densities exceeding 18000 stems ha^{-1}.

3 Energy Requirements

Knowledge of the electrical resistance of the soil and plants is helpful in understanding the factors affecting generator power requirements. The electrical resistance of plants varies widely among species, and the energy required for a lethal effect increases with increasing plant maturity.

When voltage is applied to a plant, the current begins to cause damage to the plant structures. As the level of damage increases, plant resistance decreases, allowing more current to flow through the plant (Diprose and Benson 1984; Diprose et al. 1980; Dykes 1980). This rise in current is not constant but occurs in two stages. The first stage is slow and linear, terminating at an inflection point. The second is characterized by an exponential increase in current. The release of steam from the plant occurs at the inflection point (Martens and Vigoureux 1983). Near the end of the first stage, the current reaches a plateau corresponding to maximum damage and death of the plant. If the current is maintained beyond that point, excessive structural damage occurs. The plant eventually breaks, thus interrupting the current (Diprose and Benson 1984; Diprose et al. 1980).

Several factors affect the amount of energy required for electrical weed control. They include species, size, life span and age of the plant, chemical composition and root structure arrangement, and soil composition and moisture content.

Multistem species were more difficult to control effectively than weeds with a single main stalk (Rasmusson et al. 1979; Vigoureux 1981).
For a constant electrode height, tall plants will remain in contact with the electrode for a longer period of time than shorter plants, and the former will absorb more energy. To ensure that shorter plants are killed, sufficient energy per plant must be provided either by increasing the voltage or decreasing the electrode height. This results in an excess of energy being provided to tall plants. Thus, the average energy applied per plant will be greater than the minimum required to kill a single plant.

4 Concept of Lethal Threshold Energy

The analysis of energy requirements for weed control in row crops is based on the mean threshold energy required to kill target plants. Diprose et al. (1984, 1985) suggested that field-grown plants require two to three times more lethal energy than plants grown in greenhouses. Tests have confirmed that the energy required for a lethal effect on field plants is five (Diprose and Benson 1984) to ten (Chandler 1978) times that required for the same effect on weeds grown indoors. Under experimental conditions, the lethal threshold energy varied from 4 to 111 kJ plant^{-1} (Table 1). In contrast, the energy required for a lethal effect on field-grown weeds often exceeds the minimum thresholds (Table 2).

Table 1. Lethal energy applied (kJ plant^{-1}) under experimental greenhouse conditions. The voltage were applied until the plant severs (Diprose et al. 1978).

Species	Plant height (m)	Voltage (kV)	Contact time contact (s)	Lethal energy (kJ plant^{-1})
Beta vulgaris L.	-[a]	1.0	33.7	4
		2.5	32.2	32
		5.0	8.5	28
Chrysanthemum segetum L.	-	2.0	37.2	65
		3.0	8.5	21
		5.0	6.9	23
Sinapis arvensis L.		2.0	147.8	86
Large	1.0	3.0	52.0	111
		4.0	13.2	71
Small	0.2-0.6	2.0	16.2	6
		3.0	8.9	7
		4.0	4.8	17

[a] Information not available.

Table 2. Lethal energy (kJ plant^{-1}) required to kill weeds under experimental field conditions (Vigneault et al. 1990).

Species	Age of plant	Voltage (kV)	Electrode contact time	Lethal threshold energy (kJ plant^{-1})
Abutilon theophrasti	3 weeks	3.0	0.116	0.365
Medik.	4 weeks	4.0	0.175	1.908
Beta vulgaris L.	Bolting	4.0	21.8	104[a]
	"	5.2	11.6	137
	"	6.2	9.9	139
	"	7.2	8.1	156
	"	7.6	8.7	207
	"	8.4	4.3	104
Brassica campestis L.	Fruiting	14.4	0.41	2.13
	"	14.4	0.27	0.47
Chenopodium album L.	Fruiting	14.4	0.65	3.20
	"	12.0	0.62	2.93
	"	12.0	0.20	0.53
	"	12.0	0.23	1.81
Xanthium strumarium L.	3 weeks	1.0	0.110	0.040
	4 weeks	2.0	0.094	0.254

[a] Voltage was applied until the sugar beet severed.

A machine applying electric voltage ranging from 12.0 to 14.4 kV destroyed lamb's quarters (*Chenopodium album* L.) and birdsrape mustard (*Brassica campestris* L.) with an average of 2.12 and 1.30 kJ applied per plant respectively (Table 2). This indicates that it would take an average of 1.71 kJ plant^{-1} to kill mature weeds. These values are much lower than those previously reported by Diprose et al. (1980) for the control of bolting beets (Table 2), since the values they mentioned represented the lethal energy required to sever the plants.

Table 3. Effect of tractor speed on the energy required for a lethal effect on *Chenopodium album*[a] (Vigneault et al. 1990).

Tractor speed (km h^{-1})	Voltage (kV)	Electrode contact time (s)	Lethal threshold energy (kJ plant^{-1})	Mortality (%)
2.72	14.4	0.65	3.20	100
4.12	14.4	0.49	1.16	50
6.17	14.4	0.34	0.79	28

[a] Average plant height: 85 cm.

Table 4. Effect of voltage and plant height on the energy required for a lethal effect on *Chenopodium album*[a] (Drolet and Rioux 1983).

Voltage (kV)	Average plant height (cm)	Electrode contact time (s)	Lethal threshold energy (kJ plant^{-1})	Mortality (%)
12.0	28	0.20	0.53	100
12.0	39	0.23	1.81	100
12.0	79	0.62	2.93	100
14.4	81	0.65	3.20	100

[a] Average tractor speed: 2.80 km h^{-1}.

Drolet and Rioux (1983) studied the relationship between mean lethal threshold energy, tractor speed and applied voltage. For a constant voltage, the electrode contact time and the energy applied per plant decreased as tractor speed increased (Table 3). The percent mortality decreased in direct proportion to the energy applied per plant (Table 3). Increasing the voltage from 12 to 14.4 kV increased the energy applied per plant but had no effect on percent mortality (Table 4). Percent mortality remained unchanged even when the height difference between the plant was greater, electrode contact time was longer and greater mean energy was applied per plant (Table 4). The relationship between applied voltage and contact time is not linear (Diprose et al. 1978). Doubling the voltage decreases the required contact time by four (Diprose et al. 1980).

Where weed density is very high and the weeds occur in patches, the largest plants can shield the smaller plants from contact with the electrode (Diprose et al. 1985; Drolet and Rioux 1983; Rasmusson et al. 1979), thereby reducing the overall efficiency of the treatment. Drolet and Rioux (1983) have shown that in lamb's quarters, the first plants in a patch had higher percent mortality than the last plants (Table 5). When two passes were made in opposite directions, the total energy applied per plant was doubled, but more uniform mortality was obtained throughout the patch.

5 Comparison of Energy Requirements and Cost of Weed Control Techniques

Electricity can be used either as the primary method of weed control, or to destroy the weeds remaining after one or more conventional treatments. The density of the target weed population will vary, depending on whether the electrical treatment is applied to the initial population or to the residual population. Population densities

Table 5. Effect of the number of passes[a] on average mortality of a group of plants[b] and the lethal threshold energy applied per plant (Drolet and Rioux 1983).

Number of passes	Total number of plants treated	No. of plants/patch	Mean distance between 1st and 3rd plant	Average mortality 1st plant	2nd plant	3rd plant	Average mortality/patch	Lethal threshold energy
			(cm)		(%)		(%)	(kJ $plant^{-1}$)
1	30	3	21.2	80	60	50	63.3	1.32
2[c]	33	3	14.7	64	64	64	64	2.21

[a] Electrical machine characteristics Tractor speed: 2.79 km h^{-1}, Voltage: 14.4 kV, Electrode contact time: 0.65 s.
[b] Average plant height: 82.5 cm.
[c] Each pass was made in the opposite direction.

are also affected by treatment time, soil type, field history and previous weed control practices. Initial weed densities of 168 plants m^{-2} in onion (*Allium cepa* L.), 569 plants m^{-2} in tobacco (*Nicotiana tabacum* L.) (Roberts 1967) and 1075 plants m^{-2} on fallow land (Drolet and Rioux 1983) have been reported prior to treatment. Residual weed densities were generally much lower, being 0.2 and 1.0 plant per linear metre of row crop (Kaufman and Schaffner 1982). Since studies of electrical control of residual plants involved different weed densities, the findings cannot be readily converted to practical advice.

5.1 Experimental Method

The energy input of chemical weed control was assessed for corn (*Zea mays* L.) and bean (*Phaseolus vulgaris* L.), crops for which herbicides are the principal method of weed suppression. Total energy requirements were calculated using herbicides at the rates recommended in Ontario (Anonymous 1996). Whereas the cost for corn was based on the use of metolachlor with atrazine and bromoxynil, the computation for beans used metobromuron and bentazon. The calculation of total energy inputs included the energy required to manufacture the herbicides (Southwell and Rothwell 1977) and the energy required to transport and apply the product in the fields. The energy input for herbicide application was determined from data provided in ASAE standards (Anonymous 1990). The calculation method used to establish the energy input of chemical weed control was similar to that reported by Sanwald and Koch (1978).

The energy requirements of mechanical weed control were calculated for an 80-kW tractor pulling three types of cultivators. Productivity was estimated based on the data obtained in cultivation tests (Anonymous 1990).

Weed densities of 5, 30 and 200 stems m^{-2} were used to calculate the energy requirements of electrical weed control. These densities correspond to light, moderate and severe infestations of emerging weed seedlings, respectively. Assuming that an average of 1.76 kJ^{-1} plant is required, the transformer must supply 3,520 MJ ha^{-1} for an infestation density of 200 weeds m^{-2}. Total energy consumption in the form of tractor fuel will be considerably higher since substantial energy is lost due to inefficiencies of the tractor engine, drive train, generator, transformer and rolling resistance. According to Drolet and Rioux (1983), overall tractor efficiency, including the energy required to move and operate both the tractor and the machine, was 21.4%. The power available for the electrode represented 54 to 59% of the power output from the tractor power takeoff (Drolet and Rioux 1983; Kaufman and Schaffner 1980).

The Lasco electrical weed control system consists of a low voltage AC generator driven by the tractor power takeoff, a stepup transformer to modulate the voltage, an insulator-mounted electrode suspended above the crop and a rolling coulter ground contact device to complete the circuit (Fig. 3). The horizontally positioned

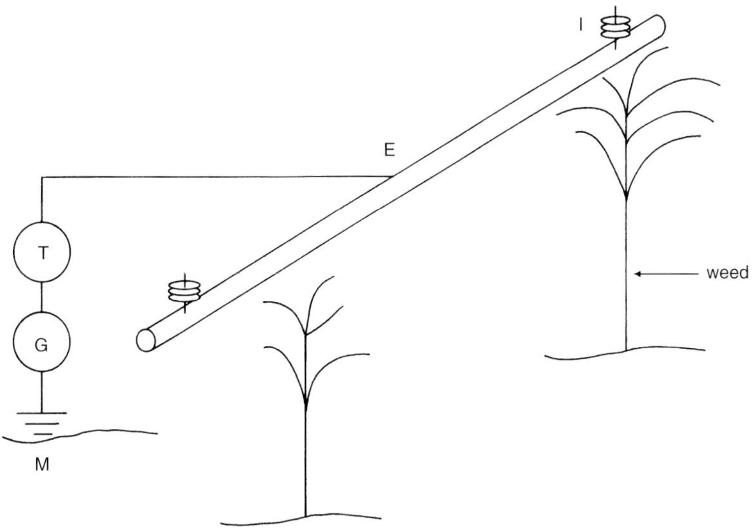

Fig. 3. Diagram of an electrical weed control system. G Current generator. T Transformer. E Electrode. I Insulator. M Grounding using a rolling coulter.

electrode conducts the current to the plants and is protected by a plastic shield to prevent accidental contact.
An 80-kW tractor with a 50-kW generator operating in a field with a weed density of 200 plants m^{-2} can supply an average of 1.76 kJ $plant^{-1}$ and will consume 16.5 GJ ha^{-1}. The full operating capacity of the electrical weeder is 0.047 ha h^{-1} [Eq. 9]. The actual capacity will be somewhat lower, however, because of time lost in headlands and because the generator is rarely used at full capacity.

5.2 Comparison of Three Weed Control Methods

There are some advantages to using electricity to control escaped weeds in a conventional weed control program. For weed densities of five plants m^{-2}, chemical and electrical weed control call for approximately the same amount of energy (Table 6). The mechanical method requires much less energy, since cultivators offer little resistance and are lighter to pull. Cultivation breaks the surface crust, generally operates at the surface and does not alter soil structure. Although mechanical cultivation does not adversely affect soil conditions, it can cause physical damage to crops and reduce crop productivity slightly. The use of electricity does not disturb the soil surface and therefore does not promote erosion or soil moisture loss. There is no danger of physical damage to crops, which is an important advantage for fruit and vegetable crops near harvest time. Chemical weed control, may entail a risk of chemical residues being left on the crops.
However, electricity is not suitable as the primary method of weed control at densities of 200 stems m^{-2} and over. Under such conditions, electrical weed control

Table 6. Comparison of total energy input for electrical, mechanical and chemical weed control techniques.

Weed control technique	Weed density (no m^{-2})	No. of passes	Application time	Formulation[a]	Rate[b] (kg ha^{-1})	Average total energy input (MJ ha^{-1})
Electrical	5	1	POE[c]	-[d]	-	418
	30	1	POE	-	-	2460
	200	1	POE	-	-	16500
Mechanical						
Corn						
Rigid tine harrow	-	2	PRE,POE	-	-	44
Danish S-tine cultivator	-	1	POE	-	-	34
Bean						
Rotary hoe	-	3	PRE,POE	-	-	23
Danish S-tine cultivator	-	1	POE	-	-	34
Chemical						
Corn	-	1	PRE	EC+SU	1.9-2.6+1.0-1.5	615
		1	POE	EC	0.3	45
Bean	-	1	PRE	SU	1.0-1.7	203
	-	1	POE	SN	0.8-1.1	144

[a] EC = emulsifiable concentrate; SU = suspension; SN = solution.

[b] Rates recommended in Ontario, active ingredient basis.

[c] POE = postemergence; PRE = preemergence.

[d] Not available.

requires approximately 20 times more energy and takes 50 times longer than chemical methods (Vigneault et al. 1990). Even at a low weed density of 15 plants m^{-2}, the electrical method requires approximately twice as much energy and takes five times longer than the chemical method. Furthermore, the electrical method, like its mechanical counterpart, requires several passes to obtain an acceptable level of weed control, compared to only one treatment for chemical weed control.

6. Conclusion

If electrical weed control is to be cost-competitive with conventional methods, several improvements need to be considered (Vigneault et al. 1990). An improved electrode design having at least two diagonally opposite contact points would increase the efficiency of current flow between the electrode and the plant by up to 13% (Martens and Vigoureux 1983). The efficiency of the transfer of available energy from the tractor to the electrode could also be improved. Tractor speed could be increased to meet the requirements of an electrical machine with a better hourly operating capacity. The effect of the electric current frequency used for plants at different stages of growth should be analysed to determine which frequency is the quickest and most energy efficient (Martens and Vigoureux 1983). To avoid wasting energy, a monitoring system could be devised to identify the point at which current flow should be stopped–essentially the lethal energy threshold–because sufficient damage has already been done (Diprose et al. 1980). Further research is needed in order to quantify the resistance of different species in terms of their phenological stage and the mean lethal energy to be applied (LD_{50}), as well as to identify associated factors.

The electrical method is slower, less efficient and requires more energy to destroy small weed seedlings than conventional control techniques. It may, nevertheless, be justified for weed suppression in crops of high commercial value, such as herbs. This method is also cost-effective in fields where the area to be treated is small, in locations where persistent herbicides are not acceptable, and in regions where there is a risk of erosion or where a soil management program is in effect. In addition, electricity can be used to destroy weeds in certain situations associated with crop rotation, such as sunflowers (*Helianthus annuus* L.) growing in soybean fields (*Glycine max* L. Merr), and weed beet or bolting beet in sugar beet crops. The electrical method will not damage the crop plants given the substantial height difference; hence it represents an efficient approach in such contexts.

References

Anonymous, (1990). Agricultural machinery management data. Agricultural Engineers Yearbook. Am. Soc. Agric. Eng., St. Joseph, MI. ASAE D. 497 p.

Anonymous, (1996). Guide to Weed Control. Ontario Ministry of Agriculture, Food and Rural Affairs. Publ. 75. 264 p.

Chandler J.M., (1978). Crops and weed response to electrical discharge. Proc. South. Weed Sci. Soc. 31:63.

Diprose M.F., Benson F.A., (1984). Electrical methods of killing plants. J. Agric. Eng. Res. 30:197-209.

Diprose M.F., Benson F.A., Hackam R., (1980). Electrothermal control of weed beet and bolting sugar beet. Weed Res. 20: 311-322.

Diprose M.F., Benson F.A., Willis A.J., (1984). The effect of externally applied electrostatic fields, microwave radiation and electric currents on plants and other organisms, with special reference to weed control. Bot. Rev. 50:171-223.

Diprose M.F., Fletcher R., Longden P.C., Champion J., (1985). Use of electricity to control bolters in sugar beet (*Beta vulgaris* L.): a comparison of the electrothermal with chemical and mechanical cutting methods. Weed Res. 25: 53-60.

Diprose M.F., Hackam R., Benson F.S., (1978). Weed control by high voltage electric shocks. Proc. Br. Crop Prot. Conf. - Weeds 2: 443-450.

Drolet C., Rioux R., (1983). Evaluation d'une rampe utilisant un courant électrique pour le contrôle des mauvaises herbes. Res. Branch, Agric. Can., Ottawa. ERDAF Rep. No. 345Z.01843-1-EC24. 66 p.

Dykes W.G., (1980). Principles and practices of electrical weed control in row crops. Am. Soc. Agric. Eng., St. Joseph, MI. Paper 80-1007. 9 p.

Guiraud D., Givelet M., (1981). Destruction électrothermique des betteraves montées. Compte rendu, II° conférence du COLUMA 1: 54-62.

Hackam R., (1985). Development of a commercial scale high voltage machine for weed control in row crops. Res. Branch, Agric. Can., Ottawa. ERDAF Rep. No. 03SU.01916-2-EC35. 122 p.

Kaufman K.R., Schaffner L.W., (1980). Energy requirements and economic analysis of electrical weed control. Am. Soc. Agric. Eng., San Antonio, TX. Paper 80-1007. 10 p.

Kaufman K.R., Schaffner L.W., (1982). Energy and economics of electrical weed control. Trans. Am. Soc. Agric. Eng. 25: 297-300.

Lutman P.J.W., (1980). A review of techniques that utilize height differences between crops and weeds to achieve selectivity. Pages 291-317 in J. O. Walker, ed. Spraying Systems for the 1980s. Crop Prot. Counc. Monogr. No. 24.

Rasmusson D.D., Dexter A.G., Warren H., (1979). The use of electricity to control weeds. Proc. North Cent. Weed Control Conf. 34: 66.

Roberts H.A., (1967). The problem of weed seeds in the soil in crop production in a weed free environment. Oxford Symp., Br. Weed Control Counc. No. 2. Blackwell Scientific Publ. Pages 73-82.

Sanwald E., Koch W., (1978). Physical methods of weed control. Proc. Br. Crop. Prot. Conf. - Weeds 3: 977-986.

Southwell P.H., Rothwell T.M., (1977). Analysis of output/input energy ratios of food production in Ontario. Res. Branch. Agric. Can., Ottawa. Contract No. OSW76-00048. 419 p.

Uvarov E.B., Chapman D.R., Isaacs A., (1971). A Dictionary of Science. 4th ed. Penguin Books, Inc., Baltimore, MD. 443 p.

Vigneault C., (1985). Use of electrocution as the primary means of weed control in row crops. Am. Soc. Agric. Eng., St. Joseph, MI. Paper 85-1507. 8 p.

Vigoureux A., (1981). Results of trials carried out in Belgium in 1980 about killing weed beets by electric discharge. Meded. Fac. Landbouww. Ryksuniv. Gent. 46: 163-172.

Vigoureux A., (1982). Mécanisation de la destruction des montées à graines en culture betteravière. Publ. Trimest. Inst. R. Belge L'Amélior. Betterave. 50: 3-36.

Mechanical control

Mechanical Weed Control in Agriculture

Daniel C. Cloutier and Maryse L. Leblanc

1 Introduction

Weeds are plants that are considered undesirable in a crop at a given time. Weeds are harmful for a number of reasons. They reduce crop yields, interfere with the harvest, support pathogens and insect pests and contaminate seeds.

Weed control is as old as farming itself. However, progress in mechanized weed management did not begin until the early 18th century, when Jethro Tull invented a seed planter for row crops, which allowed weeds between the rows to be killed by cultivation.

Physical control was the main method used against weeds until herbicides appeared in the mid-20th century (Wicks et al. 1995). Mechanical weed control is a proven technique that kept fields free of weeds long before the advent of herbicides. This technique has experienced somewhat of a rebirth in the past few years. The objective of this chapter is to present the principles behind mechanical weeding and to give some examples to illustrate them.

2 Overview of Physical Control Methods

Physical weed management is based on several different techniques: manual weed removal, pulling, mowing, smothering with mulch, thermal methods (electricity, heat or cold), flooding and tillage.

Manual weed removal and pulling are common around the world, and it has been estimated that 50–70% of the world's farmers control weeds with these methods (Hill 1982; Wicks et al. 1995). Pulling is usually done by hand, although a mechanical weed puller has been developed in the United States to remove tall weeds from a shorter crop (Wicks et al. 1995). In tropical areas, manual removal is often the principal method used, since labour is abundant and cheap. Hand pulling is used for valuable crops for which high labour costs are justified. It can also be employed to eliminate weed escapes (preventing an uncommon weed from becoming widespread), to prevent surviving weeds from producing seed, to destroy a

pocket of infestation or to ensure compliance with seed certification programs. Using a hand tool is easier and faster than weeding strictly by hand; tools include various types of hoes and weeding spuds (Clément 1981; Habault 1983).
Mowing and cutting are other weed management methods commonly used in orchards or in establishing forage crops, in the latter case allowing the crop to overgrow the weeds and become better established. These techniques are also used to control the height of weeds and minimize competition with the crop plants in orchards and vineyards (Kempen and Greil 1985). Mowing and cutting are also used to prevent weed seed production (Ross and Lembi 1985).
Mulching is also widely used to control weeds. Mulches are applied before or after the crop has become established. Mulches can be divided into two categories: natural and synthetic mulches. Natural materials used for mulching include straw, sawdust, plant residues and crushed stones, while plastic, paper, cardboard and synthetic fibres are among the synthetic materials used. Mulches prevent weeds from emerging by forming a physical barrier and excluding light. In hot, sunny regions, the use of plastic mulches can also destroy weeds by solarization (Braun et al. 1988; Silveira et al. 1993). This technique consists in leaving plastic mulch, usually clear, on the soil for several weeks during a period of high solar radiation to sterilize the soil and destroy weeds by heat.
Thermal methods may be based on high or low temperatures to kill weeds. Heat-based methods include electricity (Vigneault and Benoit, Chap. 12), infrared radiation, microwaves, hot water, steam and flaming (Ascard 1995). Farmers can choose between nonselective flaming, done over the entire field, or selective flaming, which is done by directing the flamer on the weeds to avoid damaging the crop.
Flooding is another weed control method, often used in rice (*Oryza sativa* L.) and cranberry (*Vaccinium macrocarpon* Ait.) cultivation. Completely submerging the weeds smothers them. When feasible, this technique is very effective (Schlesselman et al. 1985).
After hand weeding, cultivation is the most common mechanical method of weed control.

3 Cultivation

Mechanical cultivation methods for weed control are often divided into three categories: primary, secondary and tertiary tillage (Wicks et al. 1995; Hahn and Rosentreter 1989).
Primary tillage is used to break, turn over, loosen or stir the soil and is the first in a series of treatments to prepare the soil for planting. Primary methods are aggressive. This cultivation is carried out at a considerable depth, leaving an uneven soil surface. Primary tillage destroys any actively growing vegetation, eliminates residues and loosens the soil. This type of tillage controls weeds by burying the seeds (thus preventing emergence) or burying the vegetative reproductive structures of perennial weeds more deeply (destroying or damaging them by exposure to the

cold and air). Primary tillage tools comprise mouldboard ploughs, disk ploughs, chisel ploughs and field cultivators.
In secondary tillage, the soil is not worked as deeply as in primary tillage. This type of cultivation is used after ploughing to level the ground, prepare the seedbed and incorporate such substances as fertilizer, lime and manure into the soil. Secondary tillage is also used to destroy weeds before seeding, notably in the stale seedbed technique (Leblanc and Cloutier 1996). This method consists of preparing the seedbed and leaving it for 2 or 3 weeks to encourage the weed seeds to germinate. Then, germinated weeds are cultivated. Secondary tillage equipment includes various types of harrows, cultivators and tillers.
Tertiary tillage consists of hoeing and cultivating during the growing season. Hoeing designates the breaking of the surface crust and loosening of soil around crop plants (Clément 1981). Hoeing aerates the soil, destroys weeds and breaks up soil capillaries, thus preventing the evaporation of water from the soil. In hoeing, the soil is worked at a shallow depth. The purpose of cultivating is to destroy weeds. Hoeing and cultivating are often confused since they involve the same tools and their effects are similar.

4 Cultivators

The term cultivator is used here in a generic sense to refer to machines and tools used for mechanical weed control.

4.1 Mode of Action

Mechanical cultivation destroys weeds in several different ways. After a cultivator passes over a field, the main cause of mortality in weeds is smothering by complete or partial burial (Rasmussen 1992). Cultivators also uproot weeds, exposing the roots at the soil surface (Weber and Meyer 1993). Other actions include breaking the contact between the roots and soil, tearing the plant and depleting the weed seedbank (annuals) or propagules (perennials) (Ross and Lembi 1985). Cultivation is more effective in dry soils. When rainfall occurs right after cultivation, its effec tiveness may be reduced because the weeds can reroot. Cultivating in wet soil also causes clods and soil compaction and encourages the spread of perennial weeds.

4.2 Types of Cultivators

Cultivators can be divided into three types: those operated between the crop rows (inter-row cultivators), near the crop rows (near-row cultivators) and across the rows (broadcast or intra-row cultivators). The optimum working depth is generally around 3 cm; however, this depth can be adjusted on most cultivators by means of the tractor's three-point hitch or the depth wheels attached to the cultivator frame. Except in blind (broadcast) cultivation, the cultivator must cover the same number of rows (or whole fraction thereof) as the seeder or transplanter, since two adjacent passes of the seeder or transplanter are never exactly parallel (nor is the spacing between two passes equal to that between two rows done by the seeder). Farmers can avoid the problem by choosing a cultivator with the same width as the seeder or transplanter.

4.3 Inter-Row Cultivation

Inter-row cultivation is by far the most widely used and effective approach. There is minimal risk to the crop and weed control is excellent. With the proper equipment, even shrubs or tree saplings can be destroyed. The only constraints are crop height and growth stage. It is best to perform inter-row cultivation early in the season. Although one late-season pass is feasible, the risk of cutting the crop roots is usually too high. On the other hand, early treatment promotes deep rooting of the crop. When weeds are too well developed, the implements will get clogged with weeds and damage the crop.

Shields can be used early in the season to keep the dirt thrown up by the cultivator from burying or breaking the crop seedlings. Cultivator shields come in the shape of tunnels, protective screens, disks or toothed wheels.

Automatic guidance systems (mechanical or electronic) allow cultivation to be done at greater speeds and reduce the risk of crop damage. In mechanical systems, guide wheels follow the ridges or shallow or deep furrows created during seeding. Electronic systems use sensors or crop-sensing wands that determine the position of the crop row and automatically position the cultivator.

The effectiveness of an inter-row cultivator depends on the proportion of the field that is treated. Since these cultivators cannot go too near the crop row, they can only cover 50 to 70% of the total area. Inter-row cultivators use mainly two types of shanks: vibrating and rigid shanks. Each cultivator consists of a shank, usually long and narrow, which ends in a sweep. The shank connects the sweep to the frame.

4.3.1 Vibrating Shanks

Cultivators using this type of shank are called light-duty cultivators. The most commonly used shanks are Danish S-tines. These shanks are S-shaped and support various types of sweeps, including goosefoot sweeps, uprooting sweeps, and sweeps that are narrow and straight, narrow and curved, or rounded. The shanks vibrate in all directions, pulverizing the soil and killing weeds less than 15 cm high. There are several types of vibrating shanks.

4.3.2 Rigid Shanks

Heavy cultivators use this type of shank. The most effective models use a rigid shank to which wide, sharp sweeps are attached. This type of cultivator is effective against shrubs and tree saplings as well as in fields with heavy crop residues, as are common in no-till and ridge-till production systems. However, a more powerful tractor is needed than for other cultivators. The rigid shank type of cultivator is not used for horticultural crops.

4.4 Near-Row Cultivators

Other cultivators are designed to get as close as possible to the crop row using brushes, disks or special blades, which are mounted on one or more toolbars. They are used mainly for horticultural crops. The accuracy required for operating near the row is obtained either by having a second person seated on a rear-mounted cultivator to steer it, or by using an open frame design with one or more toolbars mounted in front of the driver. In the second case, the tractor seat and engine are offset, allowing the driver to see the crop row.

4.4.1 Brush Weeders

Brush weeders have a number of common characteristics. The rotating brush are driven by the tractor PTO. The bristles on these rolls are made of fibreglass and are

Fig. 1. Horizontal-axis brush weeder manufactured by Bärtschi-Fobro (Switzerland).

Fig. 2. Vertical-axis brush weeder manufactured by Svensk Ekologimaskin AB (Sweden).

flexible. A second person is needed beside the tractor driver to steer the brushes, so the weeder can go as close as possible to the crop without damaging it.

The brush weeder manufactured by the Swiss company Bärtschi-Fobro uses horizontal-axis rotary brushes, each measuring 50 to 76 cm in diameter. Weeds are killed by uprooting, burial or breaking. The implement has little effect on the crop since its action is focused on the inter-row. The crop plants are covered by a protective tunnel, which keeps them from getting buried (Fig. 1).

Some models such as the Thermec Brush Weeder (Svensk Ekologimaskin AB, Sweden) use vertical-axis brushes. The angle, rpm and rotating direction of the brushes can all be adjusted. The brushes can weed very close to the row or in the row. They can also be used to earth up the crop or remove soil and weeds from the row, depending on the direction in which the brushes rotate (Fig. 2).

4.4.2 Disks

Disks are also used on near-row cultivators. Steered by a second person, they cut the soil very close to the crop plants. They can be operated in well-crusted soils. Shanks are used to weed the inter-row portion.

4.4.3 Blades

Many models of cultivators use various types of blade to control weeds near the crop row. Some require a second person to steer the cultivator manually. One of the first models to be developed has a semi-mounted frame that is steered laterally

Fig. 3. Flexible blade cultivator, manufactured by Bezzerides (United States).

using pedals that turn the wheels, allowing the cultivator to pass as close as possible to the crop row. The shanks are either fixed or steered by hand.

Systems without manual steering have a toolbar which allows the farmer to adjust the spacing on the cultivator to match the production system in use. The tools used are either traditional shanks or specialised blades. Manufacturers offer a wide variety of implements. Shanks may consist of very sharp blades or square or rounded metal rods ranging from several millimetres to several centimetres thick. The blades can be rigid or curved and highly flexible. Some farmers have created their own blades. Shanks work at ground level or a few centimetres below the soil surface, cutting the weeds as they pass. These cultivators are mounted on tractors with an open front frame design or with offset seats and engines (in the latter case, the engine is offset laterally from the central axis of the tractor, allowing the driver to see the crop row and the position of the cultivator in relation to the crop row so that any necessary adjustments can be made quickly). One or more rows can be cultivated at the same time (Fig. 3).

4.4.4 Rotary Cultivators

Rotary cultivators are driven by the tractor's PTO and have a vertical, horizontal or oblique axis. They are equipped with blades, points or knives that turn, and pulverize the soil. Used for inter-row weed control, rotary cultivators are very effective and can operate as close to the crop row as horizontal brushes and disks. These weeders are also called rotary tillers or rototillers.

4.4.5 Basket Weeders

Fig. 4. Basket weeder, manufactured by Buddingh (United States).

These cultivators have basket weeders made of quarter-inch spring wire, which rotate at a speed determined by the speed of the tractor as they are pulled over the ground. The Buddingh Model K (United States) (Fig. 4) has two horizontal axles perpendicular to the tractor axis. The axles bear the baskets and are connected to a chain drive that turns the rear rotor faster than the front rotor. The first set of baskets loosens the soil and the second pulverizes it, uprooting the young weed seedlings. The German company Kress also manufactures a basket weeder.

Fig. 5. Rotary hoe, various manufacturers (United States).

4.5 Broadcast or Intra-Row Cultivators

Cultivation in the row can be done before or after crop emergence.

4.5.1 Pre-Emergence Cultivation

Broadcast, or blind, cultivation is so named because the cultivator passes over the crop as well as the weeds. The most commonly used cultivators for this type of operation consist of rotary hoes, rigid-tine harrows, flex-tine harrows and chain harrows. The rotary hoe is a harrow with spiked wheels which is pulled by a tractor (Fig. 5). [In France the term is used to designate a rototiller (Habault 1983), i.e. a PTO-driven transverse horizontal rotary cultivator (Clément 1981).] The tool leaves small dents in hard soil. It consists of a set of rolling wheels on a single horizontal axle. Each wheel is mounted on a spring-loaded arm. The arms are attached to the toolbar in such a way that there are two sets of ground driven wheels: one in front that projects the soil and another in back that pulls out and buries the remaining weeds. Each wheel consists of roughly 16 spokes with a spoon-shaped tip (Fig. 6), although the number may vary from one model to another. The implement is usually between 4.6 and 9.5 m wide.

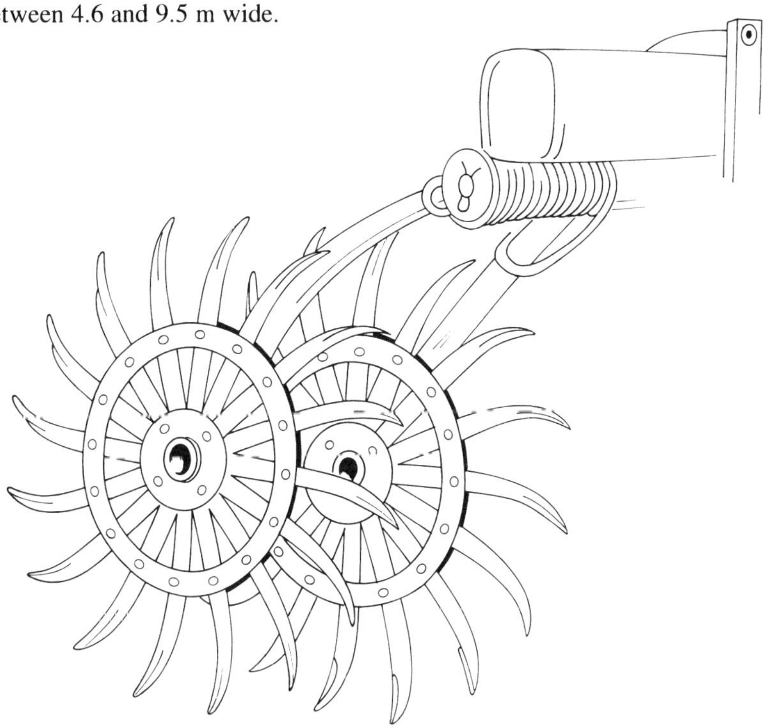

Fig. 6. Drawing of rotary hoe.

Fig. 7. Flex-tine harrow, various manufacturers (Europe).

Chain harrows, or flex-tine cultivators, have short shanks fitted on chains rather than a rigid frame so that they hug the ground. They are especially effective on light soils. Other harrows have rigid frames and a variety of shanks. Flex-tine cultivators have fine, flexible teeth that destroy weeds by vibrating in all directions (Fig. 7). Rigid-tine harrows, best for heavy soils, consist of several sets of rods or rigid blades angled at the tip (Fig. 8) that are mounted on a rigid frame or floating sections. The rods or blades vibrate perpendicularly to the direction in which the tractor is moving. The tension on the shanks can be adjusted individually or collectively, depending on the model, allowing the farmer to choose the intensity of the treatment. The working width ranges from 4.5 to over 20 m. Cultivation depth

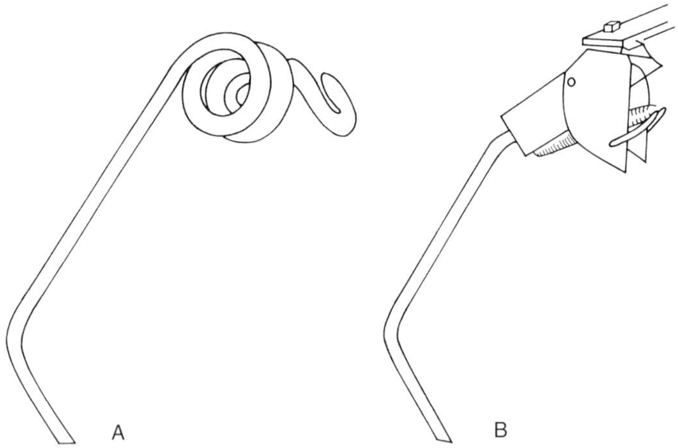

Fig. 8 A.B. Harrow shanks: **A** Flex-tine shank, manufactured by Hatzenbichler (Europe). **B** Rigid-tine shank, manufactured by Rabe Werk (Europe).

is adjusted by depth wheels on the harrow or by the tractor's hydraulic system. Rotary hoes and harrows are produced by a large number of manufacturers.
Pre-emergence cultivation is selective because the crop seeds are planted more deeply than the weed seeds or are larger than the weed seeds, and are therefore not affected or only slightly affected by cultivation. This is a gentle treatment that destroys only white-stage weeds, dicot seedlings before the two-leaf stage and monocots at the one-leaf stage. Pre-emergence cultivation is employed in large seeded crops such as corn, cereals, sugar beets, beans and peas.

4.5.2 Post-Emergence Cultivation

Post-emergence broadcast cultivation can be done with a rotary hoe or the various types of harrows described in the previous section. Treatment is selective given the fact that the crop is better rooted than the weeds. Since the crop has larger seeds (and therefore more energy reserves) or is transplanted, it becomes established faster than the weeds. This type of cultivation works well with weeds that have germinated but have not emerged or weeds that are at the one-or two-leaf stage. In this case, rotary hoes can be operated at high speeds, but harrows are pulled slowly to avoid damaging the crop.
The finger weeder, the Buddingh Model C (United States), has weeders shaped like sawed-off cones (Fig. 9). Each cone has rubber fingers that point outwards (horizontally) and metal ones that point downward (vertically). Each pair of cones is used for a single row, with one cone positioned on either side of the row. The cones are ground-driven by the metal fingers which are set back from the row to avoid damaging the crop. They work the soil like a vertical-axis rotary hoe. The rubber fingers mesh together over the row, pulling out the weeds in the process. The dis-

Fig. 9. Intra-row cultivator, Model C, made by Buddingh (United States).

tance between the cones can be adjusted according to the growth stage of the crop. This type of cultivator is effective only against young weed seedlings and is gentle to the crop. Kress, a German company, manufactures this type of weeder too.
Although most cultivators are designed to minimize the amount of soil thrown onto the crop row, specialized models have been developed solely for that purpose, that is, to ridge or hill the soil. Ridging is used as a production method, while hilling consists of piling soil around the crop. Ridge-tilled corn is an important production system in North America. The principle of the ridger or hiller is the same as that of inter-row cultivators. As the tool passes between the rows, the soil is projected into the row by a hilling sweep, disk or specialized sweep, killing weeds by smothering them. Repeated passes may be made. This technique is used for crops that can tolerate partial covering. Hilling is used for potatoes, corn, leeks, asparagus, artichokes and other crops. Weed control is enhanced when weeds are no taller than 50 cm.

5 Conclusion

Before the advent of herbicides, mechanical methods of weed control were used successfully for several centuries. Cultivator technology continued to evolve even after the development of herbicides, and these implements are efficient and versatile. In some cases, they are the only weed control tools available and they are often a cost-effective alternative to herbicides.
Cultivators are agricultural implements that require careful adjustment to ensure optimal performance. To kill a maximum number of weeds, cultivators should be operated as close to crop rows as possible without injuring the crop. The effective use of cultivators requires a fair amount of experience and careful observation, which may explain why research teams arrive at widely varying conclusions in similar situations.
The effectiveness of cultivation is directly influenced by cultivation depth and degree of soil moisture. Cultivation that is too shallow may spare weeds and cultivation that is too deep increases the risk of crop damage. Working depth can be adjusted by means of wheels attached to the frame or the three-point hitch. The use of weights and reduction of tractor speed can also increase the operating depth. Cultivating when the soil is too wet leads to clod formation and may not destroy weeds. The optimal level of soil moisture depends on the cultivator type, with rotary hoes and harrows being best suited to moist soils.
In general, two types of cultivators are required for effective weed control: one at pre-emergence or early post-emergence and a second later in the season. The vulnerability of crops to damage depends on their growth stage. For example, legumes are most vulnerable at the hook stage, when cultivation can reduce yield. In Quebec, two to seven cultivations are usually carried out, depending on the crop and the degree of weed infestation. Tractor speeds range from 3 to 20 km h^{-1} depending on the cultivator type and the growth stage of the crop.

Cultivation is not only effective in controlling weeds; it also benefits the crop by breaking up the surface crust, aerating the soil, stimulating the activity of soil microflora, reducing the evaporation of soil moisture and facilitating the infiltration of rainwater.
Cultivator selection is only one component of an effective weed control program. Technical mastery of the cultivator is critical, as is weeding at the appropriate growth stage of the weeds and the crop. Delaying treatment for a few days may significantly reduce the effectiveness of a cultivation operation. The timing of treatment is probably more critical in successful weed control than the choice of cultivator.

Acknowledgements. We would like to thank Pierre Jobin and Yvon Douville of the Centre de Développement d'Agrobiologie (Agrobiology Research Centre) at St. Élisabeth de Warwick and Pierre Lachance of the Montérégie (Eastern Sector) regional directorate of the Quebec Department of Agriculture, Fisheries and Food. We would also like to thank all the farmers who shared their knowledge with us.

References

Ascard J., (1995). Thermal weed control by flaming: Biological and technical aspects. Report no. 200, Swedish University of Agricultural Sciences, Department of Agricultural Engineering, Alnarp. 189 p.

Braun M., Koch W., Mussa H.H., Stiefvater M., (1998). Solarization for weed and pest control - Possibilities and limitations, pp. 169-178, in R. Cavalloro and A. El Titi (eds). Weed control in vegetable production, Proc. EC Experts Group, Stuttgart, 28-31 Oct. 1986, A.A. Balkema, Rotterdam, 303 p.

Clément J.-M., (1981). Larousse Agricole. Librairie Larousse, Paris, 1208 p.

Habault P., (1983). Lexique de termes agricoles et horticoles: termes scientifiques, techniques et économiques. J. B. Baillière, Paris, 151 p.

Hahn R.H., Rosentreter E.E. (Eds.), (1989). ASAE Standards 1989: Standards, engineering practices and data developed and adopted by the American Society of Agricultural Engineers, 36th edition. ASAE, St-Joseph, 659 p.

Hill G.D., (1982). Impact of weed science and agricultural chemicals on farm productivity in the 1980's. Weed. Sci. 30:426-429.

Kempen H.M., Greil J., (1985). Chapter 4: Mechanical Control Methods, pp. 51-62, in E.A. Kurtz and F.O. Colbert (eds.). Principles of Weed Control in California, Thomson Publications, Fresno, CA, 474 p.

Leblanc M.L., Cloutier D.C., (1996). Effet de la technique du faux semis sur la levée des adventices annuelles. Xième Colloque International sur la Biologie des Mauvaises Herbes à Dijon (11-13 septembre 1996), Annales de l'Association Nationale pour la Protection des Plantes 10: 29-34.

Rasmussen J., (1992). Testing harrows for mechanical control of annual weeds in agricultural crops. Weed Research 32:267-274.

Ross M.A., Lembi C.A., (1985). Applied Weed Science. Burgess Publ. Co., Minneapolis, Mn, 340 p.

Schlesselman J.T., Ritenour G.L., Hile M.M.S., (1985). Chapter 3: Cultural and Physical Control Methods, pp. 35-49, in E.A. Kurtz and F.O. Colbert (eds.). Principles of Weed Control in California, Thomson Publications, Fresno, CA, 474 p.

Silveira H.L., Caixinhas M.L., Gomes R., (1993). Solarisation du sol, mauvaises herbes et production, pp. 6-13, in J.-M. Thomas (ed.). Maîtrise des adventices par voie non chimique; errata et complément, Thème 3, Comm. Fourth International Conf. of the International Federation of the Organic Agriculture Movement, ÉNITA, Quétigny (France), 38 p.

Weber H., Meyer J., (1993). Mechanical weed control with a brush hoe, pp. 89-92, in J.-M. Thomas (ed.). Maîtrise des adventices par voie non chimique, 4th International Conf. of the International Federation of the Organic Agriculture Movement, ÉNITA, Quétigny (France), 393 p.
Wicks G.A., Burnside O.C., Felton W.L., (1995). Mechanical Weed Management, pp. 51-99, in A.E. Smith (ed.). Handbook of Weed Management Systems, Marcel Dekker Inc., New York, 741 p.

Mechanical Weed Control in Corn (*Zea mays* L.)

Maryse L. LEBLANC and Daniel C. CLOUTIER

1 Introduction

Mechanical weed control in corn was practised as early as the 19th century. Over the past 30 years, however, effective selective herbicides have more or less replaced mechanical cultivation (Lampkin 1990). Although cultivation, or tillage, is still done because of the benefits to the soil, weed control is performed through an early-season application of herbicide. Tillage not only controls weeds but loosens the soil and breaks the surface crust, a common problem in corn growing. Crusts tend to form in silty clay soils after a period of rain followed by hot, windy weather. The crust slows oxygen diffusion and reduces heat transfer, makes emergence difficult for corn seedlings, and has a negative impact on crop uniformity. Removing the surface crust by cultivation also promotes mineralization of the nutrients required by corn (Souty and Rode 1994). In addition, cultivation helps to preserve soil moisture needed for plant growth, since the layer of loosened soil limits the capillary rise of moisture. This function of cultivation is most effective in regions with a dry climate and when the corn root system is not very well developed. When the roots are well distributed throughout the soil or the foliage provides shade, little moisture is lost even if the field has not been cultivated.
The exclusively mechanical weed control strategy explained here requires two types of treatments: an early pass over the entire field (broadcast cultivation), followed by a later pass between the crop rows (inter-row cultivation).

2 Broadcast Cultivation

The main problem in using cultivation for weed control in corn is removing the weeds from the crop row. Farmers have traditionally avoided passing over the rows, for fear of causing irreversible damage to the corn plants and seeing their yields decline. Today, tools are available that are kinder to crops. They pull out the weeds while they are still at an early stage of growth and have relatively poor root systems, but do little damage to the corn plants whose root systems are better developed. These implements can be pulled at a speed at least twice as fast as conventional cultivators and are generally used at the pre-emergence or early post-emergence stage.

Box 1. How Does Cultivation Destroy Weeds?
C. LaHovary

Cultivation can pull up, cut or bury a seedling. Although both dicots and monocots can be killed by pulling them up, only dicots can be easily destroyed just by cutting. Grasses (monocots) are more difficult to control by cutting than dicots because their meristem is protected by the sheath and remains intact when cut. However, as monocot growth nears the shooting stage, the meristem is increasingly exposed and vulnerable. Burying or smothering weeds is an effective complement to cutting or pulling them out, particularly when weeds are young. Dicots can be controlled by burial alone with 50% success, whereas grasses are more difficult to bury because of their upright growth habit. If the soil becomes wet after burial, this will benefit the weed since the seedling that has been pulled out will take root and begin to grow again. Furthermore, abundant rainfall reduces the effectiveness of burial by scattering the soil. Pulling up seedlings seems to be the best way of controlling weeds since it breaks the soil-root bond. Cutting slows growth but may leave meristem areas intact, particularly in grasses. One of the most important elements in cultivation is burial of seedlings, whether they have been pulled up or cut. Pulling up followed by burial is the optimum method of destroying weeds mechanically.

2.1 Cultivation Tools Used

The rotary hoe (Cloutier and Leblanc, Chap. 13) is used at depths between 2 and 5 cm, depending on the soil type and moisture conditions. Weights can be added to improve soil penetration when the crust offers excessive resistance. As the tractor moves forward, it drags the spiked wheels along with it, making them turn. The minimum speed required for effective weeding, is 10 km h^{-1}, although speeds of 20 km h^{-1} are common in the field. A wide variety of models are available on the market, including high ground clearance models for fields with considerable residues.
Harrows (rigid-tine and flex-tine) are more aggressive than the rotary hoe. The tractor speed used depends on the growth stage of the crop. In pre-emergence crops, the harrow can be operated at speeds of up to 15 km h^{-1}, compared with half this speed in a post-emergence situation. Reducing the tension on the tines allows faster cultivation. Several test runs are required to adjust the implement properly so as to prevent crop damage and control the weeds effectively. There is no need to work the soil at depths greater than 5 cm. This type of harrow does not work well in residues.

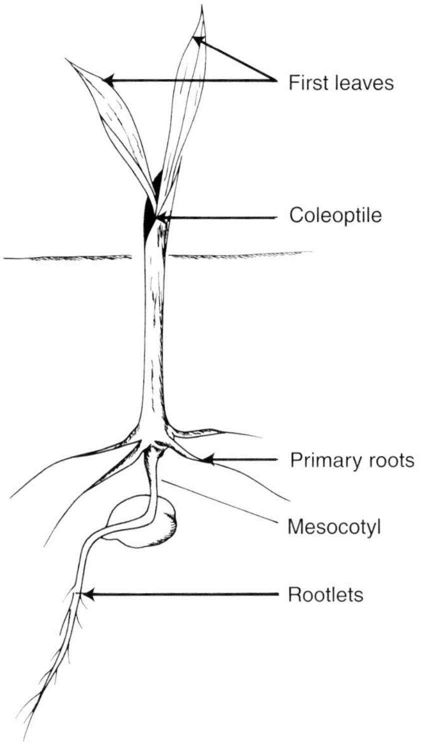

Fig. 1. Corn seedling.

2.2 Damage to Crop

Pre-emergence cultivation can be done faster and more aggressively than post-emergence cultivation, without harming the crop. The ideal time is roughly 24 h before emergence. Corn takes roughly a week to come up, but this period can be longer or shorter depending on the region and climatic conditions. Emergence can also be deliberately prolonged by seeding at a greater depth, providing farmers with greater leeway for synchronizing cultivation with weed seed germination. Planting at depths of 5 cm or more may, however, reduce seedling vigor.

In corn, the establishment of the seminal root system, formation of the mesocotyl (first internode) and tillering node and the elongation of the mesocotyl and coleoptile occur between germination and emergence (Fig. 1). The coleoptile, a sheath enclosing the young leaves, is pushed upward by the elongation of the mesocotyl until its base reaches the soil surface. There, inhibited by light, its edges spread apart at the tip, releasing the first foliage leaf. The length of the coleoptile is generally invariable in a given variety, while the length of the mesocotyl depends on the depth of planting, ranging from very short to up ten centimetres in the case of a deeply planted seed (Gay 1984). When the seed is too deep, the plant expends most of its reserves in producing the mesocotyl and therefore emerges in a much weakened state.

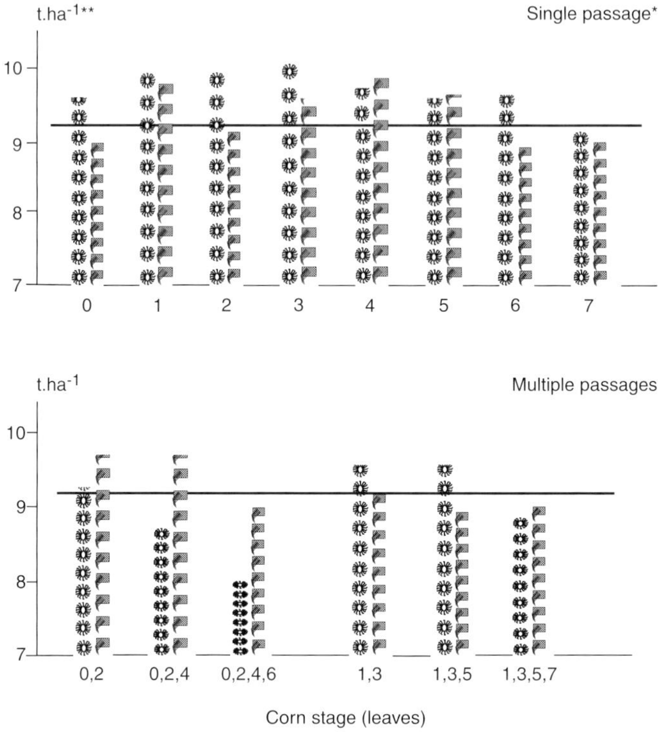

Fig. 2. Results of a 1994-95 study in Quebec to determine potential damage to different stages of corn from single and multiple passes of a rotary hoe and rigid-tine harrow. Soil types: clay and sandy loams; cultivators: Yetter rotary hoe and RabeWerk rigid-tine harrow (tines at low tension); tillage depth: 4 cm; speed: 15-18 km h^{-1}.
* In order to study the effect of weeders only and to avoid confounding their physical effects with the competing effects of weeds on corn yield, herbicides were applied on all surfaces (1 kg ha^{-1} atrazine, 1.9 kg ha^{-1} metolachlor). ** Field corn yield at 15% humidity.

After emergence, broadcast cultivation options are more limited. Both harrows and hoes have little negative impact on yield, provided they are used before the six-leaf stage no more than three times a season (Fig. 2). Rigid-tine harrows are generally more aggressive than hoes, usually bending down the corn as they pass over it (the plants are flexible, however, and will spring back after a few days). Around the end of the seven-leaf stage, cultivation is risky because the apex is at the soil surface and the tassel has begun to develop (Gay 1984). Any damage at this stage channels resources to the foliage and roots for repair purposes, resulting in decreased resources for cob production.

2.3 Effectiveness in Weed Control

Table 1. Relationship between growth stage of *Chenopodium album* L. and effectiveness of weed control using a rotary hoe (Douville et al. 1995).

Stage	Control (%)
Cotyledon	90-100
2-leaf	65
4-leaf	35

The younger the weeds, the more effective cultivation is. The best time for cultivation is when weed seeds have germinated but have not yet emerged. (To determine if this has occurred, take a handful of soil and examine the weed seeds to see if they have begun to sprout.) The young shoot is fragile and easily destroyed by the teeth of the hoe or harrow. Once weeds have emerged, cultivation should be done immediately since, according to recent studies, the effectiveness of tillage decreases rapidly as weed seedlings develop, becoming almost nil when weeds reach the three-to four-leaf stage (Table 1). Rotary hoes are excellent for controlling weeds at the cotyledon stage, while flex-tine and rigid-tine harrows, which are slightly more aggressive, are effective until the one-leaf stage (Douville et al. 1995). As the season progresses, all these implements become less effective, since the weeds are better rooted. Therefore, cultivation regime cannot be based strictly on calendar date, and the growth stage of the weeds must be taken into account in deciding when to cultivate. Some weed species (e.g. *Ambrosia artemisiifolia* L.) are more vigorous and establish root systems more quickly than others; hence, they must be treated at an earlier stage of growth. Other species germinate at greater soil depths (e.g. *Avena fatua* L.) and are more difficult to destroy with a rotary hoe or harrow. Readers should also note that rotary hoes and harrows are not effective in controlling perennial weeds.

3 Inter-Row Cultivation

When corn has reached a certain height, a second type of cultivator is used strictly in the inter-row area, generally a goosefoot sweep or a Danish S-tine. Inter-row cultivation on its own controls weeds mainly between the rows and there is very little control in the crop row, allowing weed populations to take over and compete with the crop. However, when combined with an early broadcast treatment using a rotary hoe or a rigid-tine or flex tine harrow, inter-row cultivation allows weed control to be extended later in the season. It can also be used to supplement band application of herbicides (Leblanc et al. 1995). Cultivation must be done on the same number of rows that were seeded in one pass, since adjacent seeder passes might result in divergent or convergent rows. (If not, the accuracy of cultivation will be reduced since it will be necessary to cultivate farther from the row to avoid damage to the crop.)

3.1 Cultivation Tools Used

Several models of inter-row cultivators are available on the market. The operating principle of these devices is very simple: cut, pull out and bury. The tools come in various shapes which reinforce these actions, providing more effective weeding. Inter-row cultivators are commonly divided into light and heavy cultivators.

3.1.1 Light Cultivators

Most farmers are familiar with these implements and their operating principles, adjusting and modifying them to suit their needs. Light cultivators, operated at speeds of 5 km h^{-1} on average, generally consist of a tool bar to which different types of tines or shanks can be attached. The shape of the tools selected depends on soil type and the treatment to be performed. Some models allow the working depth to be adjusted by means of a simple lever. Often, several rows of shanks are

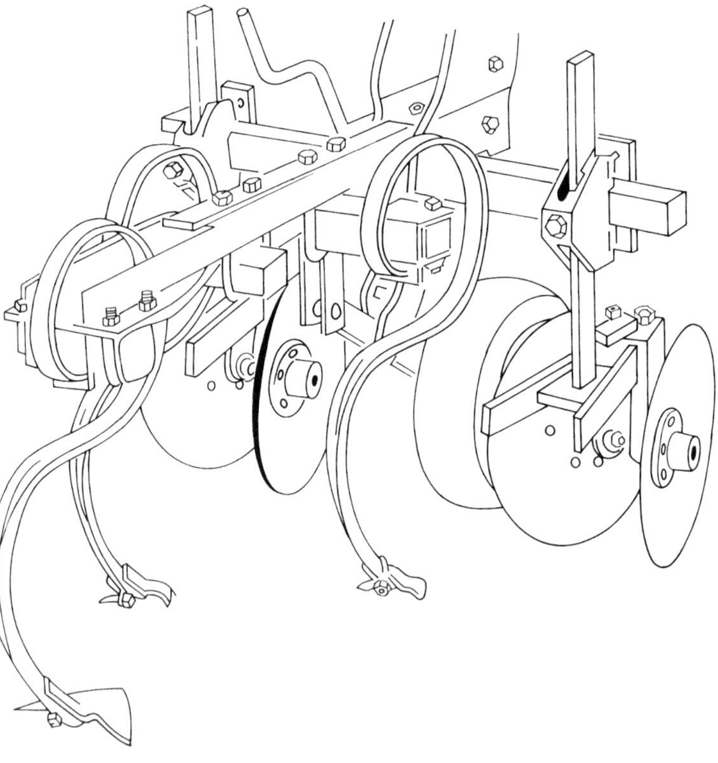

Fig. 3. Combination of Danish S-tines and disks.

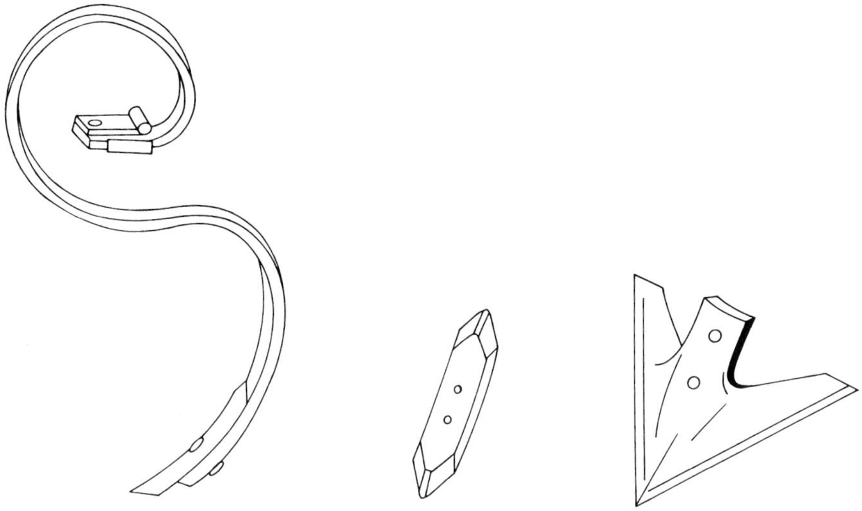

A B C

Fig. 4. A Danish S-tine. **B** Narrow sweep. **C** Goosefoot sweep.

arranged in a triangle to provide coverage of the entire inter-row (Fig. 3). These cultivators offer good clearance beneath the tool bar (>50 cm), allowing them to be used on fairly tall crops. There are several types of rigid and vibrating shanks. To obtain maximum weeding intensity, shanks that can move in all directions are best. Danish tines (vibrating shank), which are shaped like an S, are the most widely used tool; they vibrate extensively and dislodge weeds from the soil effectively (Fig. 4). Various types of sweeps can be installed on the shanks. Goosefoot sweeps (25-45 cm) work along their entire width, loosening the soil and undercutting weeds by their roots. Narrow sweeps (6 cm), which stir up the soil, are used for deeper work. Disk hillers and rolling shields (Cloutier and Leblanc, Chap. 13) are also used for inter-row cultivation—sometimes in combination with shanks, sometimes alone (Fig. 3). These implements cut through the soil and are adjustable. Depending on their orientation to the row, they throw the soil into the row or the inter-row.

3.1.2 Heavy Cultivators

Heavy cultivators, which are more recent, are mainly employed by corn growers using no-till or ridge-till production systems. The technology is designed to work with heavy crop residues. As their name implies, they are heavier, requiring 20 to 50% more tractor power for cultivation (St-Pierre 1993). These machines are also more expensive than light cultivators. However, they allow the work to be completed in the same amount of time. Several models of ridge-tillers are also available (some with only one goosefoot sweep per inter-row), in which ridging wings attached to the goosefoot sweeps carry out the ridging operation.

3.2 Damage to Crop

Inter-row cultivation is safe for crops as long as the goosefoot sweeps do not go too close to the corn rows or too far into the soil, which may cut the corn roots. Using cultivator shields is advisable when the corn is at the one- or two-leaf stage, to avoid damaging or smothering the young plants. The number of rows cultivated per pass must correspond to the number of rows seeded per pass, to ensure that the sweep does not destroy rows that are not exactly equidistant. Cultivation can be carried out until the corn has reached the ten-leaf stage (roughly 50 cm tall), or around 6 to 7 weeks after it emerges. However, late-season cultivation is not always beneficial since it may cause serious root system damage in corn. As early as a month after planting, the roots of the corn already extend into the middle of the inter-row, many in the first 4 cm from the soil surface (Mengel and Barber 1974). Fairly shallow cultivation is recommended. According to Weaver (1926), cultivating at a depth of 10 cm deep reduces the yield, whereas an operating depth of 4 cm will not. The first treatment in the season can be deeper to loosen the soil and allow it to dry and warm up, which also makes later cultivations easier. The optimum cultivation depth can be defined as that which is deep enough to kill weeds but shallow enough to minimize injury to corn roots. Earthing up (ridging up) during the last cultivation may also be beneficial, providing mechanical support to the corn stem and promoting the development of brace roots, which prevent stalk lodging during strong winds.

3.3 Effectiveness in Weed Control

Cultivators with goosefoot sweeps are much more aggressive than hoes or harrows, killing weeds at a more advanced stage of development (four to five-leaf stage). Due to their rapid growth, weeds can cover a field in a very short time, quickly reducing the chances of success of a weed control program. Cultivation depth can be adjusted, but going deeper than 5 cm deep is unnecessary since most weeds germinate at or near the surface. However, goosefoot sweeps can be used only for inter-row weeding. This allows weeds to enjoy uninterrupted growth in the row, where they can absorb fertilizer applied during seeding (indeed, weed biomass in the row may be 20 times higher than in the inter-row). Weed management can be greatly improved, therefore, by carrying out broadcast tillage early in the season. Later in the season, the cultivator shanks can be adjusted to throw the soil onto the row (ridging up) and thereby smother any late-germinating weeds that escaped preceding treatments. Late cultivation also prevents inter-row weeds from reaching maturity or from growing too tall. Ridge-tillers are much more aggressive than light cultivators and kill most weeds. Since they remove the top layer of soil and cut the weeds, their effectiveness is less influenced by the growth stage of the weeds.

4 Cultivation Conditions

Although cultivation for weed control in corn can be carried out under a wide range of soil conditions, it appears to be less effective in very wet or very dry soils. According to Lovely et al. (1958), who evaluated several weed control methods, the presence of wet soil before or after cultivation significantly reduces its efficiency. Peters et al. (1959) demonstrated, however, that within certain limits, proper timing of cultivation in relation to weed emergence and weed growth stage has a greater effect on efficiency than does soil moisture. Cultivation is ineffective when the soil is so dry that it inhibits weed germination. Optimum conditions consist of lightly crusted soil and weeds that have germinated but have not yet emerged (Coleman 1954; Lovely et al. 1958; Peters et al. 1959; Rea 1955).

5 Cost-Effectiveness of Cultivation

According to an economic study done in Quebec, mechanical weed control is just as cost-effective as conventional chemical methods (St-Pierre 1993). A mechanical weed management program usually entails three or four passes a season. According to the study, the cost of four cultivations is less than that of one herbicide application (Table 2). The rotary hoe, although its cost falls in the mid-range, is the least expensive tool to operate per hectare because it can be used at high speeds. These findings are specific to the region where the study was carried out and cannot be generalized.

Table 2. Cost of weed control in Quebec in 1993.

Type of control	Average price (4.6 m swath)		Cost[a] ha^{-1} $pass^{-1}$	
	FF	($Can)	FF	($Can)
Rotary hoe	15 750	(3500)	31	(9)
Rigid-tine harrow	24 335	(6950)	55	(16)
Light cultivator	12 255	(3500)	45	(13)
Heavy cultivator	36 765	(10500)	82	(24)
Herbicides[b]			320	(91)

[a] Including labor.

[b] DUAL (metolachlor) + MARKSMAN (atrazine and dicamba). The cost of using the sprayer was estimated at $8 per ha (or 28 FF per ha).

6 Conclusion

Using two types of cultivators in combination provides synergy by harnessing the best features of each one. The main problem farmers face is the wide variation in weed emergence times. Setting cultivation periods and the number of passes based on the calendar alone may result in ineffective or unnecessary cultivation. Instead, treatments must be based on the growth stage of the weeds. Risks of reduced yields increase when broadcast cultivation is done in corn beyond the six-leaf stage. Late cultivation is not always beneficial since it may result in significant root damage to corn. To sum up, growing corn without using herbicides is not only feasible but cost-effective. However, careful attention must be paid to the conditions under which cultivation is done (soil moisture, speed and depth of cultivation, type of soil, compaction). In short, timely treatments must be combined with the optimal use and adjustment of mechanical equipment.

Acknowledgements. We would like to thank Yvon Douville and Pierre Jobin of the Centre de Développement d'Agrobiologie (Agrobiology Research Centre), Pierre Lachance, agricultural crop protection advisor for the Quebec Department of Agriculture, Fisheries and Food, and the association of corn producers who farm without herbicides and who have contributed to the development of mechanical weed control in Quebec.

References

Coleman F., (1954). The control of weeds by tillage. J. Inst. Brit. Agr. Eng. 10:3-12.

Douville,Y., Jobin P., Leblanc M., Cloutier D., (1995). La houe rotative dans la culture du maïs. Prod. Plus 4(7):43-45.

Gay J.P., (1984). Le cycle du maïs, pp. 1-11 in Physiologie du maïs, INRA, Paris, 574p.

Lampkin N., (1990). Organic farming. Farming Press Book, Ipswick, U.K., 701 p.

Leblanc M.L., Cloutier D.C., Leroux G.D., (1995). Réduction de l'utilisation des herbicides dans le maïs-grain par une application d'herbicides en bandes combinée à des sarclages mécaniques. Weed Res. 35:511-522.

Lovely W.G., Weber C.R., Staniforth D.W., (1958). Effectiveness of the rotary hoe for weed control in soybean. Agron. J. 50:621-625.

Mengel D.B., Barber S.A., (1974). Development and distribution of the corn root system under field condition. Agron. J. 66:341-344.

Parish S., (1990). A review of non-chemical weed control techniques. Biol. Agric. Hort. 7:117-137.

Peters E.J., Klingman D.L., Larson R.E., (1959). Rotary hoeing in combination with herbicides and other cultivations for weed control in soybean. Weeds 7:449-458.

Rea H.E., (1955). The control of early weeds in cotton. Southern Weed Conf. Proc. 8:57-60.

St-Pierre H., (1993). Le sarclage et les sarcloirs, quelques coûts, pp. 37-46 in La culture du maïs sans herbicide, Direction régionale du Richelieu-Saint-Hyacinthe, M.A.P.A.Q., Saint-Hyacinthe, Québec, 51 p.

Souty N., Rode C., (1994). La levée des plantules au champ: un problème mécanique? Sécheresse 5:13-22.

Weaver J.E., (1926). Chapter IX: Root habits of corn or maize, pp. 180-191 in Root Development of Field Crops, McGraw-Hill Book Company Inc., New York, 291 p.

Mulching and Plasticulture

Serge BÉGIN, Sylvain L. DUBÉ and Joe CALANDRIELLO

1 Introduction

Although the main objective of plasticulture in horticultural crop production is to reach harvest earlier and improve productivity, mechanical control of pests is another, albeit less widespread, application. Two techniques are employed: mulches and floating row covers. Mulching involves covering the soil at the base of cultivated plants with a layer of protective material such as straw, thoroughly decomposed manure, peat moss, bark chips, sawdust, paper or plastic sheets. Floating row covering constitute a semiforcing technique using perforated plastic films that are spread over both the soil and plants (Gerst 1992).

Mulches smother weeds, conserve water, prevent the movement of salts to the soil surface, add organic materials to the topsoil and reduce leaching of fertilizer. The use of traditional plant fibre mulches to control weeds has largely been abandoned in favour of polyethylene, which is better suited to intensive farming. Polyethylene mulches have been used for mulching since the early 1950s (Lamont 1993).

The use of plastic mulches in fruit and vegetable production enhances productivity (Downes and Wooley 1966), fruit quality (Downes and Wooley 1966; Decoteau et al. 1990), root system development (Jones et al. 1977), and controls weeds effectively (Smith 1968). Plastic mulches appear to provide a good return on investment (Sanders et al. 1986). The beneficial effects of mulch can be attributed mainly to increased soil temperatures (Taber 1983; Bégin et al. 1994), but also to water conservation (Jones et al. 1977), microclimate change (Sanders 1986) and reduced leaching of nutrients (Locascio 1985).

Plastic materials modify the spectrum of incident light, alter the temperature and serve as a physical barrier, all of which are useful for weed suppression. This chapter will examine the use of mulches and plasticulture for physical control of pests in horticultural crops.

2 Weed Control

2.1 Effects on Photosynthesis

Both black and clear polyethylene are used as mulch for horticultural crops. Black plastic controls weeds by blocking the light, and preventing their photosynthetic activity. Under clear plastic, weeds proliferate but are eventually crushed by the film or are killed by the high temperatures. Since weeds may grow through the openings intended for the crop, herbicides are still required. However, by limiting leaching, plastic mulches reduce the dose of herbicides needed to obtain an equivalent level of control (Brun 1992).

Photoselective mulches (Fig. 1), that combine the advantages of both black and clear films, have been available for a number of years. They control weed growth while accelerating soil warming. For example, some photoselective plastic materials absorb or reflect photosynthetically active radiation (wavelengths of 400–700 nm), but transmit wavelengths between 700 and 2500 nm (Bégin et al. 1995).

Opaque mulches made of cellulose fibre are effective for controlling weeds, although they may degrade quickly. Some materials currently being tested may provide similar yields to those obtained with plastic films on sweet corn and peppers.

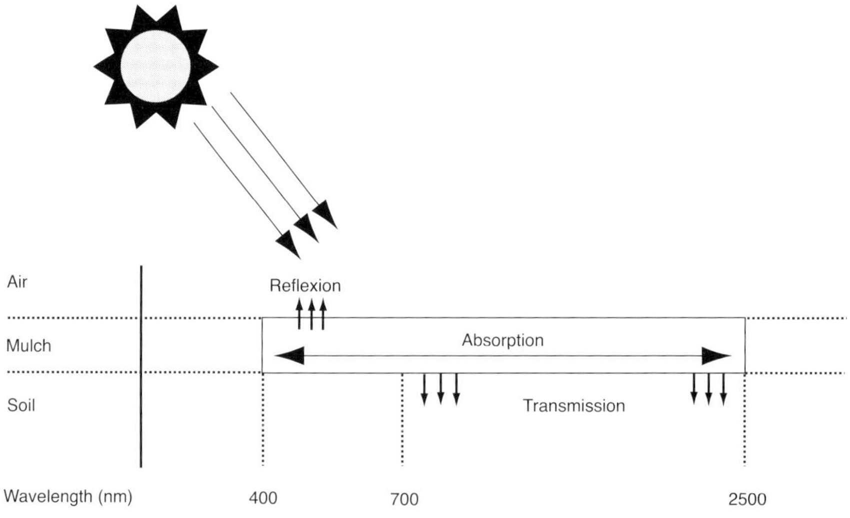

Fig. 1. Behavior of ideal photoselective film at various wavelengths of incident solar radiation (400–2500 nm).

2.2 Photomorphogenesis

Light is the source of energy for many light-dependent phenomena that are independent of or complementary to photosynthesis. Some of these phenomena affect plants by modifying their growth or morphology (hence the name photomorphogenesis). Weed germination and growth can be hindered by covering weeds with materials that filter out certain wavelengths. In addition, by promoting the reflection of specific wavelengths, crop growth can be influenced.

Phytochrome is a wavelenght sensitive pigment that is often associated with photomorphogenesis. It is a photoreversible pigment that exhibit peak absorption in the near infrared (660 nm) and far-red (730 nm) ranges respectively (Fig. 2). Phytochrome is present in all green plants and is particularly abundant in buds, subapical zones, meristem and storage organs. The two forms of phytochrome are called Pr (red-absorbing phytochrome) and Pfr (far-red-absorbing phytochrome). Phytochrome is synthesized in its Pr form (Jose and Schafer 1978). The seeds of some weeds must be stimulated by near-infrared light to germinate, meaning that the Pfr form must be active. The active form (Pfr) returns to its inactive form (Pr) in a photoconversion reaction induced by the absorption of far-red wavelengths. In the dark, Pfr may revert spontaneously to Pr through enzymatic degradation. A photostationary state may occur when the proportion of Pfr to Pr is stable (the destruction of Pfr always equalling the production of Pfr) (Heller 1985).

When dicots or conifers grow under a canopy of other plants, they receive wavelengths that tend to be concentrated in the far-red band, causing the stems to become elongated or etiolated. Under these conditions, some species delay the development of lateral stems and channel their energy into height growth, in order to

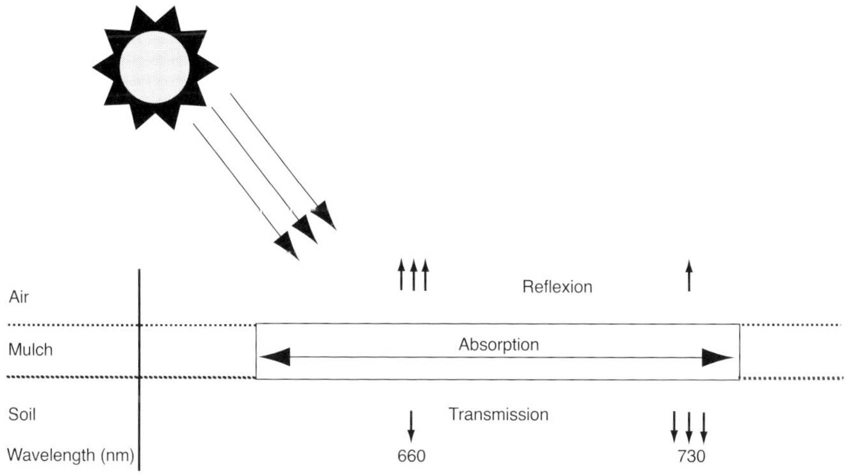

Fig. 2. Behavior of an ideal photoselective film for controlling wavelengths affecting morphogenesis.

bring the apex above the plant cover. This is typical of row crops, with the plants at the ends of the rows being shorter and stouter, with more lateral stems than the plants in the middle of the field.

Light quality can be quantified by the ratio of red to far-red (R/FR ratio), or the ratio of radiant flux measured in the 655–665 nm band to that measured in the 725–735 nm band (Smith 1982). Although the R/FR ratio of light reflected from white plastic mulch (1.00) and black plastic mulch (0.94) is very similar, according to Anderson et al. (1985), the difference is significant enough to affect the height of tobacco plants in the field, indicating that the use of black mulch is preferable in tobacco.

According to Decoteau et al. (1988), the differences in yield found between tomatoes produced with white plastic mulch and black plastic mulch can be attributed to the greater proportion of radiant flux in the blue band (400 nm) reflected by white mulch. This plays a role in regulating the growth of tomatoes, causing them to have shorter stems and increased growth in axillary stems (Tanada 1984). A similar tendency has been found in soybeans and peppers: plants that receive reflected light with a low R/FR ratio (<1) from red or black mulch exhibit 51% more elongation than plants that are exposed to reflected light from white or yellow mulch (Decoteau et al. 1990). The latter mulches produce reflected light with a higher R/FR ratio (>1) and reflects more light in the 400–700 nm band. Photoselective materials for weed control should therefore favour wavelengths that do not promote weed growth but favour heat accumulation in the rhizosphere. Additional objectives should be the promotion of optimal photosynthetic activity in the crop above the mulch and optimal control of photomorphogenesis to enhance growth and development.

3 Controlling Insects with Colour

The high reflectance of aluminized films deters some insects and can be an effective method of physical pest control. Insects ability to locate their host plants is impaired by the reflected light, although the repellent properties of the film decrease as the crop grows and covers the exposed plastic surfaces. Highly reflective sheeting and mulches help to prevent insect invasions and reduce crop damage.

In tomato and squash production, reflective materials have been found to repel some insects but attract others (Wolfenbarger et al. 1968). Bees, important pollinators, are attracted by aluminized material (Schalk et al. 1979). Thrips are attracted by blue (Csizinszky et al. 1990) and black and white (Brown 1989); aphids by yellow (Black 1980) and blue (Csizinszky et al. 1990); and whitefly by red (Csizinszky et al. 1990). The tomato fruitworm (*Heliothis zea*) and tomato pinworm (*Keiferia lycopersicella*) are attracted by reflective or aluminized materials (Schalk et al. 1987). In contrast, some species of leafminers (Wolfenbarger et al. 1968) and aphids (Schalk et al. 1987) are repelled and disoriented by these materials, which also slow the transmission of viruses in squash and watermelon (Black 1980) and lettuce (Nawrocka et al. 1975) and hinder the development of whitefly eggs and nymphs on broccoli (Chu et al. 1994).

The repellent properties of aluminized mulches are probably due to the increased reflection of light, particularly in the ultraviolet band at wavelengths less than 390 nm (Kring and Schuster 1992). Although white mulches emit a similar radiant flux in the visible wavelengths, they emit very little in the ultraviolet range. The density of the aluminium pigments used is crucial in controlling aphids (Wolfenbarger et al. 1968), and at least 50% of the soil surface must be reflective to repel insects (Garnaud 1992).

4 Controlling Diseases with Colour

Light requirements for sporulation in fungi often vary with environmental conditions and with particular species. Ultraviolet light affects sporulation in many pathogenic fungi. Continuous diffused light and total darkness also inhibit sporulation.

Blue light induces sporulation in *Trichoderma viride* and *Verticillium agaricinum*, but inhibits it in *Alternaria tomato*, *A. cichorii* and *Helminthosporium oryzae*. In the case of *Botrytis cinerea* Pers. ex Fr., UV-B (280-320 nm) induces sporulation, while blue light inhibits it (Reuveni et al. 1989).

Fungal diseases caused by *Botrytis*, *Sclerotinia*, *Alternaria* and *Stemphyllium* can be controlled by using UV-absorbing vinyl films as a greenhouse cover. Grey mould is effectively controlled by using photoselective films (Reuveni et al. 1989). A high ratio of blue light to ultraviolet light is required to prevent sporulation and a high radiant flux of blue light is required to provide long-term inhibition. Inhibition of sporulation in *B. cinerea* by blue light is initiated by the conversion of a sporulation-promoting form of mycochrome to a sporulation-inhibiting form (Tan 1974).

5 Controlling Crop Pests with High Temperatures

The solarization or sterilization of soil with plastic row covers or mulches is a promising alternative to chemical disinfectants. This process involves covering moist soil for several weeks with clear plastic. Through the greenhouse effect, heat builds up under the plastic, increasing the soil temperature to a level that significantly decreases populations of harmful organisms. Although numerous studies have been carried out on this form of physical control for various crops and pathogens at southern latitudes (Martin 1992), more recent experiments show that this technique also has potential at northerly latitudes. Most harmful soil microorganisms are more sensitive to temperatures around 40 °C than useful soil microorganisms. Thus, maintaining the soil temperature at 40 °C or above for an extended period of time eliminates most fungi, bacteria, weed seeds and nematodes.

Some strains of *Verticillium dahliae* are controlled after less than 30 min of exposure to temperatures of 50 °C. At 37 °C, exposure for 26 to 29 days is required to

Table 1. Plant diseases controlled by solarization.

Common name	Scientific name
Seedling blight	*Rhizoctonia* and *Pythiacees*
Wilt	*Fusarium* and *Verticillium*
Tomato corky root rot	*Pyrenochaeta lycopersici*
Clubroot	*Plasmodiophora brassicae*
Soil-borne fungi	*Sclerotinia minor* and *S. rolfsii*

obtain the same result (Pullman et al. 1981). The effectiveness of solarization depends on the soil temperature, length of exposure (Martin 1992) and the depth of soil heating (Lopez-Herrera et al. 1994). A solarization period of 6 or 10 days is required to eradicate the sclerotia of *Botrytis cinerea* at depths of 15 and 25 cm respectively. Total eradication of *Verticillium dahliae* is achieved when soil is solarized to a depth of 25 cm, while only 63% control of *Fusarium oxysporum* is achieved at the same depth (Katan et al. 1976).

Using a polyethylene covering on the soil affects soil microflora by promoting anaerobic activity, leading to increases in populations of pectinolytic bacteria among other things (Stapelton and deVay 1984). Solarization increases the volatile ammonia concentration (two to five times) and organic sulphur compounds in the soil. These two volatile compounds have fungistatic effects on fusarium rot and fungicidal effects on pea root rot. Studies suggest that soil amendments promoting the release of these volatiles should be used in conjunction with impermeable films. This also promotes the growth of populations of antagonists, which have better heat tolerance. For example, *Trichoderma* spp. have been found to reduce the virulence of *Fusarium* spp. (Foury 1995). Sublethal temperatures are also useful, since they weaken pathogens and make them more susceptible to their natural enemies.

Stevens (1989a) reported more than 90% weed control for 10 months after solarization using a preplant treatment with plastic mulch. This author also observed an appreciable decrease in nematodes (Stevens 1989b). Solarization also allows the control of some economically important diseases (Table 1).

6 Controlling Crop Pests with Physical Barriers

6.1 Floating row cover

Floating row covers are being used increasingly to provide mechanical crop protection. In Europe, over 16 500 ha of vegetables are grown under these covers (Thicoipe 1992). Unwoven sheets of polyamide, polyethylene, polyester or polypropylene are laid directly over the crop or installed on stand wires, forming a barrier for insects. Floating row covers are mainly used for high value-added crops such as peppers, tomatoes and cucurbits. They control cabbage maggot and carrot fly just as well as chemical insecticides. They are also effective against flea beetles, flies and potato beetles (Gerst 1992) as well as cabbage maggot on broccoli (Purser 1990). In addition, row covers provide an effective barrier against virus transmission by aphids (Gomez 1989). Mesh size must be taken into account. For example, 300-μm mesh is required to control thrips, while 500-μm is required for aphids. Plastic floating row covering are considered to be more environmentally friendly than many pesticides (South 1991). It also holds promise for the production of virus-free potatoe seeds (Hemphill 1988).

6.2 Mulch

The use of black plastic mulch in orchard crops has been found to reduce the population of one species of nematode (*Pratylenchus hamatus*), but increase the numbers of another species (*Meloidogyne incognita*) (Duncan et al. 1994). The latter pest probably tolerates higher temperatures better and is able to continue its development. Plastic mulch may also be used as a barrier against leafminer larvae, preventing them from reaching the plants to pupate (Chalfant et al. 1977).
Mulching the soil with plastic creates a physical barrier that hinders the transmission of some soil pathogens to the foliage, such as *Sclerotium rolfsii* in peppers (Garnaud 1992). Generally, when combined with soil sterilization, plastic mulch significantly reduces disease caused by soil fungi such as *Fusarium*, *Pseudomonas*, *Phoma*, *Verticillium*, *Sclerotinia*, *Alternaria* and *Pythium*. In France, mulch-grown shallots have been very successful due mainly to the fungicidal effect of plastic film on white rot (*Sclerotium cepivorum*). The use of black plastic mulch and sheeting reduces the incidence of early blight of tomato (*Alternaria solani*), although not as effectively as fungicides (Stevens et al. 1993). *Pythium* (*Pythium* spp.) is a major problem in tomato crops when the fruits lie on the ground. Fungal diseases in fruit are also caused by other fungi such as *Phytophthora* and *Rhizoctonia solani*. Polyethylene film prevents rot, serving as a physical barrier to infection (Duncan 1993).

7 Conclusion

Plasticulture can be successfully incorporated into integrated pest management programs and organic farming methods, while we await the revolutionary changes promised by proponents of biotechnology. Plasticulture could be of importance for high value-added crops, particularly in cold climates. The problem of disposing of the plastic afterwards has not been resolved, however. Farmers have two choices: developing a cost-effective system with obligatory recycling or lobbying manufacturers to produce photoselective plastic materials that are photobiodegradable.

References

Anderson R.A., Kasperbauer M.J., Burton H.R., (1985). Shade during growth-effects on chemical composition and leaf color of air-cured tobacco. Agron. J. 77:543-546.

Bégin S., Calandriello J., Dubé P.A., (1994). Plasticulture R&D in Quebec region. Proc. Natl. Agr. Plastic Conf. 25:7-12.

Bégin S., Calandriello J., Dubé P.A., (1995). Influence of the color of mulch on development and production of peppers. Hortscience 30:993-142.

Black L.L., (1980). 'Aluminium' mulch: less virus disease, higher vegetable yields. LA Agric. 23:16-18

Brown J.E., (1989). Effects of clear plastic solarization and chicken manure on weed control. Proc. Natl. Agric. Plast. Conf. 21:76-79.

Brun R., (1992). Les plastiques en agriculture. Co-Édition CPA et PHM-Revue Horticole, France 583 p.

Chalfant, R.B., Joworski C.A., Johnson A.W., Summer D.R., (1977). Reflective film mulches, millet borers and pesticides: effects on water melon mosaïc virus, insects, nematodes, soil borne fungi and yield of yellow summer squash. J. Amer. Soc. Hort. Sci. 102: 11-15.

Chu C.C., Henneberry T.J., (1994). The effect of reflective plastic mulches and insecticides on silverleaf whitefly populations and broccoli production. Proc. Natl. Agr. Plastic Conf. 25:137-146.

Csizinsky A.A., Schuster D.J., Kring J.B., (1990). Effect of mulch color on tomato yields and on insect vectors. Hortscience 25:1131.

Decoteau D.R., Kasperbauer M.J., Daniels D.D., Hunt P.G., (1988). Plastic mulch color effects on reflected light and tomato plant growth. Scientia Horticulturae 34:169-175.

Decoteau D.R., Kasperbauer M.J. Hunt P.G., (1990). Bell pepper plant development over mulches of diverse colors. HortScience 25:460-462.

Downes J.D., Wooley P., (1966). Comparison of five mulch-fumigation treatments on yields and quality of tomatoes and muskmelons. Proc. Natl. Agric. Plast. Conf. 7:53-55.

Duncan R.A., Stapelton J.J., (1994). Mulching young Prunus trees with black polyethylene film increases blossoming and yield, reduces needs for irrigation, eliminates weeds and controls pathogenic nematodes. Proc. Natl. Agric. Plast. Conf. 25:147-150

Duncan R.A., Stapelton J.J., May D.M., (1993). Preliminary evaluation of mulches for prevention of tomato fruits rots. Proc. Natl. Agric. Plast. Conf. 24:123-128.

Foury C., (1995). Quelques aspects de la désinfection solaire des sols. PHM 356:15-20.

Fritschen L.J., Shaw R.H., (1960). The effect of plastic mulch on the microclimate and plant development. Iowa State J. Sci. 35:59-71.

Garnaud J.-C., (1992). Défense des cultures et plasticulture. Tiré de: Les plastiques en agriculture, Coédition CPA et PHM-Revue Horticole, France. p. 543-552.

Gerst J.J., (1992). Les bâches aérées. Tiré de : Les plastiques en agriculture, Co-édition CPA et PHM-Revue Horticole, France p. 359-378.

Gomez O.C., (1989). Crop row covers exclude insects that vector virus diseases of vegetables. Proc. Natl. Agric. Plast. Conf. 21: 297-300.

Heller R., (1985). Physiologie végétale. 2. Développement. Masson. Paris. 215 p.

Hemphill D.D., (1988). Prevention of potato virus Y transmission in potato seed stock with direct covers. Plasticulture 79: 31-36.

Jones T.L., Jones U.S., Ezell D.O., (1977). Effect of nitrogen and plastic mulch on properties of troup loamy sand and yield of 'Walter' tomatoes. J. Am. Soc. Hortic. Sci. 102: 272-275.

Jose A.M., Shafer E., (1978). Distorted phytochrome action spectra in green plants. Planta. 138: 25-28.

Katan J., Greenberger A., Alon H., Grinstein A., (1976). Solar heating by polyethylene mulching for the control of diseases caused by soil-borne pathogens. Phytopathology 66: 683-688.

Kovalchuk S., (1983). Record pepper yields. Plastic mulch, fumigation, and trickle irrigation. Amer. Veg. Grower 31.44: 46-47.

Kring J.B., Schuster D.J., (1992). Management of insects on pepper and tomato with UV-reflective mulches. Florida Entomologist 75(1):119-129.

Lamont W.J., (1993). American society for plasticulture: a unique organization. Hort Technology 3:104-105.

Locascio S.J., Fiskel J.G.A., Graetz D.A., Hauck R.D., (1985). Nitrogen accumulation by pepper as influenced by mulch and time of fertilizer application. J. Amer. Soc. Hort. Sci. 110: 325-328.

Lopez-Herrera C. J., Verdu-Valiente B., (1994). Eradication of primary inoculum of *Botrytis cinerea* by soil solarization. Plant disease 78: 594-597.

Martin C., (1992). La solarisation: méthode de désinfection des sols aux perspectives nouvelles. Tiré de: Les plastiques en agriculture, Co-édition CPA et PHM-Revue Horticole, France. p. 553-566.

Nawrocka B.Z., Eckenrode C. J., Uyemoto J. K., Young D. H., (1975). Reflective mulches and foliar sprays for suppression of aphid-borne viruses in lettuce. J. Econ. Ent. 68: 694-698.

Pullman G.S., DeVay J.E., Garber R.H., (1981). Soil solarization and thermal death: a logarithmic between time and temperature for soilborne plant pathogens. Phytopathology 71: 959-964.

Purser J., Comeau M., (1990). The effect of mulches and row covers on root maggot infestation in broccoli. Demonstration and research report, University of Alaska Fairbanks Cooperative Extension Service, pp. 45-46.

Reuveni R., Raviv M, Bar R., (1989). Sporulation of *Botrytis cinerea* as affected by photoselective sheets and filters. Ann. appl. Biol. 115: 417-424.

Sanders D.C., Konsler T.R., Lamont W.J., Estes E.A., (1986). Pepper and muskmelon economics when grown with plastic mulch and trickle irrigation. Proc. Natl. Agr. Plastics Conf. 19: 302-314.

Schalk J.M., Leron Robbins M., (1987). Reflective mulches influence plant survival, production and insect control in fall tomatoes. Hortscience 22: 3032.

Schalk J.M., Creighton C.S., Fery R.L., Sitterly N.R., Davis B.W., McFadden T.L., Day A., (1979). Reflective film mulches influence insect control and yield in vegetables. J. Amer. Soc. Hort. Sci. 104(6): 759-762.

Smith D.F., (1968). Mulching systems and techniques. Proc. Natl. Agric. Plast Conf. 8: 112-118.

Smith H., (1982). Light quality, photoperception, and plant strategy. Annu. Rev. Plant Physiol. 33: 481-518.

South L., (1991). Plastics that keep the bugs out. Amer. Veg. Grower 11:14-15, 36.

Stapelton J.J., DeVay J. E., (1984). Thermal components of soil solarization as related to changes in soil and root microflora and increased plant growth response. Phytopathology 74: 255-259.

Stevens C., (1989a). Evaluation of preplant application of clear polyethylene mulch for controlling weeds in Central Alabama. Proc. Natl. Agric. Plast. Conf. 21: 65-70.

Stevens C., (1989b). The effect of preplant plastic mulch on root knot damage of sweet potato. Proc. Natl. Agric. Plast. Conf. 21: 80-82.

Stevens C., V., Khan A., Brown J.E., Ploper L.D., Collins D.J., Wilson M.A., Rodriguez-Kabana R., Curl E.A., (1993). The influence of soil solarization on reducing foliage diseases as related to changes in the soil rhizosphere microflora. Proc. Natl. Agric. Plast. Conf. 24: 170-188.

Taber H.G., (1983). Effect of plastic soil and plant covers on Iowa tomato and muskmelon production. Proc. Natl. Agric. Plast. Conf. 17: 37-45.

Tan K.K., (1974). Blue-light inhibition of sporulation in *Botrytis cinerea*. Journal of General Microbiology 82: 201-202.

Tanada T., (1984). Interaction of green and red light with blue light on the dark closure of *Albizzia pinnules*. Physiol. Plant. 61: 35-37.

Thicoipe J.P., (1992). Bâches au sol : protection physique contre les insectes. Fruits et légumes 95(3):41-44.

Wolfenbarger D.E., Moore W.D., (1968). Insect abundances on tomatoes and squash mulched with aluminium and plastic sheeting. J. Econ. Entomol. 61: 34-36.

Physical Barriers for the Control of Insect Pests[1]

Gilles BOITEAU and Robert S. VERNON

1 Introduction

Insecticides have become such an important part of insect pest management that they are the first line of defence considered whenever a problem develops. As a result, producers and consumers have become quite dependent on insecticides for the production of inexpensive and esthetically pleasing foods (Chagnon and Payette 1990; Storch 1996). The availability, relatively low cost, high efficiency and ease of application of insecticides continue to ensure that they will remain the control method of choice, despite the known deleterious effects on human and environmental health. However, increasing public and media awareness of the negative side effects, erosion of the existing pesticide arsenal, and the steady development of insecticide resistance have encouraged and led to a renewal of interest in alternative control methods (e.g. Duchesne and Boiteau 1996).

Physical and mechanical control methods are but one of a group of alternatives considered for integration with, or even replacement of the use of synthetic chemical insecticides in pest management. Physical and mechanical control methods can be defined as the alteration of the environment by physical or mechanical means to make it hostile or inaccessible to insect pests (Banks 1976). Mechanical methods involve the operation of machinery developed specifically for pest management purposes (Metcalf and Metcalf 1993). Physical methods manipulate the physical properties of the environment to the detriment of pest populations (Metcalf and Metcalf 1993). They differ from usual cultural practices in that they would not have been developed or employed if insect pests were not a problem.

Before the advent of modern insecticides, insect control relied heavily on innovative physical and mechanical control methods (e.g. Riley 1876; Forbes 1895). Many of these approaches are becoming popular again or are being reexamined for further development and improvement (Duchesne and Boiteau 1996). Among the main factors shaping this revival are shifts in economics and pest management philosophies, as well as growing desperation for effective control options in certain

[1] For the Department of Agriculture and Agri-Food, Government of Canada,

crops. Costs of registering, marketing and purchasing pesticides are continually increasing. For example, it is estimated that new insecticides cost about $30-40 million dollars (Cdn) in developmental and registration-related costs to reach the marketplace. The costs of developing physical or mechanical control techniques on the other hand are considerably lower, and generally bypass the rigid registration procedures that pesticides require. With the development of modern manufacturing techniques, and the ability to mass produce products at low cost, physical and mechanical approaches previously thought to be impractical can now compete economically with insecticides. With the popularity of commercial and domestic organic farming on the increase, there is an increased demand for alternative pest control strategies. In some cases, effective control options for organic growers do not presently exist. It is felt by some scientists, that the development of effective physical or mechanical control options for organic farming will also infiltrate conventional farming sectors. The scope of this infiltration will depend on efficiency, cost, ease of use, public incentives, consumer demand and desperation for control measures.

This chapter deals specifically with the utilisation of physical barriers for insect pest management. The utilisation of physical barriers by man to gather food, clothing or to protect himself accompanies him through history. A well-known example is the nets set up to intercept and capture large numbers of passenger pigeons for food (Schorger 1955). The method was so effective that it led to the rapid extinction of this species. Also, cliffs have been used as a combination barrier and trap to effectively capture or kill bison on the great plains of North America (Rorabacher 1970). Examples of insect control based on barriers are less flamboyant but can be just as effective. The common use of screen doors or window screens to prevent the entry of insect pests into our homes, or protective clothing and netting to prevent bites and stings out of doors are examples most of us can relate to. Sealed packaging is pervasive and plays an important role in protecting our food from a variety of insects and other pests.

The objective of this chapter is to outline the types of barriers used for insect control, their attributes and their past, current and future use in the different areas of entomology.

2 Definition and Attributes of Physical Barriers

A physical barrier may be defined as a structure made up of wood, metal, plastic or any other material (including living barriers) used to obstruct or close a passage or to fence in a space (adapted from Banks 1976).

Many attributes or characteristics of physical barriers have been traditionally considered negative because the attention was focused on comparing their efficiency and application methods to chemical insecticides. Now that integrated pest management (IPM) is being considered as an essential component of agricultural sustainability, physical barriers are being seen in a more positive light. They are often highly compatible with other alternative methods and insecticides. The materials and/or their installation can be costly but their utilisation is safe and they generally have a minor impact on the environment. The effect of barriers on an insect pest population is generally visible immediately and encourages their adoption by producers or consumers. However, population reduction below an economic threshold or eradication may not always be possible and is often progressive. Where results are less dramatic, barriers may be abandoned before they can prove their ability to manage pests in the context of a well-rounded integrated pest management program.

It is a misconception that barriers can only be used against the wingless or apterous life stages of insects. Barriers can be used to modify flight direction and landing location and frequency.

Barriers are especially adaptable to confined spaces, with greenhouses being the best example. Traditionally, barriers have been recommended for small land surfaces. The high cost of deploying certain barrier strategies on a large scale has been a major impediment to their use in large-scale commercial crops. The material cost of covering a crop with spunbonded polyethylene row covers to exclude certain insects, for example, doubles as the number of hectares doubles. This linear relationship also applies to the use of insecticides. However, the cost of certain physical barriers installed around a field to intercept incoming pests actually decreases per hectare as the area of the field increases. For example, if it costs $1000 to encircle a 1-ha square field with a barrier device, the cost is $1000 for that hectare. If the perimeter is doubled however, the area enclosed increases to 4 ha, reducing the cost of the barrier to only $500 h^{-1}. The relationship between fence cost and area enclosed is shown in Fig. 1. In addition, if the exclusion barrier remains functional for more than 1 year, the costs can be amortized over the life of the device, which brings the yearly per hectare costs down further.

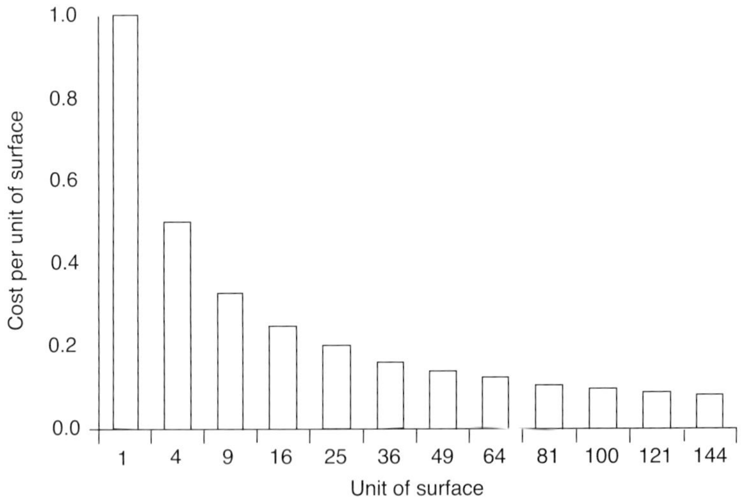

Fig. 1. Relationship between the cost ha^{-1} of a perimeter fence and the change in area of a square field.

3 Categories or Types of Physical Barriers

Table 1. List of barrier types available to manage insect pests.

Barrier Type	Examples	Target insects
Adhesive	Glues, petroleum jelly, creosote	Ants, cockroaches
Slippery surface	Fluon, Teflon, dust	Ants, cockroaches, potato beetle
Fence	Metal, plastic, soil	Chinch bug, potato beetle, cutworm
Windbreak	Net, trees	Aphid vectors
Wrapping	Burlap, aluminum	Gypsy moth, forest tent caterpillar
Organic mulch	Bark, straw	Melonworm, potato beetle
Inorganic mulch	Plastic, aluminum, whitewash	Aphid vectors, potato tuberworm
Trench	Plastic, soil	Potato beetle, chinch bug
Water	Irrigating ditch	Chinch bug
Spatial barrier	Sterile male release	Screw worm fly
Forced air	Building entrance	House flies

A list of barriers that have been used to control insects can be found in Table 1. The list is not intended to be exhaustive but provides a clear demonstration of the wide range of types of barriers.
Physical barriers have been developed for the management of insect pests in urban environments, forestry, agriculture and as experimental tools in research.

3.1 Urban Insects

3.1.1 Tent Caterpillars

The cyclical outbreaks of eastern and forest tent building caterpillars cause significant economic losses to cities and homeowners by stressing or killing ornamental trees and shrubs (Collman and Antonelli 1994). Specimen trees are protected using paper or burlap strips wrapped around tree trunks and coated with a sticky or greasy substance to prevent larvae from climbing up to the tree canopy. Plastic lined trenches covered with dust can also be used to surround shrub patches.

3.1.2 Ants, Cockroaches

The access to buildings and structures by ants and other pests such as cockroaches can be restricted by the use of barriers covered with sticky compounds, petroleum jelly or Teflon (Long 1994). These compounds are less practical for outdoor control.

3.1.3 Houseflies, Mosquitoes

The use of screen doors or window screens in homes or other buildings will prevent the entry of many insects, including wasps, houseflies and mosquitoes, and are physical controls in common use around the world. Insect netting in tents or over bedding is an effective barrier to biting or sucking pests such as mosquitoes, which in some countries transmit malaria. Strong air movement, or positive air pressure can also be used to prevent flying insects from entering buildings (Mathis et al. 1970).

3.2 Agricultural Insect Pests

Physical barriers have seen the most activity against a wide variety of insect pests in a wide variety of agricultural crops. Barriers may be as simple as collars placed around the individual stems of bean plants and other vegetable crops in small gardens to protect them from cutworms (Howard et al. 1994) or they can involve trenches, fences or windbreaks strategically placed around the perimeter of fields. Key examples are described below.

3.2.1 Mormon Cricket

The Mormon cricket (*Orthoptera: Tettigoniidae*) can be a serious insect pest in northwestern USA. This species is capable of causing significant crop losses by its periodic attacks on a wide range of field and garden crops including wheat and alfalfa (Metcalf and Metcalf 1993). Because this cricket is unable to fly, it is possible to stop its movement between fields with a 20-cm-high steel barrier or a straight edge ditch with water traps located here and there to collect them. In irrigated fields, a surfactant at the surface of some ditches can also prevent their dispersal (Metcalf and Metcalf 1993).

3.2.2 Chinch Bugs

Chinch bugs [*Blissus leucopterus leucopterus* (Say)] can be serious pests of grasses, wheat and corn in the Western and Southern States. At the time of harvest of grasses, it is possible to intercept the wingless forms of the chinch bug as it moves to corn. This is the traditional example used in textbooks describing the use of barriers in agriculture (e.g. Metcalf and Metcalf 1993). These barriers were developed by Forbes in 1895, and are constructed of a plowed ridge at the brow of which some creosote is placed on a daily basis. A barrier can also be made of strays of tarred paper 12 cm wide in width soaked in creosote. Also, strips of land 6 m wide, half in grasses and half in corn, have been sprayed with insecticides to obtain successful control of the migrating chinch bugs (Metcalf and Metcalf 1993).

3.2.3 Potato Tuberworm

Paparatti (1993) compared the utilisation of a plastic film and a natural covering of *Trifolium* to an insecticide and an insecticide-synergist combination for the control of *Phtorimea operculella* on potato in Italy. Results indicated that the plastic or the natural cover were superior to insecticides in protecting tubers from insect damage.

3.2.4 Aphid Vectors

Aphids play a critical role in the transmission of many virus plant diseases of economic importance (Howard et al. 1994). Control methods can be directed at the disease itself or to the vector. Insecticides play a secondary role in the prevention of virus transmission because of their relatively slow mode of action (e.g. Raccah 1986). Cultural and physical control methods can therefore be of great importance in the development of an integrated disease management plan for crops affected by viral diseases transmitted by aphids or other insect vectors. Mulches, nets, yellow sheets and trap crops have been used to reduce the landing of aphids in crops. Aphids are attracted to different ranges of wavelength during flight takeoff and during landing (e.g. Raccah 1986). Repelling aphids with a reflective surface covering the ground between plants (such as aluminum foils or grey mulches) was found to be effective in reducing or preventing virus infections in Florida and Israel (Moore et al. 1965; Smith and Webb 1969; Loebenstein et al. 1975). Mulching resulted in a higher quantity and quality of pepper plants (Loebenstein et al. 1975). Organic mulches were used against the aphid vectors of potato virus Y (Kemp 1978) and against whiteflies (Cohen 1982). A detailed description of the use of plastic mulches in Florida is given by Zitter and Simons (1980).
Windbreaks have also been investigated. Modification of the air movement or wind direction by windbreak barriers can significantly affect the dispersal and field distribution of insects, especially the smaller ones (Lewis and Dibley 1970). The mean speed of the wind blowing into the edge of a crop decreases quickly into the crop, and sheltered patches develop in the lee of individual plants. Wind-borne insects therefore tend to land both on the windward side of a crop, and behind individual plants. Accumulation on windward sides has been demonstrated for several species such as bean aphids (Pedgley 1982). In the case of the lettuce root aphid in a British experiment, crop damage was greatest along a line 2-3 m leeward of the windbreaks, the line thus showing where a heightening density of airborne aphids in June is likely to have occured due to sheltering by the windbreak. Similar results were found by Lewis (1966) with turnip mild yellows, a disease due to virus carried by the green peach aphid, *Myzus persicae*. Artificial windbreaks consisting of 45% slatted fences were set up. It was found that infected plants were most common behind the south-facing fence, at distances between one and four times the fence height.
Most common crop shelters on farms in temperate Europe and North America is given by hedges or rows of trees, but artificial windbreaks made of cane, straw or

netting are widely used to shelter small areas of valuable crops. Lewis (1965) demonstrated that flying insects tended to gather near a windbreak, especially on the downwind side, and that horizontal profiles of volume density are similar to those for drifting organisms such as spiders, or even for pieces of paper. Although this suggests little or no behavioral effect by many species on their concentration while wind-borne, it is likely that some do respond to the weaker wind and greater gustiness on the leeward side of a windbreak. For example, night-flying lacewings and moths tend to gather closer to windbreaks than do day-flying thrips and parasitic wasps; and so do strong fliers compared with weak ones (Lewis 1969). Volume density is affected as far downwind as ten times the height of a windbreak, but the greatest densities (up to ten times those upwind) are mostly at distances one to four times the height of the windbreak (Lewis 1969). Pedgley (1982) provides additional examples. An artificial barrier, however, may well encourage field pests to gather without being able to harbour their hedge-living enemies, but it is not yet known if windbreaks are in general harmful or useful. Each windbreak and crop must be judged individually until more is known of the ways by which they gather insects. It may be possible to take advantage of the aggregating effect of barriers to limit the application of insecticides or biological agents such as predators to the leeward areas of the barrier, thereby reducing costs of control and environmental impact in lettuce fields (Yudin et al. 1991). Many types of barriers such as nets, mesh and plastics have been used to try and reduce the spread of virus diseases by aphid vectors (Cohen and Marco 1973; Harpaz 1982).

In the case of thrips affecting lettuce production, it was postulated that barriers might be useful to control their dispersal because they were caught at an average height of 1 m on stickinyl barriers1.5 m in height. Sticky traps within the plots were used to monitor the abundance and the distribution of thrips within protected and unprotected plots. Overall, the abundance of the thrips was reduced only by 10% by the barriers. The reduction was higher in traps placed close to the barrier but catches in traps 1.3 m from the barrier were frequently superior to those in the control plots. Also, the tomato spotted wilt virus transmitted by these insects is a persistent virus. Therefore, only a small number of vectors is required to spread the virus.

A different approach was suggested by Cohen and Marco (1973), who used sticky yellow sheets positioned vertically around a pepper plot. This procedure resulted in considerable reduction of infection. The yellow sheets are now being recommended by the Extension Agricultural Advisory Board for prevention of virus in certain regions of Israel. It seems that yellow sheets are not a universal solution for all viruses since, in an attempt to protect squash, more infected plants were recorded in the sheet-surrounded plot than in the control plot (Weiss et al. 1977). Aluminum and coloured foils have been successful at repelling aphids from some crops (Adlerz and Everett 1968; Fusco and Thurston 1970) but it seems that they may attract thrips (Beckman 1972). Cohen and Melamed-Madjar (1978) reduced the spread of whitefly-transmitted tomato yellow leafcurl virus with yellow polyethylene soil mulches.

Another approach is based on interfering with the landing behaviour of aphids by placing white 2-8-mesh nets above the crop. Aphid vision is limited to 50 cm (Raccah 1986), and the plants which are immediately underneath the nets are apparently camouflaged by them. Virus incidence was reduced in pepper plots after 77 days, from 95% in the control to less than 2%. Except when aphid populations are extremely high, this approach seems effective, although not economical for low-value crops (Cohen 1981). Another interesting approach was initiated by Bar-Joseph and Fraenkel (1983), who observed fewer *Aphis citricola* van de Goot on limes which had to be sprayed with a mixture of kaolinite and montmorillonite. The number of alates on terminal twigs was reduced from 13.5 per plant in the control to no more than 2-3 on the treated plants.

Growers conventionally isolate seed potato fields with cultivated fallow borders, creating a green plant/dark soil border that is attractive to flying aphids potentially carrying potato virus Y. To eliminate this edge effect, Radcliffe et al. (1993) surrounded small seed potato fields with a 15-foot border using soybeans, sorghum and wheat. Their hypothesis was that winged aphids would land in the crop border, probe and lose their charge of virus before entering the seed potato field, reducing PVY spread. When a winged aphid is searching for a host, it alights at random on any green plant and must "taste" or "sap sample" the plant to determine if it is a suitable host. It is during these brief feeding probes that PVY is acquired and transmitted. This is why grain aphids can transmit PVY even though they do not reproduce on potatoes and is also why routine application of insecticidal sprays will not prevent transmission. PVY acquisition and transmission takes only seconds, and no insecticide kills aphids quickly enough to prevent these events from occurring. This isolation method proved to be successful. The crop borders reduced PVY in seed potato by more than 50%, with the greatest reduction in PVY occurring in the outer rows of crop-bordered fields as compared to fallow-bordered fields. Soybeans, sorghum and winter wheat worked equally well as crop borders.

3.2.5 The Colorado Potato Beetle, **Leptinotarsa decemlineata** *(Say)*

The Colorado potato beetle, *Leptinotarsa decemlineata* (Say), is the primary insect pest of potatoes in Canada (Howard et al. 1994), in the Central and Northeastern USA (Radcliffe et al. 1993) and in many European and Asian countries (Jolivet 1991). It is also common to other solanaceous plants such as tomato and eggplant, where it can also cause significant economic losses (Boiteau 1996). In recent years, potato beetles have become resistant to a wide range of synthetic insecticides, making management of this pest more difficult and costly with each growing season. Increasing the number of sprays, combining chemicals, or increasing the dosage provides only short term benefits, and greatly increases the rate of development of resistance. Once resistant to one or more insecticides, the beetle easily and rapidly becomes resistant to others with similar chemistries and mode of action.

The best way to manage the problem of insecticide resistance in the potato beetle population is to reduce the number of applications (Roush and Tingey 1994). The use of alternative insect control measures such as physical controls in an integrated pest management program would reduce the equilibrium level of potato beetle populations and thus reduce sprays. Weisz et al. (1995) have shown that although certain alternatives can be more expensive than the insecticides they replace, improvements in yield can more than compensate for the expense. In some regions of Canada in the early 1990s, the failure of pesticides to provide adequate protection against Colorado potato beetles prompted a shift in thinking among some producers to consider other alternative methods including cultural, biological, mechanical and physical methods in developing their management strategies. Fortunately there had already been considerable research in this direction, including better crop rotation strategies, new and more specific bacterial insecticides, machines to remove or burn the beetles, and barriers preventing access to the crop. Rotations should be the first line of defence. By changing the location of potato fields from year to year, fewer beetles are able to find the new fields and they also colonize much later. One or more of these techniques has been employed to varying degrees in certain areas by commercial growers desperate for new control methods.

The use of barriers to manage the Colorado potato beetle is one of the more efficacious of the alternatives developed, and is a strategy based on what is known about their basic ecology and behaviour. Adult Colorado potato beetles overwinter in the soil in potato fields and along protected headland regions surrounding the fields. Generally, headland areas with shrubs, windbreaks or woodlands are preferred as overwintering sites, where the cover of plant material and snow protects them from the harsh winter temperatures. Populations of hibernating beetles as high as 188 beetles m^{-2} of soil have been recorded in a study conducted in Massachusetts (Weber et al. 1994). As the temperature rises in the spring, adult beetles slowly emerge and start foraging for food. If the same field is replanted to potatoes, they do not have far to go. If it is not, they must find a new field. As most beetles will walk rather than fly to find new food source in the cool days of early summer or an overwintering site in the cool days of late summer throughout most of Canada and the Northeastern USA, barriers can be used effectively to prevent their entry into new plantings or into the overwintering sites. The efficiency is partly dependent on the densities of overwintered adult beetles (Weber et al. 1994). Plastic-lined trenches have been scientifically evaluated and used by commercial growers for this purpose (Boiteau et al. 1994). The trench is essentially a furrow in the soil lined with black plastic which is laid around the edges of new potato fields to trap incoming beetles (Fig. 2). The furrow is at least 25 cm deep with the sides sloping at an angle of at least 50°. When the trench is established, the plastic soon becomes covered with a fine layer of dust. As immigrating beetles fall into the trench and attempt to climb out, their tarsae, or footpads, are coated with the dust particles, which impairs their ability to escape from the trench. The beetles are also unable to fly out of the trap and soon die. Interestingly enough, many wingless insect predators with a different tarsal structure escape the trench easily and most

Fig. 2. Utilisation of plastic-lined trenches to reduce colonization of potato fields in the spring by overwintered potato beetles and to reduce adult overwintering in the fall, by trapping as they leave the fields.

winged predators fly out easily (G. Boiteau, unpubl.). To be effective, the plastic-lined trench is placed between the emerging beetles and the new field, and must be in place before beetles start emerging. The efficiency of barriers is partly dependent on the densities of overwintered adult beetles (Weber et al. 1994), and on the number of walkers versus flyers entering the field. If growers know in advance where in a field the beetles are most likely to emerge, then trenches can be restricted to those areas, thus lowering the cost of protection. Short of knowing this, the whole field should be enclosed by trenches.

At the commercial scale, plastic lined trenches are installed using a machine specially designed for this purpose. The implement, carried on a three-point hitch, opens a furrow at the required depth, unrolls the plastic over the trench and covers the edges with soil to prevent the plastic from being blown away by the wind (Misener et al. 1993). The cost of the machine, if purchased, would be in the vicinity of 3,500 $Ca. The machine has also been built by growers themselves, following detailed plans and drawings available from one of the authors (Boiteau). This particular design stresses the importance of covering any plastic surface on the shoulders of the trench with soil, or beetles may not be able to enter the trap-barrier. A modified design is commercially available from VBM, Delhi, Ontario, Canada.

The cost of the plastic, manpower and amortization of the equipment is recovered when a single application of insecticides is eliminated by placing a trench around the perimeter of an 8-ha field. It is also important to take into consideration the fact that the cost decreases with the size of the field (Fig. 1).
Trenches with walls sloping at 46 ° or more will retain all captured beetles in the laboratory and an average of 84% under field conditions. The reduction in field efficiency is caused primarily by rainfall. Small numbers of beetles may escape from field trenches during periods of rain, but the efficiency of the trench is reinstated as soon as the dust on the slopes of the trench becomes dry again. Water accumulated at the bottom of the trench does not reduce the efficiency of the trap. A potato field surrounded by plastic-lined trenches could see its populations of immigrating overwintered adult beetles reduced by 48% and its population of emigrating summer-generation adult beetles reduced by 40-90% compared with fields without trenches. Reductions in immigrating overwintered adults are also reflected in reduced numbers of egg masses subsequently deposited in the crop. Alone, plastic lined trenches are not sufficient to control Colorado potato beetle populations fully in potatoes but can easily be integrated with other control strategies such as propane burners, insect collectors plastic traps or chemical insecticides.
The effectiveness of the barrier has been evaluated experimentally and in commercial fields in New Brunswick (Boiteau et al. 1994), Ontario (M. Sears, pers. comm.), Quebec (R.-M. Duchesne, pers. comm.), Prince Edward Island (J. Stewart, pers. comm.), New York (Moyer 1995), and Massachussetts (Ferro 1996). It has been suggested that barriers are the most effective alternative to insecticides for the control of potato beetles (Ferro 1996). The effectiveness level of the barrier expressed in percentages may seem low in comparison to mechanical control methods such as propane flamers (e.g. Moyer et al. 1992). However, the effectiveness of the barrier is ongoing for the entire season whereas the effectiveness of a mechanical control method or most pesticide applications is comparatively short lived. Thus, the effectiveness of a propane flamer or insecticide might appear high against the population present at the time of the treatment, but is not sustained unless the treatment is repeated as new colonizers reach the crop.
In practice it has been found that when plastic-lined trench barriers are placed around commercial fields, they must be interrupted at one or two locations to permit access by farm machinery. The effectiveness of below ground trenches could also be less consistent in areas where rain is frequent. In answer to these concerns, an above-ground version of the trench has been developed (patent pending) in Ontario and British Columbia, Canada (R. Vernon, pers. comm.) and tested against the Colorado potato beetle and other important crawling insect pests including click beetles (the adults of wireworms) and several species of weevils including the strawberry weevil and black vine weevil. In potatoes, above-ground trenches can be especially useful in rainy areas, or as removable access barriers in combination with the below ground trench. Essentially, the above-ground trench is composed of an extruded, UV-retarded PVC plastic trough, with a patented design for allowing Colorado potato beetles and other pests to enter the device and become trapped and

killed. The device is coloured black to speed up beetle mortality through heat buildup, and is designed to release water and small wingless beneficial insects from the trough. A number of add-on devices have also been developed for the trench to separate Colorado potato beetles from the larger beneficial insects (i.e. predacious beetles), such that the Colorado potato beetles can be removed from the device without disturbing it, and the beneficials can be released where they will do the most good. The plastic trench can be pushed into ground that has been softened and flattened by a standard tractor mounted bed-shaping device, and the sections can be joined by a short coupling section to form a continuous barrier along one or more sides of a field.

Although still being evaluated, early results indicate that the above-ground trench will be as effective in intercepting and retaining beetles as the in-ground trench (Boiteau and Osborn 1999). The above-ground trench would also be comparable to the in-ground trench in cost, and less expensive than pesticides, since the lengths of trench can be reused and the costs amortized over an estimated 10 years.

Unfortunately, like many of the other physical or mechanical alternative control methods available for Colorado potato beetles, the implementation of in-ground or above-ground trenches into large-scale commercial potato operations is slow. This is partly due to the recent introduction of new chemical insecticides and genetically-modified potato cultivars (e.g. cultivars expressing *Bt* toxin in the leaves) that are currently effective and fulfill the traditional expectations of potato growers. The rate of adoption of plastic-lined trenches was increasing but the interest in such alternative methods of insect control decreased suddenly with the registration of a very effective chemical insecticide and a transgenic plant (Ferro 1996, D.W.A. Hunt, pers. comm.). For smaller-scale growers, such as backyard gardeners and organic farmers who do not accept the new chemicals and transgenic varieties, it is expected that above-ground trenches will be adopted more readily.

It is expected that full dependence on new chemical insecticides and transgenic plants will lead to a decrease in their effectiveness within a few years. In the meantime, research is continuing into the integrated use of trenches with these and other mechanical and chemical approaches.

Other barriers have been investigated. Their utilisation seems limited at this time but they may be useful in particular agroecosystems.

Straw mulch

Ng and Lashomb (1983) observed that the physical presence of wheat and the lower soil surface temperature in wheat fields restricted movement of Colorado potato beetle adults. It has since been suggested that straw may not significantly affect the dispersal of the beetles but that it increases the abundance of ground predators in the field (Cloutier et al. 1996). Zehnder and Hough-Goldstein (1990) grew potatoes in plots with and without straw mulch. They found that the numbers of overwintered adults, egg masses, and larvae were significantly lower in plots with mulch compared to those without mulch. Soil temperatures were consistently lower (2.4-3.4 °C cooler) in the mulched plots compared to those without mulch. Soil moisture potential was greater in mulched plots than in plots without mulch. When insecticide applications were integrated with mulching, the yields were higher than when only insecticides were applied to control the beetle. Moreover, mulched plots required four fewer insecticide applications. The amount of straw required for mulch in a commercial potato field would be about 500 bales ha^{-1}. The use of straw mulch may be a valuable cultural practice for incorporation into a potato pest management program; however, it may be cost prohibitive to non-organic growers (Ferro 1996).

Trap cropping

Trap cropping may be effective in intercepting newly-emerged, overwintered beetles which otherwise colonize the main crop (Wegorek 1959; Hunt and Whitfield 1996). The trap crop needs to be planted strategically between the newly planted field and overwintering sites to arrest the beetles as they walk to the field. Using a mark-release-recapture technique Ferro (1996) found that 70% of colonizing beetles remained within the three-row potato barrier for less than 24 h before they moved into the potato field. This means that the attraction of these plants was not sufficient to prevent the beetles from colonizing the rest of the field.

3.2.6 Root Maggots of Vegetable Crops

In many vegetable crops, including onions, carrots and cole crops (i.e. cabbage, broccoli, cauliflower, Brussels sprouts), root-feeding insects are often the main cause of economic loss to farmers. Well-known examples are the larvae, or maggots of several fly species, including the onion fly, *Delia antiqua*, the carrot rust fly, *Psila rosae*, and the cabbage fly, *Delia radicum*. Presently, control of these and other root-feeding pests depends almost totally on the use of insecticides applied as granules to the soil, or as sprays and drenches applied to the crop during the growing season. The continuing dependence on insecticides to control root feeding pests places the North American vegetable growing industry at considerable risk. The arsenal of granular and drench insecticides is seriously limited at the present time, is dwindling, and does not appear sustainable in the long run. In Canada and the USA, there is only one granular insecticide remaining for onion maggot and cabbage maggot control and resistance is already building to the remaining product. If the ability to control root maggots by chemical means is lost, there are presently few, if any, economically viable alternatives for control, especially for large-scale production operations.

Alternatives to insecticides for the management of root feeding pests primarily involve cultural and physical methods, and are most often deployed by small-scale organic growers and backyard gardeners. Cultural controls might entail planting certain crops at strategic times (Ellis and Hardman 1988; Finch 1993), or planting in lower risk areas along with strategic crop rotations and the use of trap crops to avoid pest populations (Kettunen et al. 1988). Planting resistant varieties of certain crops such as carrots will also reduce damage (Ellis and Hardman 1988). A small number of physical control approaches have been developed for the management of root maggots, all of which focus on restricting the access of the adult flies to egg-laying sites on the host. The use of various spunbonded polyester row covers, for example, has been shown to successfully exclude carrot rust flies and cabbage flies from their respective crops (Haseli and Konrad 1987; Folster 1989; Antill et al. 1990; Hough-Goldstein 1987; Millar and Isman 1988). In addition, plant collars placed around individual plants in brassica crops will retard egg laying and damage (Schoene 1914; Wheatley 1975; Skinner and Finch 1986; Havukkala 1988). Although some crops benefit in terms of higher yields and shorter time to harvest under row covers, others are adversely affected by the modified microclimate. In carrots grown under row covers for example, yields either increased (Folster 1989; Haseli and Konrad 1987; Ellis and Hardman 1988), or decreased (Davies et al. 1993; Antill et al. 1990), depending on the materials used. Increases in certain diseases have been recorded under row covers, and weed growth can be a problem since the covers must be removed for manual or chemical weeding.

Due to the high cost of materials and the labour required for installation and maintenance of sheeting or plant collars, adoption of these physical techniques in large-scale commercial fields has been very limited. To cover 1 ha of carrots it would cost about 2500 $Ca, in material costs alone, which, of course, doubles as the area

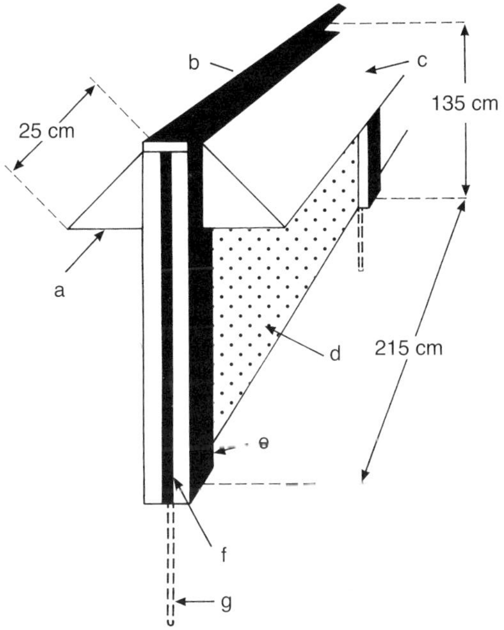

Fig. 3. Design of an experimental exclusion fence used in retarding the movement of insect pests such as cabbage flies and carrot rust flies into their host crops. The components of the fence include : **a** overhang support wing; **b** wooden fence top; **c** nylon mesh overhang; **d** nylon mesh screen panel; e hollow wooden fencepost; **f** groove in post for mesh screen panels; and **g** rebar to anchor the post and fence to the ground.

doubles. On a smaller scale, however, such as with organic farms and backyard gardens, the cost is less imposing and the use of sheeting is becoming more common. Another physical control method under development is the exclusion fence, which has been shown to prevent certain flying insect pests, including cabbage, onion and carrot rust flies from entering their respective crops (R. Vernon, pers. comm.). The concept of the exclusion fence arose as a result of studies to determine the best height for positioning sticky traps used in IPM programs for monitoring cabbage flies (Vernon 1979; Tuttle et al. 1988), carrot rust flies (Judd et al. 1985), and onion flies (Vernon 1979; Vernon et al. 1989). In these studies, it was found that these and many other vegetable pests were trapped in greatest numbers within 30 cm of the ground or crop canopy. It has also been observed that when many insects, including the above pests, encounter obstacles, such as a screen interception trap, they tend to move upward to become trapped at the apex of the structure. On the basis of these observations, it was hypothesized that fences constructed of fine nylon window-screening material with a screen overhang at the top (Fig. 3) might impede the movement of several common insect pests into their host crops.

When placed around small plantings of rutabaga, exclusion fences 1 m in height typically excluded about 80% of female cabbage flies from entering the enclosures, with corresponding reductions in maggot damage to rutabagas also occuring (R. Vernon, pers. comm.). The effectiveness of the fence shown in Fig. 3 declined as fence height was reduced from 1 m, and as the length of the overhang at the top was reduced from 24 cm (R. Vernon, pers. comm.). Exclusion fences were just as effective in excluding cabbage flies when placed around large plantings of radish, and preliminary data has shown that the fences will work just as well for plantings of stem crucifers, such as cabbage, broccoli, Brussels sprouts and cauliflower.

In other studies, the fence also reduced populations of adult carrot rust flies entering areas of a commercial field by about 82%, with a corresponding 91% reduction in damage to enclosed carrots (R. Vernon, pers. comm.). The effectiveness of the fence in excluding female onion flies from enclosed plots in commercial onion fields ranged from 70-79%. These studies showed that exclusion fences will reduce infestation levels and damage by key root maggot species in several vegetable crops where effective pesticides are becoming seriously scarce.

Although exclusion fence technology is still in the developmental stage, it is a new pest management approach that may have considerable market potential in the future. Exclusion fences have a number of advantages over other physical control devices, such as row covers or tarpaper collars. Although expensive to use in small plantings such as backyard gardens, permanently erected fences around commercial fields become less expensive per unit area as the size of the field increases. Cost per unit area would be reduced even further for contiguous fields where part of the fence is shared (Fig. 4). In addition, the life expectancy of the fence would be at least 5 years with proper handling, which allows amortization of costs, and thus unit area cost per year to decline even further. If used as diagrammed in Fig. 4, where crops are strategically segregated by fences and properly rotated, it is also expected that the effectiveness of fences would be gradually enhanced over time.

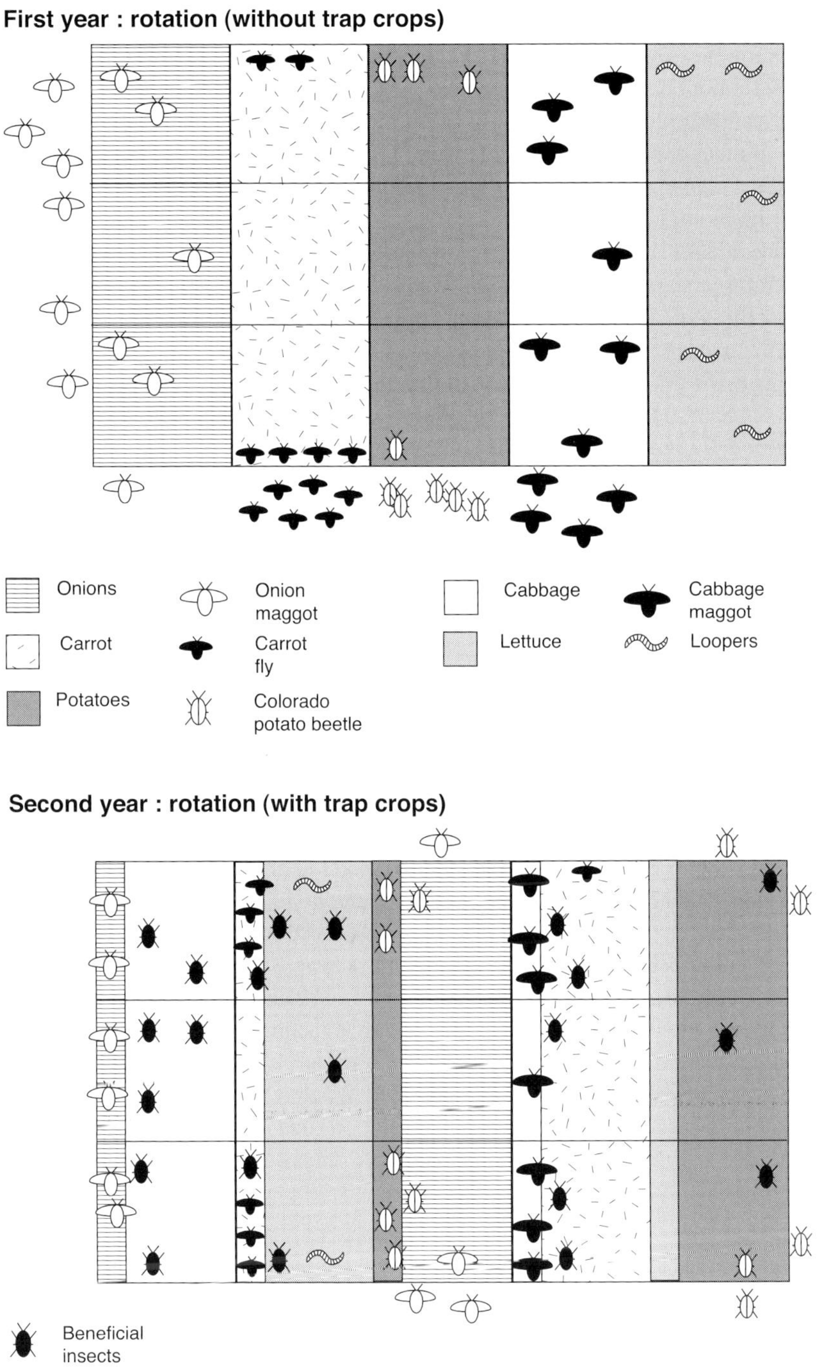

Fig. 4. Utilisation of contiguous barriers in mixed cropping

For example, rotating onions to fenced blocks far enough away would force emerging onion flies from last years blocks to cross more than one fence barrier. Planting strips of onions as trap crops in blocks where onion flies are expected to emerge might also help deter pests from locating the new plantings. This approach would apply to other root feeding pests as well. Also, if fences were to be used in lieu of traditional chemical controls for root maggots, it is expected that levels of naturally occurring biological control agents, such as parasites or predacious ground beetles, would gradually increase and have an increasing impact on root maggot populations. It has also been noted that the fences themselves will congregate populations of tiger flies and spiders which capture and feed on adult root maggot flies.
The effectiveness of exclusion fences will be reduced if surrounding plants (i.e. weed growth) become too high, or in uneven or hilly terrain where fence height is below adjacent field headland areas where pests might aggregate. In areas with high winds or drifting snow, it might also be necessary to remove and store parts of the fence, and re-erect them the following season. Designs for exclusion fences that can be rapidly assembled and dismantled are available from one of the authors (Vernon).

3.3 Forest Insect Pests

3.3.1 Spruce Budworm

Prévost (1989) suggested the use of barriers against the spruce budworm as part of a strategy based on what he calls environmental architecture. Environmental architecture exploits the fact that herbivorous insects respond to and are influenced by differences in plant texture. Two windbreaks in the pollen dilution zone are suggested as a barrier between the surrounding forests and a spruce seed orchard. Such a barrier would help prevent insect damage when the forest has a heavy component of spruce and balsam fir. The spruce cone maggot and spruce budworm, specialist herbivores of spruce, probably are attracted to spruce visually and by olfaction. A windbreak consisting of deciduous trees would mask the orchard from the insects because it would be visually and chemically different from the orchard trees. Windbreaks would also act as a physical filter for insects blown across the pollen dilution zone into the orchard. Finally, plant texture also interferes with the free flow of air and hence creates turbulence which could interfere with local dispersal of spruce budworm and spruce cone maggot.

3.3.2 Gypsy Moth

Gypsy moth [*Lymantria despar* (L.)] control remains highly dependent on chemical insecticides (Metcalf and Metcalf 1993). However, at the limit of the range of distribution of this species, it is not uncommon to rely, at least in part, on a combination of Tanglefoot and burlap band around the tree trunk to eradicate the moth from isolated foci of colonization outside its normal range of distribution. The larvae feed in the canopy during the day but descend along the tree trunk at night to seek shelter. Burlap bands provide the shelter and the Tanglefoot controls them (Metcalf and Metcalf 1993). Thorpe et al. (1994) studied the combination of sticky barriers and the release of a predator for control of the gypsy moth. The sticky barrier placed around the tree trunks, prevented the larvae of the gypsy moth from crawling up the trees to the foliage. The predator, *Podisus maculiventris*, a pentatomid, was released to prey on larvae already in the tree, thus encouraging the movement of these larvae towards the sticky barrier. The concept is interesting but results are preliminary. For now, the burlap barrier is used most often as a monitoring tool to estimate the abundance of the gypsy moth in the forest (Elkinton and Liebhold 1990).

3.4 Physical Barriers Used in Experimental Research

Researchers sometimes use physical barriers to exclude certain insects from study areas. Boiteau (1986) used polythene barriers to quantify the role of native predators, especially *Carabidae*, on the population dynamics of potato-colonizing aphids. This technique had been previously used in England for similar studies in cereals with grain aphids.

Holopainen and Varis (1986) report extensively on the effect of polythene barriers on carabid catches in pitfall traps. They have reviewed the literature and conducted experimental work on the advantages and limitations of this methodology.

In another study (Shands et al. 1972), the movement of aphid predators (*Coleoptera: Coccinellidae, Neuroptera: Chrysopidae*) was restricted using 20-cm-tall aluminum flashing placed upright in the field. The top of the flashing was bent at 90 ° over 3.5 cm and covered with a non-stick chemical. The barriers were effective at controlling the spread of the predators but not very effective as a tool to determine their effectiveness at controlling aphids. Previous trials with strips of oats up to 17 m wide, or a similar bare strip with a straight untreated aluminum flashing in the middle, were ineffective at reducing the dispersal of *Coccinellidae*. This subject is so extensive, that it alone could be the subject of a review.

4 Physical Barriers and Biological Barriers: the Interface

Technically, physical barriers are made up of physical products that are inserted or added to the agroecosystem. However, if hairs and natural glues cover the leaf surface of the plants and hamper the dispersal of insects (Boiteau and Singh 1988) should we be talking of a physical or of a biological barrier? The difference between the two types is even more difficult to maintain when the control agent is biological, but it is used according to the principles controlling the effectiveness of physical barriers.

For example, Marsula and Wissel (1994) used modeling to investigate the optimization of a spatial barrier, as part of the screwworm fly eradication program. The sterile male release technique was used to eradicate the screwworm fly from the USA. Thereafter, a barrier was established at the USA-Mexico border to prevent the seasonal reinvasion of the USA. In this case, the barrier is defined as a strip of land over which sterile males are released to prevent the reinvasion along the border. This approach has been made possible because the female mates only once and because sterile males are fully competitive with wild males. On the basis of research data, it has been possible to confirm that a random distribution model is appropriate for the dispersal of the screwworm fly. This model has been made possible by the existence of a considerable quantity of information on the biology and ecology of this insect pest. The model demonstrates, as might have been expected, that the safety or effectiveness of the barrier increases with its width. The cost increases in parallel. However, the model made it possible to consider alternatives such as multiple barriers. Effectiveness can be increased by using two or three parallel strips. The first strip has a high density of sterile males and controls most of the wild flies. The second strip has a lower density of sterile males and controls the remaining wild population. A single wide strip with a high density of sterile males would have been more expensive than the multiple strips because of the high production costs of sterile males. The conclusions of this particular system may not extend directly to other barrier-pest systems, but they warrant consideration. In most field crops, the value of the land area itself rarely allows for the multiplication of barriers around fields. However, in the same way that barriers can be spatial as well as physical structures, multiple barriers do not all have to be of the same design. For example, Prévost (1989) suggested that multiple windbreaks of different tree species around the perimeter of seed cone nurseries could provide both barrier and filter against insect pests. The mathematical model provides renewed evidence that the combination of control measures can be advantageous. For example, it may be possible to reduce the adult population of CPB with trenches and to control the flyers' progeny with beetle collectors or bacterial sprays.

5 Conclusion

Screens, double door entrance systems, etc. continue to be incorporated into the design of greenhouses and food storages as a first line of defence in preventing insects and other pests from entering into these confined spaces. Containers and constantly improving packaging protect our food supply from significant yield losses to insects throughout the world. The extension of physical methods of insect management to agriculture and forestry has, however, been hampered by four main factors. As often alluded to above, a major impediment to the implementation of many physical control methods is the cost involved. The up-front cost of certain exclusion barriers, such as row covers or exclusion fences, can be relatively high in contrast to other popular control approaches (i.e. pesticides). Another obstacle is the level of difficulty involved with implementing certain physical control strategies. A common complaint of organic producers about row covers is the amount of handling required, which might include initial placement, removal for weeding, fertilizing and harvesting, and final storage or disposal of the covers. Other barriers alone are often not sufficient for managing pests at the levels expected of traditional pesticide control programs and must therefore be used in combination with other alternative control methods, thus increasing the complexity of the management system. Even something as simple as establishing a live windbreak requires long-term planning and several years of patience before the strategy will be useful. These first two factors, cost and management complexity, become less important, however, when traditional control methods (such as pesticides) can no longer ensure the sustainability of food and fibre production. The use of plastic-lined trenches, for example, increased in use by commercial potato growers when the Colorado potato beetle developed resistance to all available insecticides in certain regions of Canada.

The third obstacle to the adoption of physical barriers has been the general availability of the products required. Previous barrier designs called for expensive and often heavy materials that made them too labour-intensive to install and maintain. New materials and processes for manufacturing and installing these materials and devices have made previously impractical methods economically feasible. Some recent examples are spunbonded polyester row covers and plastic-lined trench technology.

The final obstacle to the adoption of physical barriers, especially in conventional agricultural systems, is the chronic dependence of growers on pesticides for pest control. This dependence on pesticides has developed as a natural consequence of the low cost, high efficiency and ease of use this technology has offered growers for half a century. It is an unfortunate fact of life that, given the choice, growers will generally choose pesticides over alternative strategies, even when the pesticide strategies are obviously non-sustainable. Once again, the Colorado potato beetle provides us with a good example. For years, the main line of defence for the management of this pest was the routine application of pesticides. In Canada, despite the growing availability of a number of alternative physical control methods, growers continued to rely on pesticides even after resistance to the existing pesticides had occurred. It was not until the number of pesticide applications became cost-prohibitive and their control effectiveness collapsed, that some growers desperately adopted physical control strategies such as the plastic-lined trenches mentioned above. Nevertheless, when a new insecticide for Colorado potato beetle control was eventually released, growers immediately abandoned the alternative strategies once again in favour of routine spraying.

The attitude that pesticides are the "best" and, in the minds of many, the "only" method for controlling pests, is gradually changing. This can be attributed to the gradual implementation of well-researched integrated pest management programs throughout North America, and to the ongoing development and demonstration of cost-effective alternative control methods. In addition, each year brings a number of pest organisms closer to defeating the pesticide approach, either through resistance buildup, or through the corporate- and regulatory-based attrition of the pesticide arsenal. Add to this the improved availability, cost and efficacy of physical control strategies. It is therefore expected that these alternatives will become important tools in the IPM programs of the future.

References

Adlerz W.C., Everett P.H., (1968). Aluminum foil and white polythene mulches to repel aphids and control watermelon mosaic (*Citrullus lanatus*). J. Econ. Entomol. 61: 1276-1279.

Antill D.N., Senior D., Blood-Smyth J., Davies, Emmett B., (1990). Crop covers and mulches to prevent pest damage to field vegetables. Brighton Crop Protection Conference, Pests and Diseases. Vol. 1, 355-360.

Banks H.J., (1976). Physical control of insects - Recent developments. J. Aust. Entomol. Soc. 15: 89-100.

Bar-Joseph M., Fraenkel H., (1983). Spraying citrus plants with kaolin suspensions reduces colonization by the spiraea aphid (*Aphis citricola* van der Goot). Crop Prot. 2 (3): 371- 374.

Beckman R.L, (1972). Colour preference and flight habits of thrips associated with cotton. J. Econ. Entomol. 65: 650-654.

Boiteau G., (1986). Native predators and the control of potato aphids. Can. Entomol. 118: 117-1183.

Boiteau G., (1996). La pomme de terre au Canada. pp. 541-542, in P. Rousselle, Y. Robert et J. C. Crosnier, (eds.), La pomme de terre : production, amélioration, ennemis et maladies, utilisations. INRA Editions, Versailles, 607 p.

Boiteau G., Osborn W.P.L., (1999). Comparison of plastic-lined trenches and extruded plastic traps for controlling *Leptinotarsa decemlineata* (Coleoptera: Chrysomelidae). Can. Entomol. 13: 567-572.

Boiteau G., Singh R.P., (1988). Resistance to the greenhouse whitefly, *Trialeurodes vaporariorum* (Westwood) (Homoptera: Aleyrodidae), in a clone of the wild potato *Solanum berthaultii* Hawkes. Ann. Entomol. Soc. Am. 81: 428-431.

Boiteau G., Pelletier Y., Misener G.C., Bernard G., (1994). Development and evaluation of a plastic trench barrier for protection of potato from walking adult Colorado potato beetles (Coleoptera: Chrysomelidae). J. Econ. Entomol. 87: 1325-1331.

Chagnon M., Payette A., (1990). Modes alternatives de répression des insectes dans les agro-écosystèmes québécois, tome 1: Document synthèse. Ministère de l'Environnement et Centre québécois de valorisation de la biomasse, Québec, 81 p.

Cloutier C., Jean C., Bauduin F., (1996). More biological control for a sustainable potato pest management strategy, pp. 15-52 in R.-M. Duchesne and G. Boiteau (eds.). Lutte aux insectes nuisibles de la pomme de terre. Développement d'une approche durable. Agriculture et Agroalimentaire Canada, L'Union des producteurs agricoles et Gouvernement du Québec Ministère de l'Agriculture, des Pêcheries et de l'Alimentation, Québec, 204 p.

Cohen S., (1981). Reducing the spread of aphid-transmitted viruses in peppers by coarse-net cover Cucumber mosaic virus and potato virus Y. Phytoparasitica 9: 69-76.

Cohen S., (1982). Control of whitefly vectors of viruses by color mulches *Bemisia tabaci*, pp. 45-56 in K.F. Harris and K. Maramorosch (eds.). Pathogens, vectors and plant diseases: approaches to control, Academic Press, New York.

Cohen S., Marco S., (1973). Reducing the spread of aphid-transmitted viruses in peppers by trapping the aphids on sticky yellow polyethylene sheets. Phytopathology 63: 1207-1209.

Cohen S., Melamed-Madjar V., (1978). Prevention by soil mulching of the spread of tomato yellow leaf curl virus transmitted by *Bemisia tabaci* (Gennadius) (Hemiptera: Aleyrodidae) in Israel. Bull. Entomol. Res. 68: 465-470.

Collman S.J., Antonelli A., (1994). Biology and control of tent caterpillars. Washington State University. Cooperative Extension. Extension Bulletin 1106, 4 pp.

Davies J.S., Hembry J.K., Williams G.H., (1993). Novel production systems to minimize pesticide use in intensive field vegetable crops. Crop protection in Northern Britain 1993: Proceedings of a Conference, Dundee University, 23-25 March 1993, 195-200.

Duchesne R.-M., Boiteau G. (eds.), (1996). Lutte aux insectes nuisibles de la pomme de terre. Agriculture et Agroalimentaire Canada, L'Union des producteurs agricoles et Gouvernement du Québec Ministère de l'Agriculture, des Pêcheries et de l'Alimentation, Québec, 204 p.

Elkinton J.S., Liebhold A.M., (1990). Population dynamics of Gypsy moth in North America. Annu. Rev. Entomol. 35: 571-96.

Ellis P.R., Hardman J.A., (1988). Non-insecticidal contributions to an integrated programme for the protection of carrots against carrot fly. Bulletin - SROP. 1988, 9: 1, 33-39; In working group, integrated control in field vegetable crops, Denmark, 21-23 September 1987.

Ferro D., (1996). Mechanical and physical control of the Colorado potato beetle and aphids, pp. 53-67 in R.-M. Duchesne and G. Boiteau (eds.). Lutte aux insectes nuisibles de la pomme de terre. Développement d'une approche durable. Agriculture et Agroalimentaire Canada, L'Union des producteurs agricoles et Gouvernement du Québec Ministère de l'Agriculture, des Pêcheries et de l'Alimentation, Québec, 204 p.

Finch S., (1993). Integrated pest management of the cabbage root fly and the carrot fly. Crop Prot. 12: 423-430

Folster E., (1989). Pay attention to crop rotation when using netting. Deutscher-Gartenbau. 43: 11, 688.

Forbes, S.A., (1895). Experiments with the muscardine disease of the chinch-bug, and with the trap and barrier method for the destruction of that insect. University of Illinois Agricultural Experimental Station, Urbana. Bulletin 38: 25-86.

Fusco R.A., Thurston R., (1970). Effect of coloured foils on green peach aphid infestations of burley tobacco. Tob. Sci. 14: 126-127.

Harpaz I., (1982). Nonpesticidal control of vector-borne viruses, in Pathogens, Vectors and Plant Diseases: Approaches to Control (K. F. Harris and K. Maramorosch, eds.), Academic Press, New York, pp. 1-22.

Haseli A, Konrad P., (1987). An alternative for plant protection in vegetables. Pest attack control with nets. Gemuse-Munchen. 23: 7, 320-324.

Havukkala I., (1988). Non-chemical control methods against cabbage root flies *Delia radicum* and *Delia floralis* (Anthomyiidae). Ann. Agr. Fenn. 27: 271-279.

Holopainen J.K., Varis A.-L., (1986). Effects of a mechanical barrier and formalin preservative on pitfall catches of carabid beetles (Coleoptera, Carabidae) in arable fields. J. Appl. Entomol. 102: 440-445.

Hough-Goldstein J.A., (1987). Tests of a spun polyester row cover as a barrier against seedcorn maggot (Diptera: Anthomyiidae) and cabbage pest infestations. J. Econ. Entomol. 80: 768-772.

Howard R.J., Garland J.A., Seaman W.L. (Eds.), (1994). Diseases and pests of vegetable crops in Canada. The Canadian Phytopathological Society and Entomological Society of Canada, Ottawa, 554 p.

Hunt D.W.A., Whitfield G., (1996). Potato trap crops for control of Colorado potato beetle (Coleoptera: Chrysomelidae) in tomatoes. Can. Entomol. 128: 407-412.

Jolivet P., (1991). Le doryphore menace l'Asie, *Leptinotarsa decemlineata* Say 1824 (Col. Chrysomelidae). L'Entomologiste 47: 29-48.

Judd G.J.R., Vernon R.S., Borden J.H., (1985). Monitoring program for *Psila rosae* (F.) (Diptera: Psilidae) in southwestern British Columbia. J. Econ. Entomol. 78: 471-476.

Kemp W.G., (1978). Mulches protect peppers from viruses. Can. Agric. 23: 22-24.

Kettunen S., Havukkala I., Holopainen J.K., Knuuttila T., (1988). Non-chemical control of carrot rust fly in Finland. Ann. Agr. Fenn. 27: 2, 99-105

Lewis T., (1965). The effects of an artificial windbreak on the aerial distribution of flying insects. Ann. Appl. Biol. 55: 503-512.

Lewis T., (1966). Artificial windbreaks and the distribution of turnip mild yellows virus and *Scaptomyza apicalis* (Dip.) in a turnip crop. Ann. Appl. Biol. 58: 371-376.

Lewis T., (1969). The distribution of flying insects near a low hedgerow. J. Appl. Ecol. 6: 443- 452.

Lewis T., Dibley G.C., (1970). Air movement near windbreak and a hypothesis of the mechanism of the accumulation of airborne insects. Ann. Appl. Biol. 66: 1306-1307.

Loebenstein G., Alper M., Levy S., Palevitch D., Menagem E., (1975). Protecting peppers from aphid-borne (cucumber mosaic, potato Y) viruses with aluminium foil or plastic mulch. Phytoparasitica 3: 43-53.

Long B., (1994). Solving ant problems nonchemically. J. Pesticide Reform. 14: 22-23.

Marsula R., Wissel C., (1994). Insect pest control by a spatial barrier. Ecological Modelling. 75/76: 203-211.

Mathis W., Smith E.A., Schoof H.F., (1970). Use of air barriers to prevent entrance of house flies. J. Econ. Entomol. 63: 24-31.

Metcalf R.L., Metcalf R.A., (1993). Destructive and useful insects: Their habits and control. Fifth Edition. McGraw-Hill Inc., New York, 1094 p.
Millar K.V., Isman M.B., (1988). The effects of a spunbonded polyester row cover on cauliflower yield loss caused by insects. Can. Entomol. 120: 45-47.
Misener G.C., Boiteau G., McMillan L.P., (1993). A plastic-lining trenching device for the control of Colorado potato beetle: Beetle excluder. American Potato Journal 70: 903-908.
Moore W.D., Smith F.F., Johnson G.V. Wolfbarger D.O., (1965). Reduction of aphid populations and delayed incidence of virus infection on yellow straight neck squash by the use of aluminium foil. Proc. Fla. Sta. Horticul. Soc. 78: 187-191.
Moyer D.D., (1995). Trapping the Colorado potato beetles with plastic-lined trenches. 1994 New York State Vegetable Project Reports Relating to IPM. New York State IPM publ. #118.
Moyer D.D., Derksen R.C., McLeod M.J., (1992). Development of a propane flamer for Colorado potato beetle control. Am. Potato J. 69: 599-600.
Ng Y.S., Lashomb J.H., (1983). Orientation by the Colorado potato beetle (*Leptinotarsa decemlineata* Say). Anim. Behav. 31: 617-618.
Paparatti B., (1993). Experiments in the integrated control of *Phthorimaea operculella* (Zeller) (Lepidoptera: Gelechiidae) in northern Latium. Frustrula- Entomologica1. 6: 9-22.
Payette A., Chagnon M., (1990). Modes alternatives de répression des insectes dans les agro-écosystèmes québécois, tome 4: Techniques physiques. Ministre de l'Environnement et Centre québécois de valorisation de la biomasse, Québec, 53 p.
Pedgley D.E., (1982). Windborne pests and diseases: Meteorology of airborne organisms. Ellis Horwood Limited, Chichester, 250 p.
Prévost Y.H., (1989). Environmental architecture - Preventing loss of seed production to insects in black and white spruce seed orchards. Proceedings, Cone and Seed Pest Workshop, 1989, St. John's Newfoundland, Canada. Report N-X-274.
Raccah B., (1986). Nonpersistent viruses and control: Epidemiology and control, pp 387-429 in K. Maramorosch, F. A. Murphy et A. J. Shatkin (eds.) Academic Press, Orlando, 444 p.
Radcliffe E.B., Ragsdale D.W., Flanders K.L., (1993). Management of aphids and leafhoppers. Potato Health Management, edited by Randall C. Rowe, APS Press. pp 117-126.
Riley C.V., (1876). Potato pests. Being an illustrated account of the Colorado potato-beetle and the other insect foes of the potato in North America. With suggestions for their repression and methods for their destruction. Orange Judd Co., New York, 108 p.
Rorabacher J.A., (1970). The American buffalo in transition: a historical and economic survey of the bison in America. North Star Press. Minnesota.
Roush R.T., Tingey W.M., (1994). Strategies for the management of insect resistance to synthetic and microbial insecticides, pp. 237-254 in G.W. Zehnder, M.L. Powelson, R.K. Jansson, and K.V. Raman. (eds.). Advances in Potato Pest Biology and Management APS Press, 655 p.
Schoene W.J., (1914). The cabbage maggot in relation to the growing of early cabbage. New York Agricultural Experiment Station. Geneva, N. Y. Bulletin No. 382.
Schorger A.W., (1955). The passenger pigeon: its natural history and extinction. Norman, University of Oklahoma Press, 1973, 424 p.
Shands W.A., Simpson G.W., Storch R.H., (1972). Insect predators for controlling aphids on potatoes. 3. In small plots separated by aluminum flashing strip-coated with a chemical barrier and in small fields. J. Econ. Entomol. 65: 799-805.
Skinner G., Finch S., (1986). Reduction of cabbage root fly (*Delia radicum*) damage by protective disks. Ann. Appl. Biol. 108: 1-10.
Smith F.F., Webb R.E., (1969). Repelling aphids by reflective surfaces, A new approach to the control of insect-transmitted viruses, pp. 631-639 in K. Maramorosch (ed.). Viruses, Vectors, Vegetation, Wiley (Interscience), New York.
Storch R.H., (1996). Insect pest control on potato with conventional insecticides, pp.95-112 in R.-M. Duchesne and G. Boiteau (eds.). Lutte aux insectes nuisibles de la pomme de terre. Développement d'une approche durable. Agriculture et Agroalimentaire Canada, L'Union des producteurs agricoles et Gouvernement du Québec Ministère de l'Agriculture, des Pêcheries et de l'Alimentation, Québec, 204 p.

Thorpe K.W., Webb R.E., Aldrich J.R., Tatman K.M., (1994). Effects of spined soldier bug (Hemiptera: Pentatomidae) augmentation and sticky barrier bands on gypsy moth (Lepidoptera: Lymantriidae) density in oak canopies. J. Entomol. Sci. 29: 339-346.

Tuttle A.F., Ferro D.N., Idoine K., (1988). Role of visual and olfactory stimuli in host finding of adult cabbage root flies, *Delia radicum*. Entomol. Exp. Applic. 47: 37-44.

Vernon R.S, (1979). Population Monitoring and Management of *Hylemya antiqua* and *Thrips tabaci* in British Columbia Onion Fields, with Observations on Other Root Maggot Populations. Master of Pest Management Professional Paper, Simon Fraser University, Burnaby, B.C.

Vernon R.S., Hall J.W., G.J.R. Judd, Bartel D.L., (1989). Improved monitoring program for *Delia antiqua* (Diptera: Anthomyiidae). J. Econ. Entomol. 82: 251-258.

Weber D.C., Ferro D.N., Buonaccorsi J., Hazzard R.V., (1994). Disrupting spring colonization of Colorado potato beetle to nonrotated potato fields. Entomol. Exp. Applic. 73: 39-50.

Wegorek W., (1959). The Colorado Potato Beetle (*Leptinotarsa decemlineata* Say). (Translated from Polish). Prace Naukowe Instytutu Ochrony Roslin. Vol. 1, No. 2.

Weiss M., Cohen S., Marco S., Harpaz I., (1977). Failure of sticky yellow traps to reduce virus infection in squash (*Myzus persicae, Aphis gossypii, Aphis fabae*). Hassadeh (Tel-Aviv) 58(1): 75-78.

Weisz R., Smilowitz Z., Saunders M., Christ B., (1995). Integrated Pest Management for Potatoes. College of Agricultural Science Cooperative Extension, PennState, State College. 16 pp.

Wheatley G.A., (1975). Physical barriers for controlling cabbage root fly. Report of the National Vegetable Research Station for 1974. p. 97.

Yudin L.S., Tabashnik, B.E., Mitchell, W.C., Cho J. J., (1991). Effects of mechanical barriers on distribution of Thrips (Thysanoptera: Thripidae) in lettuce. J. Econ. Entomol. 84: 136-139.

Zehnder G.W., Hough-Goldstein J., (1990). Colorado Potato Beetle (Coleoptera: Chrysolelidae) population development and effects on yield of potatoes with and without straw mulch. J. Econ. Entomol. 83: 1982-1987.

Zitter T.A., Simons J.N., (1980). Management of plant viruses by alteration of vector efficiency and by cultural practices. Annu. Rev. Phytopathol. 18: 289-310.

Control of Insects in Post-Harvest: Inert Dusts and Mechanical Means

Paul FIELDS, Zlatko KORUNIC and Francis FLEURAT-LESSARD

1 Introduction

Residual chemicals are currently used to control stored-product insects pests that are found in granaries and food-processing facilities. Long-term use of these chemicals has resulted in the development of insect populations that are resistant to the common insecticides (e.g. malathion, chlorpyrifos-methyl, pirimiphos-methyl, fenitrothion or deltamethrin) (White and Leesch 1995). Inert dusts are used in a fashion similar to the residual chemicals, and hence offer a convenient alternative for the control of insect pests in grain-and food-processing industry installations such as flour mills, food warehouses and retail outlets.

Inert dusts have been used for thousands of years (Ebeling 1971). Today, diatomaceous earth and silica gel are the predominant inert dusts used commercially, both of which are composed of silicon dioxide. Over the past 15 years, research has been directed at new dust formulations based on silicon dioxide. Formulations that are effective at 1000 ppm (or a lower rate) have been developed, and are now used around the world (Fields and Muir 1995; Banks and Fields 1995; Golob 1997; Korunic 1997; 1998). A recent workshop dealt with all aspects of diatomaceous earth as a protectant for stored grain (Fields 1998).

Mechanical control is another ancient method of controlling stored-product insects and it is still in use today (Fields and Muir 1995; Banks and Fields, 1995). This approach is used extensively in flour mills, and may become more widespread due to the impending ban of methyl bromide, a fumigant that is widely used in flour mills. Methyl bromide is slated to be banned in most parts of North America and Europe by 2005 because it is an ozone-depleting substance. Wheat that is destined to be milled into flour, or durum and corn that is to be milled into semolina, can also be disinfested by impact machines that hit grain hard enough to kill insects within the kernel. In some cases, pneumatic conveying of grain is sufficient to control certain insects in certain life stages. Some mites and lepidopteran larvae can be controlled simply by moving the grain.

2 Inert Dusts

2.1 As Traditional Grain Protectant

Ashes have been used as an insecticide for thousands of years by aboriginal peoples in North America and Africa (Ebeling 1971). Modern research on inert dusts to protect stored grain began in the 1920s (Ebeling 1971; Fields and Muir 1995; Golob 1997; Korunic 1997, 1998). The main advantage of inert dusts is their low toxicity to mammals. In Canada and the USA, diatomaceous earth is registered as an animal feed additive, and silicon dioxide is registered as a human food additive. Silica gel has an acute rat LD_{50} of 3160 mg kg^{-1} (Ebeling 1971). Other advantages of inert dusts are that they are effective for long periods and do not affect end use quality of the product (Desmarchelier and Dines 1987 ; Korunic et al. 1996).
There are four types of inert dusts: (1) powdered clay, sand and earth, (2) diatomaceous earth, (3) silica gel, and (4) non-silica dusts such as rock phosphate and lime. Powdered clay, sand and earth have been used as thick layers of dust on the top surface of a grain bulk. Diatomaceous earth is made up of the fossilized skeletal remains of diatoms, single-celled algae that are found in fresh and salt water. Diatoms are microscopic and have a fine skeleton made up of amorphous silica (SiO_2 + n H_2O). The accumulation of diatom skeletons over thousands of years produces the sedimentary rock, diatomaceous earth. The diatomaceous earth deposits currently mined are millions of years old, and certain deposits are hundreds of meters thick. The major constituent of diatomaceous earth is amorphous silicon dioxide (SiO_2) with minor amounts other minerals (aluminum, iron oxide, calcium hydroxide, magnesium and sodium). Its insecticidal properties depend upon the geological origin of the diatomaceous earth, some types being 20 times more effective than others (Korunic 1997). The diatomaceous earths that are the most effective in insect control have SiO_2 content above 80%, a pH below 8.5, and a tapped density below 300 g l^{-1} (Korunic 1997). Silica gels, which are produced by drying aqueous solutions of sodium silicate, are very light hydrophobic powders (Quarles 1992). Non-silica dusts, such as rock phosphate, have been used in Egypt. Tests have shown that lime (calcium oxide) has some insecticidal activity (Fields and Muir 1995).

2.2 Mode of Action of Inert Dusts

Insects die when they have lost approximately 60% of their water or about 30% of their body weight (Ebeling 1971). Since stored-product insects are small and have a high surface to volume ratio, it is particularly difficult for them to maintain their water balance in very dry habitats such as granaries of dry grain. Insects have several mechanisms to deal with desiccation, but the cuticle is one of the more important ones; however, it is sensitive to damage by inert dusts.

When insects move through the grain, inert dusts adsorb their cuticular waxes (Le Patourel et al. 1989). The mode of action of silicon dioxide-based inert dusts is generally accepted to be desiccation rather than abrasion. Two observations support this conclusion. First, inert dusts are more effective when the grain moisture content or the relative humidity is lower. Second, insects treated with inert dusts usually have higher rates of water loss (Fields and Muir 1995). The following species have been ranked as most to least sensitive to diatomaceous earth: *Cryptolestes* spp., *Oryzaephilus* spp., *Sitophilus* spp., *Rhyzopertha dominica* (F.) and *Tribolium* spp. (Carlson and Ball 1962; Fields and Muir 1995).

Tolerance to desiccation is usually, but not always, positively correlated with tolerance to inert dusts. There are several possible reasons for these differences. Insects that can use metabolic water would be less sensitive to inert dusts. Species that have a better reabsorption of intestinal water also would have an advantage, as well as those that have a more waterproof cuticle. Thus, the composition of the cuticular waxes can be an important factor. The role of the cuticle depends on how much diatomaceous earth it can pick up. A sensitive species such as *Cryptolestes ferrugineus* (Stephens) picks up much more than *Tribolium castaneum* (Herbst) (Fields and Korunic 1999).

Behaviour may also play a role. Insects that move extensively through the grain, such as *C. ferrugineus*, may be more damaged than insects that are more sedentary, such as *R. dominica*. Insects are repelled by diatomaceous earth, which is an important consideration when only part of the grain bulk is treated, such as a top dressing. Finally, the type of grain also determines the effectiveness. Of the following grains, milled rice requires the greatest amount of diatomaceous earth for control of its insect pest, followed by sunflower, maize, paddy rice, oats, barley, wheat and durum, in that order.

Despite the numerous advantages of inert dusts, they have seen limited use. However, there has been an increase in use in the past decade due to concerns over the effects of chemical insecticides on worker safety, food safety and development of resistance by insect populations. Diatomaceous earth is registered as a grain protectant or for treating storage structures in Australia, Canada, USA, Croatia, Germany, China and some other Asian countries. In Australia, diatomaceous earth is used principally as a treatment for empty silos. However, it can also be used to treat the entire mass of feed grain. In all other countries, diatomaceous earth can also be used to treat grain aimed for human consumption. In India, during the 1960s, 70% of the grain was treated with activated kaolin clay. Egypt used rock phosphate and sulphur (Ebeling 1971).

The main problem with the use of diatomaceous earth is that it reduces grain bulk density and grain fluidity (Korunic et al. 1996; Fields 1998). Adding diatomaceous earth at 2 kg t^{-1} reduces grain bulk density by approximately 4 kg hl^{-1} for maize and by 6 kg hl^{-1} for wheat. Another limitation is that grain must be dry for the diatomaceous earth to cause enough desiccation to control insect populations.

Application of inert dusts can be undesirable because of the dust generated. Diatomaceous earth can be used as a mild abrasive, and there is concern over increased wear on grain-handling machinery. However, diatomaceous earth is relatively soft, its Moh's hardness index being 2, which is softer than that of gold (2.5-3), copper (2.5-3), nickel iron (5), quartz (7) and diamond (10) (Glover 1997). It is necessary to determine to what extent diatomaceous earth increases wear on grain-handling and milling equipment.

Depending upon the source and processing method, diatomaceous earth can contain from 0.1 to 60% crystalline silica. Diatomaceous earths registered as insecticides generally have less than 7% crystalline silica. For other uses, diatomaceous earth is heated or burned and the crystalline silica content can increase up to 60%. Crystalline silica has been shown to be carcinogenic if inhaled (IARC 1997). Proper masks should therefore be used when handling diatomaceous earth of high crystalline silica content.

One solution to the airborne dust problem is to apply the diatomaceous earth in a water suspension. This method of application is widely used in Australia to treat empty structures (Bridgeman 1998). However, slurry application does somewhat reduce their effectiveness (Maceljski and Korunic 1971). On the farm, diatomaceous earth can be blown into empty granaries through the aeration ducts.

Another way to reduce the problems associated with diatomaceous earth is to lower the concentration needed to achieve control. A mixture of diatomaceous earth (90%) and silica gel (10%) is twice as effective as diatomaceous earth used alone (Korunic and Fields 1995). Silica gels are very effective, but they greatly reduce grain bulk density and are so light that they are difficult to apply to grain. This is another reason why it is better to use them in a mix with diatomaceous earth.

Insect resistance to residual insecticides has been one of the factors motivating the search for alternatives to chemical insecticides. Laboratory experiments have shown that the susceptibility of *T. castaneum*, *C. ferrugineus* and *R. dominica* can drop significantly when exposed to diatomaceous earth for five to seven generations (Korunic and Ormesher 1998). Although there are no reported cases of insects developing resistance to diatomaceous earth under field conditions, these results suggest that it will be necessary to use resistance management strategies to prevent widespread resistance to diatomaceous earth products.

Diatomaceous earth can be combined with other treatments to increase the effectiveness (Bridgeman 1998). In Australia, diatomaceous earth is used as a top dressing in conjunction with long duration, low dose phosphine fumigation (SiroFlo). The major limitation of SiroFlo used alone is that phosphine concentrations are too low to obtain control at the surface of the grain bulk. Diatomaceous earth serves a dual function: it controls insects directly, and provides a physical barrier to retain the phosphine. Diatomaceous earth is also used as a top dressing, in conjunction with cooling of the grain mass by ambient air or refrigerated air aeration. Low temperatures slow or stop the development of insects, but it is difficult to maintain the grain surfaces at low temperature. Diatomaceous earth can also be used in conjunction with heat. One method to control insects in food-processing facilities is to raise the air temperature of the building to 50 °C for 24 h. In one experiment, it was found that control of *Tribolium confusum* J. du Val was obtained at 41°C when diatomaceous earth was also used, whereas a temperature of 47 °C was needed in the absence of the dust (Doudy and Fields, in press). Diatomaceous earth could make heat sterilization more acceptable to the industry because it would require lower temperatures to obtain control. As the use of methyl bromide as fumigant declines, new methods such as diatomaceous earth and heat will become more widely used.

3 Mechanical Means

Stored-product insects are very sensitive to mechanical shock. Even the effect of the movement of grain during conveying can reduce the fecundity of the insect and prevent a normal population increase (Banks 1986). However, the only mode of grain transportation that destroys significant numbers of individuals, including internal stages, are pneumatic conveyers (Fleurat-Lessard 1985, White et al. 1997, Paliwal et al. 1999).

3.1 Traditional Uses in Stored Grain

In 19th century France, mechanical mixing of grains was seen as a replacement for the traditional hand shoveling of grain that was usually stored on a wooden plat-

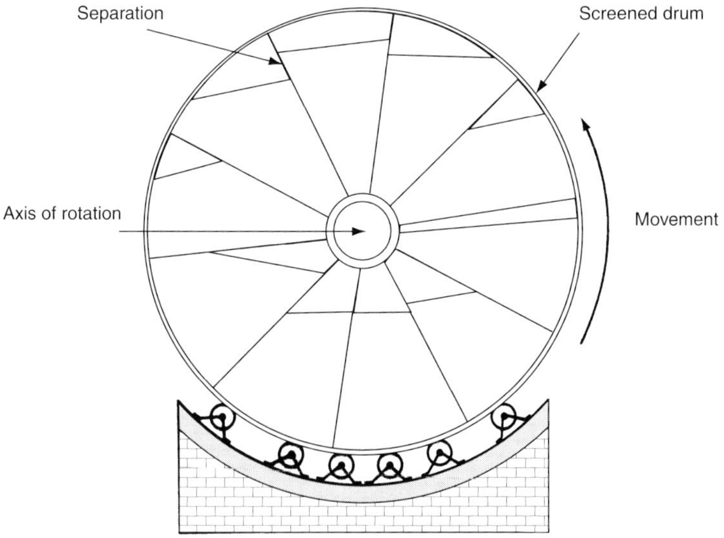

Fig. 1. Cross-sectional view of a mobile granary invented in the first half of the 19th century to prevent insect infestation in grain stored for long durations. It was made up of a rotating cylinder with internal baffles (Vallery 1839).

form in layers of thickness 30 to 50 cm (Sigaut 1978). The first real attempt at mechanical mixing was a rotating granary (Vallery 1839). It consisted of a large horizontal drum with internal baffles and an axle running through the centre (Fig. 1). The concept of keeping the grain in motion, one which was used later for the design of certain grain dryers, offered the advantage of reducing the number seeds infested by pests that must bore holes into the seed to lay their eggs or to begin their development (e.g. *Sitophilus* spp., *R. dominica*, and bruchids). This also reduced feeding and mating. More recent studies have shown that disturbing the grain, either continuously or periodically, reduces the reproductive potential of insects significantly (Joffe 1963; Bahr 1990). Joffe (1963) showed that turning maize every 2 weeks reduced *Sitophilus* spp., *Tribolium* spp. and *Cryptolestes* spp. by 87, 75 and 89%, respectively. Pneumatic conveyers can reduce internal stages of *S. oryzae* by up to 85% (Fleurat-Lessard 1985), and external insects, such as *Oryzaephilus surinamensis* (L.) and *C. ferrugineus* and *T. castaneum* by over 90% (Bahr 1990; Paliwal et al. 1999). Bruchid larvae, such as *Acanthoscelides obtectus* (Say), can be effectively controlled by moving the beans every 8 h. Eggs are laid outside the bean, and first instar larvae take about 24 h to bore through the seed

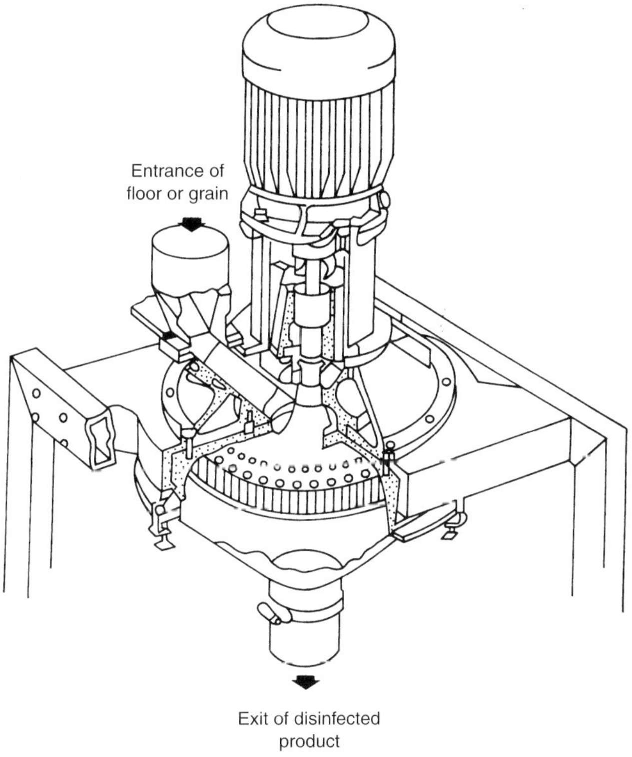

Fig. 2. Impact device (entoleter) used to destroy insects in whole grain or flour by the centrifugal forces.

coat. Moving the beans causes the larvae to lose the pit that they have started to bore in the seed coat, and they eventually die from starvation or crushing during movement (Quentin et al. 1991).

An indirect method of mechanical control that is widely used in mills is the exclusion of insects by sieving. Most cleaning sieves have a mesh size of 2 mm that separate the adult insects and other impurities from the flour. High-volume centrifugal separators can eliminate over 80% of external stages with a single pass. It is recommended that flour be passed through sieves with an opening of 0.6 to 0.8 mm, just before bulk storage or bagging of the finished product. This eliminates adult insects, but does allow the passage of insect fragments.

Another common physical control is obtained by impact, using devices called entoleters. Flour is projected onto a spinning disk equipped with pins. All insect stages are killed as they hit the pins or the walls of the entoleter (Fig. 2). The use of the entoleters is one of the final processing steps before packaging or bulk storage.

3.2 Integration into the Preventive Control Strategies for the Cereal Industry

Originally, the impact desinfestor were only used to destroy insects in flour (Stratil et al. 1987). They are now used to disinfest grain just before it enters the first grinding mill. It was noted that grain with internal feeders, such as *Sitophilus* spp., would break apart more easily than uninfested grain. Impact experiments showed that over 90% of the internal stages of *S. granarius* were destroyed when the kernels reached a speed of 33 m s^{-1}. However, under these conditions, 4-12% of uninfested kernels is also broken (Bailey 1962). These studies served as a starting point for using flour entoleters as a method of breaking infested kernels so that sound kernels can be separated from infested ones before milling (Fleurat-Lessard 1989), thus reducing insect fragment counts in the flour. The entoleters should be running at 1500 rpm^{-1} to maximize the control of internal insects stages in grain, whereas a speed of 3000 rpm^{-1} is needed to control insects in flour. If the entolelers are used at an appropriate stage of processing, one can ensure that flour or milled products are free of live insects or insect fragments (Fleurat-Lessard 1989). French mills that export a large percentage of their product must have entolelers to insure that the flour will meet export standards (Fusillier 1986; Lagarde 1986) (Fig. 3). Another inconvenience of impact desinfestor is that they can change the end use quality. For example, couscous semolina will be broken into smaller particles. Flour is not altered significantly by impact devices, but flour that becomes filled with air during treatment can have rheological properties slightly different from untreated flour.

Pneumatic conveyers can also reduce stored-grain insect populations. High throughput (38 t h^{-1}) pneumatic conveyers used at ports to unload ships, effectively control the adults of *Sitophilus oryzae* (L.), *R. dominica* and *Cryptolestes* spp. Insect mortalities of 48 to 95% have been observed immediately after pneumatic

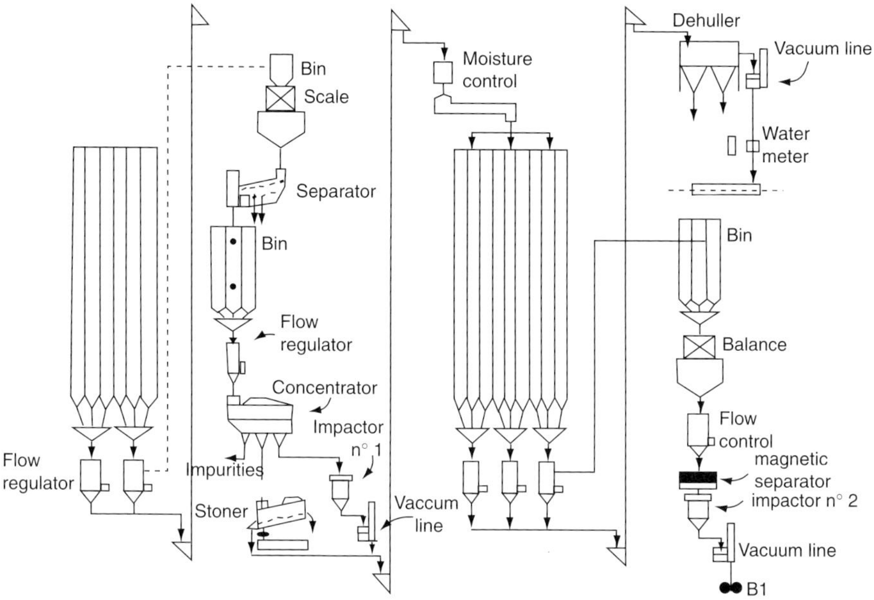

Fig. 3. Flow chart of a cleaning processes of wheat and flour in a flour mill used to reduce insects and insect parts in the final product using separators and impact devices (After a Buhler-Miag diagram).

conveying, whereas the mortality was 99% 1 week later (Bahr 1990). There are pneumatic conveyers designed for farm use that convey grain at a rate of 3 to 5 t h^{-1}. Moving wheat with a small pneumatic conveyer gave 100% control of *C. ferrugineus* and *T. castaneum* larvae and adults. Moving grain with the conventional auger caused 89% mortality of *T. castaneum* and 94% mortality of *C. ferrugineus* (White et al. 1997; Paliwal et al. 1999). Pneumatic samplers used in large grain-handling facilities can themselves cause mortality and give a false impression of live insect populaitons. Insect-infested grain fed into a pneumatic sampling system at a terminal elevator that carried grain at 11 m s^{-1}, caused 73, 65, 65 and 22% mortality of *C. ferrugineus*, *O. surinamensis*, *T. castaneum* and *Sitophilus granarius* (L.), respectively (Bryan and Elvidge 1977).

Stored-product mite populations are greatly reduced when grain is moved pneumatically or by more conventional methods. However, mites can rapidly reestablish themselves if the moisture and temperature conditions are favourable. Survival of mites was 2 to 5% following movement of the grain by an auger, whereas there was no survival when pneumatic unloading was used (White et al. 1997). Moving grain within a terminal elevator using belt conveyers, bucket elevators and a free fall from 2 to 32 m reduced psocids (*Liposcelis* spp.) by 0 to 70% and *Cryptolestes* spp. by 0 to 96% (Rees et al. 1994). Hence, the turning over of grain stocks within an elevator should reduce populations, but it does not eliminate them.

4 Conclusion

Mechanical methods were probably one the first means of protecting stored grain from insects. These are undergoing a renaissance now that the dangers of chemical fumigants are better known. Their integration into modern grain storage techniques will be aided by the applying these techniques (e.g. entoleters, pneumatic conveyers) as a regular step of grain handling, and in some cases the level of control is similar to chemical methods. The disadvantage of most physical control methods, with the exception of inert dusts, is that similar to fumigants, they offer no residual protection, which contact insecticides can provide. This is balanced by the low probability that insect populations could develop resistance to physical methods, as they have to the contact insecticides. Another advantage is that they leave no residues on the grain and can be used by organic growers and distributors.

References

Bahr I., (1990). Reduction of stored product insects during pneumatic unloading of ship cargoes. pp. 1135-1144, in F. Fleurat-Lessard and P. Ducom (eds.) Proceedings 5th Int. Working Conf. Stored Product Protection. INRA/SPV Bordeaux, 2066 p.

Bailey S.W., (1962). The effects of percussion on insect pests in grain. J. Econ. Entomol. 55: 301-304.

Banks H.J., (1986). Impact, physical removal and exclusion for insect control in stored products. pp. 165-184, in E.J. Donahaye and S. Navarro. Proceedings 4th Int. Working Conf. Stored Products Protection, Tel Aviv, 668 p.

Banks H.J., Fields P.G., (1995). Physical methods for insect control in stored grain ecosytems. pp 353-409 in D.S. Jayas, N.D.G. White, and W.E. Muir (eds) Stored-grain ecosystems. Marcel Dekker, Inc., New York, 757 p.

Bridgeman, B., (1998). Application Technology and Usage Patterns of Diatomaceous Earth In Stored Product Protection. in Z. Jin, Q. Liang, Y. Liang, X. Tan and D. Guan (eds), Proceedings 7th Int. Working Conf. Stored Products Protection, Beijing, Sichuan Publishing House of Science and Technology, Chengdu, China, pp. 785-789.

Bryan J.M., Elvidge J., (1977). Mortality of adult grain beetles in sample delivery systems used in terminal grain elevators. Can. Entomol. 109: 209-213.

Carlson S.D., Ball H.J., (1962). Mode of action and insecticidal value of a diatomaceous earth as a grain protectant. J. Econ. Entomol. 55: 964-969.

Desmarchelier J.M., Dines J.C., (1987). Dryacide treatment of stored wheat: its efficacy against insects, and after processing. Aus. J. Exper. Agr. 27: 309-312.

Dowdy A., Fields P.G. Heat combined with diatomaceous earth to control the confused flour beetle (Coleopthera: Tenebrionidae in a flour mill. J. Stored. Prod. Res., in press.

Ebeling W., (1971). Sorptive dusts for pest control. Ann. Rev. Entomol. 16: 123-158.

Fields P.G., (1998). Diatomaceous earth: Advantages and limitations. in Z. Jin, Q. Liang, Y. Liang, X. Tan and D. Guan (eds), Proceedings 7th Int. Working Conf. Stored Products Protection, Beijing, Sichuan Publishing House of Science and Technology, Chengdu, China, pp. 781-784.

Fields P.G., Korunic Z., (2000). The effect of grain moisture content and temperature on the efficacy of diatomaceous earths from different geographical locations against stored-product beetles. J. Stored. Prod. Res., 36:1-13.

Fields P.G., Muir W.E., (1995). Physical Control. pp. 195-222, in B. Subramanyam and D.W. Hagstrum (eds.) Integrated Pest Management of Insects in Stored Products, Marcel-Dekker Inc., New York, 426 p.

Fleurat-Lessard F., (1985). Les traitements thermiques de désinfestation des céréales et des produits céréaliers : possibilité d'utilisation pratique et domaine d'application. Bull. Org. Eur. Prot. Plantes 15: 109-118.

Fleurat-Lessard F., (1989). La désinsectisation des stocks de farine. Ind. Céréales 60: 17-22.

Fusillier A., (1986). Forte diminution de la teneur en " filth " par une nouvelle conception du diagramme de nettoyage. Ind. Céréales 39: 21-25.

Glover T.J., (1997). Pocket Ref. Sequoia Publishing, Inc. Littleton, CO, USA, 554 p.

Golob P., (1997). Current status and future perspectives for inert dusts for control of stored product insects. J. Stored Prod. Res. 33: 69-79.

International Agency for Research on Cancer (IARC), (1997). Silica, some silicates, coal dust and para-aramid fibrils. Volume 68, IARC Working Group on the Evaluation of Carcinogenic Risks to Humans, Lyon, 506 p.

Joffe A., (1963). The effect of physical distrubance or " turning " of stored maize on the development of insect infestations I. Grain elevator studies. S. Afr. J. Agric. Sci. 6: 55-68.

Korunic Z., (1997). Rapid assessment of the insecticidal value of diatomaceous earths without conducting bioassays. J. Stored Prod. Research, 33: 219-229

Korunic Z., (1998). Diatomaceous earths, a group of natural insecticides. J. Stored Prod. Res. 34:87-97.

Korunic Z., Fields P.G., (1995). Diatomaceous earth insecticidal composition. US patent 5,773,017.

Korunic Z., Fields P.G., Kovacs M.I.P., Noll J.S., Lukow O., Demianyk C.J., Shibley K. J., (1996). The effect of diatomaceous earth on grain quality. Postharvest Biol. Technol. 9: 373-387.

Korunic Z., Ormesher P., (1998). Evaluation and standardised testing of diatomaceous earth. In Z. Jin, Q. Liang, Y. Liang, X. Tan and D. Guan (eds), Proceedings 7th Int. Working Conf. Stored Products Protection, Beijing, Sichuan Publishing House of Science and Technology, Chengdu, China, pp. 738-744.

Lagarde A., (1986). Limitation de l'infestation dans les moulins. Incidence sur la conception des machines et de l'installation. Ind. Céréales 39: 13-18.

Le Patourel G.N.J., Shawir M., Moustafa F.I., (1989). Accumulation of mineral dusts from wheat by *Sitophilus oryzae* (L.) (Coleoptera: Curculionidae). J. Stored Prod. Res. 25: 65-72.

Maceljski, M., Korunic Z., (1971). Trials of inert dusts in water suspension for controlling stored-product pests. Zastita Bilja (Plant Protection), Belgrade, 22:119-128.

Paliwal J., Jayas D.S., Muir W.E., White N.D.G., (1999). Effect of pneumatic conveying of wheat on mortality of insects. App. Eng. Agr. in press.

Quarles W., (1992). Silica gel for pest control. IPM Practioner, 14:1-11.

Quentin N.E., Spencer J.L., Milles J.R., (1991). Bean tumbling as a control measure for the common bean weevil *Acanthoscelides obtectus*. Entomol. Exp. Appl. 60: 105-109.

Rees D., van Gerwen T., Hillier T., (1994). The effect of grain movement on *Liposcelis decolor* (Pearman), *Liposcelis bostrychophila* Badonnel (Psocoptera: Liposcelidae) and *Cryptolestes ferrugineus* (Stephens) (Coleopter: Cucujidae) infesting bulk-stored barely. pp. 1214-1219, in E. Highley, E.J. Wright, H.J. Banks and B.R. Champ (eds.) Proceedings 6th Int. Working Conf. Stored Product Protection. CABI International, Wallingford, U.K., 1274 p

Sigaut F., (1978). Les réserves de grain à long terme. Techniques de conservation et fonctions sociales dans l'histoire. Maison des Sciences de l'Homme, Paris et Université de Lille III, 202 p.

Stratil H., Wohlgemuth R., Bolling H., Zwingelberg H., (1987). Optimization of the impact machine method of killing and removing insect pests from foods, with particular reference to quality of flour products. Getreide, Mehl Und Brot 1987; 41:294-302.

Vallery, (1839). Description de l'appareil à conserver les grains, dit grenier mobile. Bull. Soc. Enrouragement Indust. Natle. 38: 123-131.

White N.D.G., Leesch J.G., (1995). Chemical Control pp. 287-330 in Subramanyam, B, and Hagstrum, D, Editors, Integrated management of insects in stored products: Marcel Dekker Inc., New York, 425 p.

White N.D.G., Jayas D.S., Demianyk C.J., (1997). Movement of grain to control stored-product insects and mites. Phytoprotection 78: 75-84.

Pneumatical Control

Pneumatic Control of Insects in Plant Protection

Mohamed KHELIFI, Claude LAGUË, Benoît LACASSE

1 Operating Principle of Pneumatic Systems

The pneumatic control technique consists of using moving air to eliminate undesirable insects from crops. Pneumatic energy could be used in different modes: suction, blowing, or a combination of both. Pneumatic control systems are often referred to as vacuums. The mode used depends on the species of insect to be controlled and on the characteristics of the crops to be protected. The use of vacuums is limited to some crops and some insects. Hence, insects that easily fly as soon as they feel vibrations or when they are disturbed in their environment such as, the Whitefly (*Aleurodidae*), are best controlled by vacuuming. The Colorado potato beetle (CPB) [*Leptinotarsa decemlineata* (Say)] and insects such as the European corn borer (*Ostrinia nubilalis* (Hbn.)) or the cabbage maggot [*Delia radicum* (L.)] cannot be dislodged easily by this mode. Their behavior is completely different and requires more elaborate methods. The CPB beetle holds on to the plants firmly. Furthermore, the foliage of the host potato plants becomes denser as growth advances and the generated vacuuming force rapidly dissipates. It is therefore very difficult to suck up CPBs that are deep in the plants or those that are firmly gripped to the foliage. On the other hand, the European corn borer attacks corn inside the stems. In this case, it is impossible to remove this insect by a pneumatic system.

2 Background

The idea of using pneumatic systems dates back to the 1950s. The first uses to control some insect pests occured in cotton fields of the United States (deVries 1987; Bédard 1991). However, the gradual introduction of efficient and economic chemical pesticides made this technique less attractive to growers. Currently, factors such as growing demand for food products free of pesticide residues, problems of resistance developed by some insect species to many chemicals, and risks of environmental contamination by pesticides have revived the interest in alternatives to the chemical means, and in particular, to the pneumatic control technique (Misener and Boiteau 1993a).

Box 1. Aerodynamics of the Pneumatic Control

To determine the main air suction or blowing parameters allowing for a better control of insect pests, it is also important to know the drag coefficient of the insect. This non-dimensional coefficient has the advantage of allowing the force induced to the insect by an airstream of a known speed to be computed and the terminal velocity of the insect which represents the equilibrium speed between the drag force and the weight of the insect to be determined. The drag coefficient is variable according to the flow regime. It can be expressed as:

$$C_D = 2\,\frac{mg}{V_t^2\,\rho_a\,A}$$

where:
C_D = drag coefficient,
m = mass of the insect, kg,
g = gravity acceleration, m s^{-2},
V_t = terminal velocity, m s^{-1},
ρ_a = air density, kg m^{-3},
A = projected surface of the insect, m^2.

Although the shape of insects is very irregular, it is possible to estimate the drag coefficient by analogy to other objects of known shapes such as a sphere taking into account a correction factor (Misener and Boiteau 1993b). DeVries (1987) estimated that the drag coefficient of a CPB adult is about 0.35 whereas its terminal velocity is about 12.5 m s^{-1}. CPB adults reach a constant velocity after falling 10 m against 10, 9, and 8 m for the fourth, third, and second instar larvae, respectively (Misener and Boiteau 1993b). The respective terminal velocities are 9.4, 9.5, 7.3, and 5.9 m s^{-1}. This shows that larvae of the third and second instars are more susceptible to being carried by a vertical airstream (air suction case) than fourth instar larvae and adults.

At the beginning of the 1990s, at least eight companies worldwide had tried to market pneumatic systems to control insect pests (Khelifi 1996). The effectiveness of the developed machines is variable. The major drawback of these pneumatic systems is their contribution to soil compaction, destruction of beneficial insects, and mechanical damage to the crops (Moore 1990).

3 Examples of Pneumatic Systems

3.1 Beetle Eater

The Beetle Eater was designed by James Syznal Sr. in the Northeast of the USA (Florence, Massachussetts) and the patent was bought by Thomas Equipment Ltd.

of New Brunswick, Canada. The approximate cost of this vacuum machine is 170 970 FF (41 700 $Can). The system requires a 45-kW tractor. The operating speed is 8 to 9.6 km h^{-1}. Two fans (71 cm diameter) respectively generate the blowing and suction airstreams. The system has four combined blowing/suction hoods. The first fan blows air through a rectangular slot 25.4 cm long by 7.6 cm wide on each unit. The second fan vacuums the air at the summit of the units through a 30.5-cm diameter flexible plastic tube. Airflow speed in the main duct is 480 km h^{-1}. The effectiveness of this machine in controlling the CPB varies from 15 to 50% (Bédard 1991). The machine has to be modified regularly to maintain its effectiveness since it is affected by the height of the plants, the growing stage of the insects, and the roughness of the ground surface. A picture of the Beetle Eater as well as a discussion of the effects of this particular pneumatic system are presented in the chapter of Vincent and Boiteau (Chap. 19).

3.2 Bio-Collector

The Bio-Collector (Bio-land Technik, Mühlhausen, Germany) can collect CPBs over 1 ha in 1 h, at a speed of 6 km h^{-1} with a four-row machine. Models having two, three, or four rows are available. Reynaud (1990) reported an effectiveness of 80 to almost 100% for CPB adults and of 75 to 80% for larvae. The cost of this machine is about 57 000 FF (14 000 $Can).
This system was tested in Quebec in commercial and experimental plots by Dr. Raymond-Marie Duchesne, entomologist at the Ministry of Agriculture, Fisheries and Food of Quebec. It was more effective against small populations of CPBs, but success was highly variable, averaging 50% removal of all the stages. As a result, the Bio-Collector was judged not efficient for controlling the CPB in potato crops in the particular Quebec agricultural context. Some adjustments are needed to adapt this machine to potato-cropping techniques and consequently improve its overall effectiveness.

3.3 Bug-Vac

The design of the Bug-Vac by E. Show (Driscoll, Watsonville, California) dates from 1986 (Wilcox 1988). This huge vacuum was successfully tested in 1987 and its success in controlling CPB was comparable to that of chemical pesticides. The number of beneficial insects in the plots treated with pesticides was 50 to 70% lower than that in the plots controlled with the Bug-Vac. Inman (1990) indicated that this vacuum is relatively ineffective against *Lygus hesperus* at the larval stage. Pickel et al. (1995) conducted many trials with the three models of vacuums used by growers in California to control *L. hesperus* Knight in strawberry fields. These vacuums can treat one, two, or three strawberry rows. The three-row vacuum costs about 342 000 FF (83 000 $Can.), and its control is comparable to that of chemical pesticides and higher than that obtained with the two other models. However, Pickel et al. (1995) considered that the levels of damage caused to strawberry fields are economically unacceptable despite the level of insect control that was obtained with either of the models.

3.4 Bug-Buster

The Bug-Buster was designed by Industrial Air Products Inc., Phillips, Wisconsin. It was intended for insect pests that are flying or lightly gripped to plants. Models with two, three, or four centrifugal fans are available. Each model requires a power of about 15 kW per fan. There are three possible configurations of the suction hoods, so that they can be adapted for control in strawberry, potato or lettuce crops. Three ha could be swept in 1 h by this vacuum. Many trials conducted at the University of Wisconsin (Madison) revealed that potato plant defoliation by CPB larvae could be maintained at an acceptable level with weekly treatments (Puttré 1992). Chemical and mechanical control of leafhoppers (*Cicadellidae*) in carrot and soybean crops gave comparable results.

The Bio-Vac (Premier Tech, Rivière du Loup, Québec) and the Pash Tash (Israël) are respectively described in the chapters of Vincent and Boiteau (Chap. 19) and Weintraub and Horowitz (Chap. 21).

4 Design Criteria

The design of a pneumatic system capable of eliminating a high (≥ 80%) proportion of insect pests involves analysis of the growing stage and the physical characteristics of the insect to control, as well as the growing stage, the geometry, and the resistance to airstreams of the plants to be protected.

A fundamental study of the design parameters, namely the airflow rate and speed, the shape of the control units, i.e. the hoods, and the orientation of the airstreams is thus necessary. With the exception of the study by DeVries (1987), this important aspect has often been neglected. Khelifi (1996) demonstrated the importance of these parameters in designing efficient pneumatic control systems.

5 Airstream Effects on the Plants

Plants can resist the mechanical effects of an air stream up to a threshold above which they experience considerable damages. To determine the airflow velocity threshold that can induce such damages, DeVries (1987) carried out a limited number of tests in a wind tunnel. He exposed a single 46-day-old (from planting) and 76-cm-tall potato plant (cultivar Rosa) to different airstreams for 30 s. The plant was visually checked to assess the damage. There was little or no foliar damage up to air speeds of 12.5 m s^{-1}.

Similarly, Khelifi et al. (1995a) conducted many laboratory tests on three varieties of potato plants (Superior, Norland, and Kennebec). Three growing stages (26, 40, and 58 cm) and seven levels of airspeeds (from 12.5 to 31 m s^{-1}) were considered. Potato plants were placed inside a test bench (Khelifi et al. 1992) and then exposed to a horizontal airstream for 20 s. The degree of damage was visually evaluated. Results showed that the resistance of potato plants to airstreams is mainly related to the airspeed and the growing stage. The variety of potato plants is not important. Plants having less than 12 leaves and a mean height less than 40 cm could be exposed to air speeds as high as 27.5 m s^{-1} (at the foliage level) without incurring any visible damage.

6 Forces Required to Dislodge Insects from the Foliage

A pneumatic system to control insect pests designed without taking into account insect physical characteristics and behavior is likely to fail. Indeed, it is from such appropriate knowledge that one can consider the possibility of using a pneumatic system and define its mode of action. For example, the use of a pneumatic system to control the CPB gave results that are in general variable and sometimes contradictory. This was mainly related to the lack of information about the mobility of CPBs and their capabilities of grasping to the foliage at different levels of their growing stages. The reader willing to learn more about the forces required to dislodge the CPB can consult Vincent and Boiteau (Chap. 19).

In a similar study, DeVries (1987) indicated that CPB adults grasp more when they are at the lower surface of leaves or on their edges. The CPB uses tarsal hooks located at the end of each leg to grasp potato plants. A CPB adult can maintain its grip on the plant despite the application of forces as high as 20 times its weight. Similarly, DeVries (1987) mentioned that the CPB can resist forces of 41 mN when located on the lower surface or on the edges of the foliage as opposed to forces of 11 mN when located on the upper face of leaves.

7 Optimal Orientation of Airstreams

The optimization of the orientation of airstreams is based on insect, plant and soil factors. Knowledge of the insect behavior is used to decide whether to use a blowing or vacuuming technique and to determine the forces required to remove the insect from the foliage. The resistance of the plant to the necessary airstream should be good at the required removal force and the soil should not be easily displaced.

Khelifi et al. (1995b) conducted many laboratory tests to determine optimal airspeeds and adequate hood orientations for dislodging CPB adults and to verify the necessity of integrating a shaking system to improve the effectiveness of the control system. They used airspeeds of 20.5, 24, and 27.5 m s^{-1} at the foliage level, and six configurations beginning with simple horizontal blowing to a combination of two oblique ascending air streams and air suction from the top. Each test was conducted over five CPB adults and replicated ten times. Potato plants (cv. Superior) were at the vegetative growing stage (mean height of 25 to 30 cm with a dozen leaves). The best configurations consisted of simple horizontal blowing through the plant foliage and a combination of simultaneous air suction and blowing. Airspeeds in the range of 24 to 27.5 m s^{-1} at the foliage level dislodged the maximum number of CPBs. Based on these results, a shaking system did not seem necessary.

8 Numerical Simulation Models

Pneumatic control of insects depends on many factors including the variability in the gripping ability of the insects at different growing stages, the geometry of the plants, and the resistance of crop foliage to airstreams according to the growing stage of the plants. Success of pneumatic control of insects appears to depend upon an appropriate design of the suction or blowing hoods. The main factors that have to be considered are the geometry or the shape of the hoods, which greatly affects the airflow pattern at the plant foliage level, the dimensions of the hoods, their position relative to the plants, and obviously the airflow rate that the hoods can deliver. The selection of the optimal hood geometry also depends on many factors such as the growing stage of the plants and the mode of action planned, i.e. total or partial coverage of the plants by suction, blowing, or a combination of both.

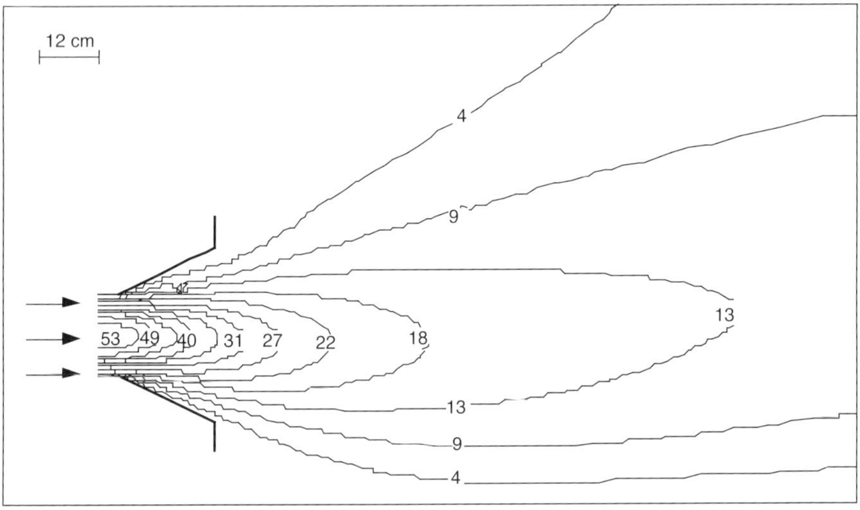

Fig. 1. Simulation for the simple air-blowing configuration (air velocities are in m s^{-1}).

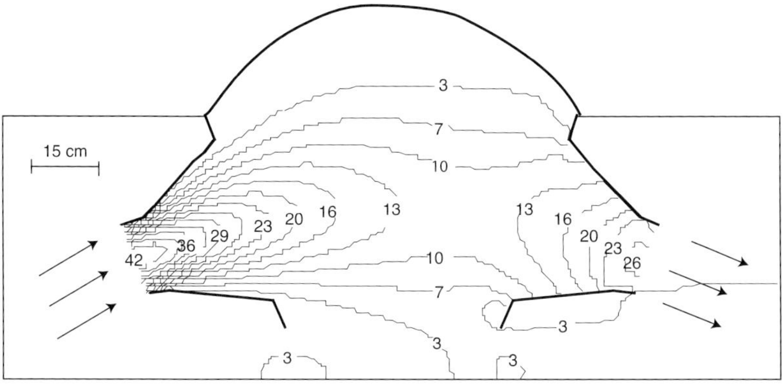

Fig. 2. Simulation for the shielded simultaneous oblique air suction and blowing configuration (air velocities are in m s^{-1}).

Because of the large number of parameters involved in the design of the hoods, the use of a numerical method of simulation, such as the finite element method (FEM), greatly facilitates the design. This has the advantage of simulating many working units and selecting the most appropriate one for pneumatic control of insect pests without extensive prototyping and testing.

Khelifi et al. (1996a) developed a finite element model to predict the airflow inside and around hoods arranged in simple geometries (air suction and blowing), and in more complicated ones (combinations of two to three hoods). This model is based on the Navier-Stokes equations and the results could be generalized to other insects. Among the eight configurations tested, two appeared promising: the simple horizontal air blowing and the shielded simultaneous oblique ascending air suction and blowing across the plant foliage (Khelifi et al. 1996b).

The airflow maintains a considerable velocity even very far from the hood outlet (Fig. 1). The flow pattern is also wide enough to cover a large section of the plant. The effectiveness of air blowing appears to be far better than that of suction because the air can penetrate more deeply into the foliage. This configuration is simple and requires less energy than the others tested. It can be efficiently used provided that an adequate catching device is used to collect the CPBs blown of the plants.

The air blowing/air suction configuration includes a semicircular shield (40 cm radius) centered at the intersection point of the y and horizontal hood axes (Fig. 2). In addition, hoods are tilted upward 20 ° from their horizontal axis. This directs the airflow more toward the upper part of the plant where the CPB tends to feed first. The advantage of this configuration is to keep the plant near the suction hood under the effect of the blown air, thus providing high suction effectiveness. This shielded simultaneous oblique ascending air suction and blowing configuration does not require a catching system because the suction hood vacuums the insects. However, the perfect semicircular shape of the shield has to be slightly modified to keep it in contact with the airstream and consequently prevent recirculation of air.

9 Conclusion

The pneumatic control of insect pests is an interesting alternative to chemical insecticides. This technique is successfully used to control insects that easily fly as soon as they are disturbed in their environment, in particular, the whitefly (*Aleurodidae*) and the tarnished plant bug [*Lygus lineolaris* (P. de B.)]. Less success was reported for other insects like the Colorado potato beetle [*Leptinotarsa decemlineata* (Say)].

When designing a pneumatic system, it is imperative to take into account the characteristics of the plants to be protected and the behavior of the insects to be controlled at different life-growing stages. The success of the control process is mainly related to the adequate design of the hoods. The designed system has to be as light as possible to avoid any excessive compaction of the soil. Also, the pneumatic control system must cause no damage to the plants.

Numerical modeling is a powerful tool for optimizing the parameters required to design pneumatic control systems. It allows many configurations of variable degrees of complexity to be simulated without the need of costly prototyping. Thus, the numerical method not only allows the design process to be accelerated, but also the time and the cost of field testing to be greatly reduced.

However, it would be wise to recall that pneumatic control does not represent a magic and complete solution to the problems caused by insect pests. This technique cannot, alone, like some other non chemical means available on the market, efficiently control the populations of insect pests. It has to be used within an integrated pest management program.

References

Bédard R., (1991). Lutte écologique contre le doryphore. Le Producteur Plus. Sept./Oct., pp. 32-34.

DeVries R.H., (1987). An investigation into a non-chemical method for controlling the Colorado potato beetle. Unpublished M. Sc. Thesis. Graduate School, Cornell University. Ithaca, NY., USA, 68 p.

Inman J.W., (1990). Western perspective: mechanical methods of pest management. American Vegetable Growers. April, pp. 54-55.

Khelifi M., (1996). Optimisation des paramètres physiques de conception d'un système pneumatique de contrôle du doryphore de la pomme de terre, *Leptinotarsa decemlineata* (Say). Unpublished Ph. D. Thesis, Department of Soil Science and Agri-Food Engineering, Université Laval, Québec, Canada, G1K 7P4, 310 p.

Khelifi M., Robert J.-L., Laguë C., (1996a). Modeling airflow inside and around hoods used for pneumatic control of pest insects, Part I: Development of a finite element model. Can. Agri. Eng. 38: 265-271.

Khelifi, M., Laguë C., Robert J.-L., (1996b). Modeling airflow inside and around hoods used for pneumatic control of pest insects, Part II: Application and validation of the model. Can. Agri. Eng. 38: 273-281.

Khelifi M., Laguë C., Lacasse B., (1995a). Potato plant damage caused by pneumatic removal of Colorado potato beetles. Can. Agri. Eng. 37: 81-83.

Khelifi, M., Laguë C., Lacasse B., (1995b). Resistance of adult Colorado potato beetles to removal under different airflow velocities and configurations. Can. Agri. Eng. 37: 85-90.

Khelifi M., Laguë C., Lacasse B., (1992). Test bench for pneumatical control of pest insects. ASAE-paper No. 92-1599. Am. Soc. Agric. Eng., St. Joseph, MI 49085.

Misener G.C., Boiteau G., (1993a). Holding capability of the Colorado potato beetle to potato leaves and plastic surfaces. Can. Agri. Eng. 35: 27-31.

Misener G.C., Boiteau G., (1993b). Suspension velocity of the Colorado potato beetle in free fall. Amer. Potato J. 70: 309-316.

Moore J., (1990). Sweeping fields controls some pests. American Vegetable Growers. March, pp. 10-11.

Pickel C., Zalom F.G., Walsh D.B., Welch N.C., (1995). Vacuums provide limited Lygus control in strawberries. California Agriculture 49: 19-22.

Puttré M., (1992). Bug-busting alternative to pesticides. Mechanical Engineering 114 (12), Dec.

Reynaud M., (1990). En bref "lutte bio contre les doryphores". Nature et progrès. Sept./Oct.

Wilcox L., (1988). Sucker for bugs. California Farm. October, pp. 46 and 50.

Pneumatic Control of Agricultural Insect Pests

Charles VINCENT and Gilles BOITEAU

1 Introduction

In the past few years, several articles have appeared in agricultural magazines on the value and usefulness of vacuum devices used to control insect pests on various crops (e.g. Hillsman 1988, Stockwin 1988, Lachance 1990, Street 1990; Pickel et al. 1995). Most of the facts in these articles are rather vague. Here, we aim at presenting the most complete and accurate information possible on the use of vacuum pest control in crops. Most of the scientific literature on the subject deals with two crop pests, the tarnished plant bug (a pest of strawberry crops) and the Colorado potato beetle. Readers are also invited to read the chapters in this book by Khelifi et al. (Chap. 18), Lacasse et al. (Chap. 20) and Weintraub and Horowitz (Chap. 21) for additional information on the pneumatic control of insects.
Pneumatic control essentially involves the use of an airstream to dislodge and kill insect pests. Insects can be dislodged from plants with negative (aspiration) or positive air pressure (blowing). Once the insects have been dislodged, they are killed by a system of turbines or are collected and killed in a dedicated system upstream of the blower. This basic design must, of course, be adapted to the specific crop in question. The first step is to assess the effectiveness of the system in the laboratory (on a test bench) and in the field. Then, the system has to be perfected and implemented in the context of the farming system used.

2 Control of Tarnished Plant Bug on Strawberries

Problems related to insect pest management in strawberry crops have been discussed in Vincent et al. (1990). Strawberries are perishable agricultural commodities with a high per-hectare value. By feeding on the fruit, the tarnished plant bug (*Lygus lineolaris* Palisot de Beauvois) diminishes both the quantity and quality of the harvest. The commercial control method currently used in northeastern North America is one or two insecticide treatments per year when populations exceed the economic threshold in Quebec of 0.25 nymph/blossom cluster (Mailloux and Bostanian 1988).

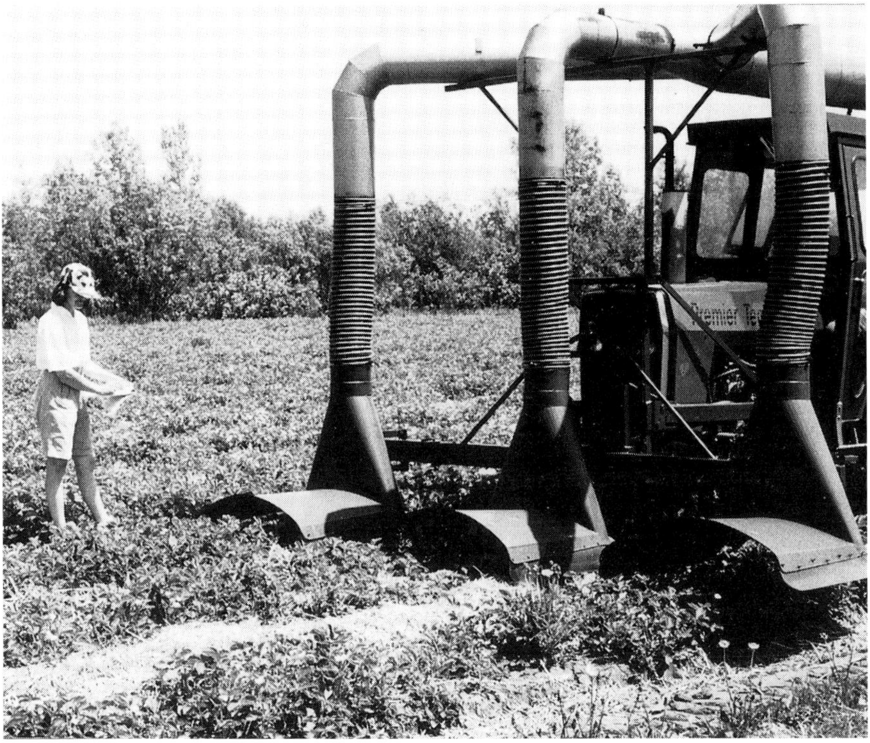

Fig. 1. Tractor-mounted Biovac in a strawberry field.

The tarnished plant bug has few natural enemies (pathogens, parasites or predators) (van Driesche and Hauschild 1987), and there are several reasons for which it is unlikely that their enemies would be effective controls. First, the adults must be killed before they oviposit. Second, the tarnished plant bug is very mobile and often inhabits areas outside strawberry fields. Finally, the bug has several host plants. Pneumatic control measures not only have the advantage of delaying the appearance of insecticide-resistant populations (Snodgrass 1996), but also appear to be a promising alternative given the poor prognosis for biological and chemical controls.
Dietrick (1961) designed the D-Vac to collect adult and nymph *Lygus hesperus* Knight in California strawberry fields. This is a backpack apparatus in which the negative airflow is provided by a small two-stroke engine. To collect insects, the inlet (roughly 0.35 m in diameter) must be held for a few seconds above the plant. Kovach (1991) used a portable vacuum collector similar to the D-Vac to successfully control *L. lineolaris* populations on a strawberry farm in New York State, but concluded that the portable device was practical only for small fields.
In California, Driscoll Associates developed the Bugvac and used it to control *L. hesperus* Knight (Glynn 1989). This is a tractor-mounted device that can treat four rows at a time. Southam (1990) reports that another device, the Ag-Vac, removed up to 80% of western plant bugs under field conditions in California. This was achieved at a speed

of 5 km h^{-1}. According to Grossman (1994), commercially satisfactory results were attained by treating California strawberry fields twice a week (no data given).
Vincent and Lachance (1993) tested a tractor-mounted vacuum collector, the Biovac (Fig. 1), on a commercial strawberry farm in Quebec, Canada. This device, which treats three rows at a time, is mounted on a tractor with a 48-kW PTO that drives a 1800-rpm turbine, creating a strong negative airflow (4.32 m^3 s^{-1}) in the main duct, i.e. 1.44 m^3 s^{-1} in each of the three inlets. Tarnished plant bug populations were evaluated on a weekly basis using the saucer pot method, and passes were made with the collector when more than 12 individuals (adults and nymphs) were captured per 100 taps. For these treatments, the hood inlets were set to barely touch the top of the plant canopy and the machine was operated at a speed of 7 km h^{-1}. Tarnished plant bug populations exceeded the threshold 15 times, and treatments with the Biovac reduced their numbers significantly 7 out of the 15 times. In 1991, populations were significantly reduced 3 out of 15 times (Fig. 2). These tests, which represented the first attempt to evaluate the performance of a tractor-mounted vacuum collector in a commercial strawberry field, showed that the Biovac was of limited use in a commercial setting when used according to the agronomic conditions described by Vincent and Lachance (1993).

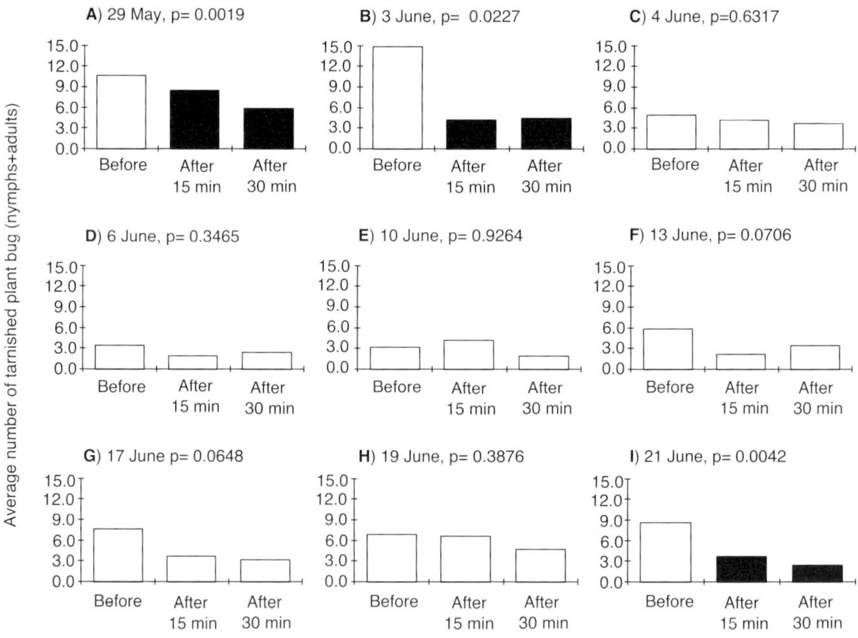

Fig. 2. Tarnished plant bug populations (nymphs and adults) before and 15 and 30 min after passes with the Biovac (Ste. Anne de la Pérade, Quebec, Canada, 1991). The average values shown by the histogram bars of the same color are not significantly different (P=0.05) according to a contrast test (after Vincent and Lachance 1993).

Zalom et al. (1993) compared the performance of three portable vacuum collectors in sampling for *L. hesperus* on California strawberry farms. The devices, which had suction capacities of 24.9, 11.9 and 8.1 m s^{-1} respectively, were used to sample populations. The densities obtained were 15.6, 5.8 and 3.9 individuals per ten plants (in the same field) respectively, suggesting that capture effectiveness increases with airflow speed (as would be expected).
Pickel et al. (1994) also compared three types of vacuum devices operated at speeds of 4 to 8 km h^{-1} and used to treat one, two or three beds per pass. The two-bed device had airflow speeds of 29.6 km h^{-1} at the hood, while the three-bed collector had airflow speeds of around 55.6 km h^{-1}. However, their results are difficult to interpret. Essentially, the three machines cannot be compared with one another, given the differences in experimental parameters. Furthermore, the authors did not specify the position of the inlets in relation to the canopy.
Treatments were done once or twice a week, regardless of the size of the tarnished plant bug populations. Treatment with one- and two-bed devices did not significantly reduce plant bug populations or damage. However, the three-bed device reduced insect damage to levels significantly lower than those observed in the control plots. In plots treated once a week, adult and nymph *L. hesperus* populations decreased respectively by 74 and 43%. However, the decrease in damage was not sufficient for the treatment to be commercially viable in California. The average percentage of damaged fruit at harvest was significantly lower in plots that received two treatments of bifenthrin, compared with plots subjected to other treat-

Fig. 3. Test bench used by Chagnon and Vincent (1996) and Vincent and Chagnon (2000).

ment combinations, including vacuuming (e.g. vacuum treatment once or twice a week, application of malathion). Pickel et al. (1994) concluded that a higher airflow speed could increase the effectiveness of vacuum treatments. However, it should be noted that, from an agronomic point of view, the situation studied by Pickel et al. (1994) was particularly challenging. The strawberry fields contained a day-neutral cultivar (i.e. Selva) which harbour high populations of phytophagous bugs (Pichette 1990), and the bifenthrin applications caused an increase in phytophagous mite populations.

To obtain a better understanding of the effects of Biovac treatments on *L. lineolaris* populations, Chagnon and Vincent (1996) set up a test bench to study the influence of various treatment parameters (such as relative height of the hood and forward velocity) and bug behaviour on removal effectiveness (Fig. 3). The test bench provided precise control of forward velocity (0 to 8 km h^{-1} range). The motors tested were rated at 7.3 and 10.9 kW, provided airflow speeds of 25.8 and 30.7 m s^{-1}, respectively (measured by Pitot tube), and generated airflow speeds of 1.20 and 1.42 m^3 s^{-1}. These parameters gave a good approximation to the operating characteristics of the Biovac. Tarnished plant bugs (nymphs or adults) were marked on the back with fluorescent powder. Most individuals (99%) were found in less than 5 min by illuminating the test area with an ultraviolet lamp. The greatest insect removal efficiency was achieved at airflow speeds of 30.7 m s^{-1}, as expected.

Optimal efficiency was achieved with the Biovac when the forward velocity (measured at the inlet) was 4 km h^{-1} and the hood was placed in the top third of the strawberry canopy. Differences in the proportion of insects vacuumed can be explained by the relative distance of the insects from the hood shortly before the machine passes; the percentage of success is inversely proportional to this distance. For example, the percentage of nymphs collected from the lower part of the plant ranged from 3% (hood just grazing top of canopy) to 26% (hood penetrating 1/3 of the way into the canopy) Vincent and Chagnon (2000). When the nymphs were in the top third of the plant before treatment, removal rates were 68% (hood penetrating 1/3 of the way into the canopy) and 61% (hood just touching the canopy). Most individuals that were not vacuumed were later found on the ground or on a lower part of the plant than before the machine passed. It is not known whether this phenomenon is due to active behaviour by the insect or mechanical dislodging of the insect when the inlet passes.

Current knowledge of the behaviour of the tarnished plant bug in the field is incomplete. According to Snodgrass (1993) and Taksdall and Sørum (1971), mirids are not very mobile on plants. They generally congregate on those plant parts that undergo rapid cell multiplication – specifically, on the shoot apexes and flower parts. However, a California study found that the position of *Lygus* spp. adults on alfalfa plants varied throughout the day (Stern 1973). Rancourt et al. (2000) has shown that tarnished plant bugs are relatively stationary on strawberry plants. A better understanding of the species' behaviour is required in order to develop commercially acceptable vacuum control methods and to determine the optimum time of day for treatment, that is, when the insects' position on the plant makes them readily accessible.

2.1 Secondary Effects on Pollination and Pollinators

To better understand the overall effectiveness of Biovac in controlling strawberry pests, its effects on pollination must be evaluated (Vincent et al. 1990). Although modern cultivars have perfect (i.e. hermaphroditic) flowers and are self-fertile, the stigmas are usually receptive before the pollen of the same flower is available, which favours cross-pollination (McGregor 1976). The spherical pollen grains, roughly 24 µm in diameter, require dispersal by gravity, wind or insect pollinators to ensure that most of the achenes are fertilized. Since fruit production is directly linked to the percentage of achenes fertilized, all pistils must be fertilized to obtain strawberries of optimal weight (Chagnon et al. 1989).

The Biovac disperses pollen by vacuuming it up and then ejecting it, but neither the distance of the recipient plant from the originating plant nor the time of day of treatment has a significant effect on dispersal (Chiasson et al. 1995). The viability of pollen dispersed by the Biovac is half that of pollen collected directly from flowers. Experiments were done in the field to determine the machine-dispersed pollen's contribution to pollination as determined by the percentage of fertilized achenes and fruit weights. In terms of both percentage of fertilized achenes and fruit weight, flowers pollinated by machine-dispersed pollen exhibited lower values than did flowers pollinated under optimal conditions (exposed to all pollinating agents). Although treatment with the Biovac increased the rate of pollination slightly, natural pollination mechanisms such as bees, wind and gravity obviously play a crucial role (Chiasson et al. 1995).

Chiasson et al. (1997) studied the behaviour of pollinating insects, particularly the honeybee (*Apis mellifera* L.), in response to Biovac passes. When the machine approached at a speed of 4 km h^{-1}, with the hoods just brushing the top of the canopy, 19% of pollinators flew away before the Biovac arrived. Of those remaining on the flowers while the Biovac passed, 61% of individuals were vacuumed and the rest remained clinging to the flowers after the machine had passed.

3 Control of Colorado Potato Beetle on Potatoes

The pest management situation for the Colorado potato beetle, *Leptinotarsa decemlineata* Say, has been described in Hare (1990) and Duchesne and Boiteau (1995). This chrysomelid beetle is one of the most intensively sprayed insects in northeastern North America. A number of insecticide-resistant populations have been reported around the world (Boiteau et al. 1987).

Boiteau et al. (1992) conducted field tests of a vacuum collector for Colorado potato beetles that uses a combined positive/negative airflow system (Fig. 4). The device, which treats four rows at a time, is tractor mounted, with the turbine driven by the tractor's PTO. The advantage of this type of design is that it takes less energy to dislodge an insect with a positive airstream and then aspirate it than it does to

A

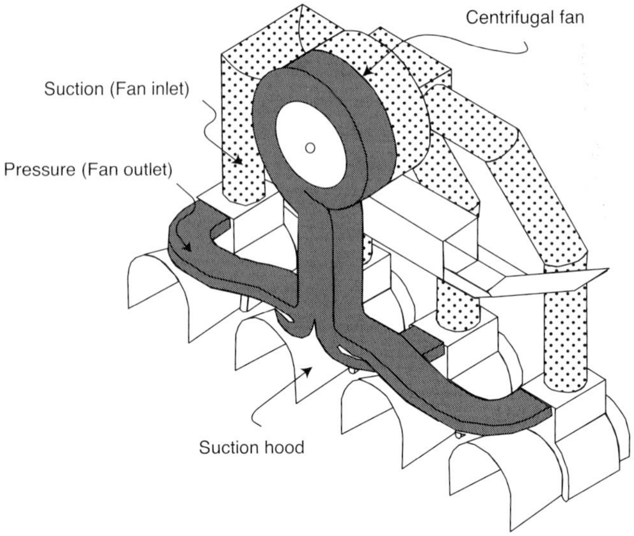

B

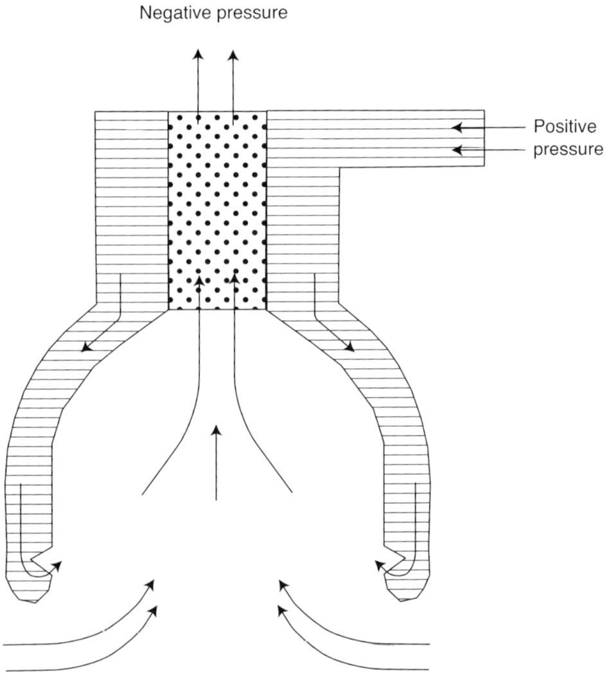

Fig. 4. General view of (**A**) Beetle Eater and (**B**) diagram of hood (After Boiteau et al. 1992).

dislodge an insect by aspiration alone. Individuals dislodged from the plants by positive pressure go into free fall and then are sucked into the turbine and killed. The Beetle Eater, tested by Boiteau et al. (1992) and Misener and Boiteau (1995), directs a positive airstream at the base of plants from above the row, shaking the plants and making the beetles fall off. Laboratory tests have shown that over 90% of adults (male and female) and larvae will fall off potato plant leaves when subjected to vibrations of over 20 Hz with an amplitude of over 0.6 mm (Boiteau and Misener 1995). Since the combined frequency and amplitude create an acceleration force that is lower than the beetles' ability to cling to the leaves, the beetles' drop-off behaviour cannot be attributed to the acceleration force produced by shaking the plants. Instead, it must be an escape mechanism allowing the beetles to avoid unusual conditions in their environment (beetles' day-to-day experiences on plants are unlikely to include being subjected to vibrations greater than 20 Hz with an amplitude of over 0.6 mm). Interestingly, the drop-off frequency was lower in insects holding onto leaf margins than for those on the leaf surface. Over 82% of the beetles dropped off within the first two seconds of the five seconds test period. This suggests that they are reacting spontaneously to vibrations of a certain frequency and amplitude by dropping off the plant, a behaviour known as thanatosis. A vacuum collector travelling too fast would have to aspirate a larger percentage of insects still on the foliage, which would probably reduce its efficiency. Hence, vacuum collectors for Colorado potato beetles can be operated at a limited range of travel speeds, particularly if their effectiveness is strongly linked to the aspiration of free-falling insects.

In the field, the Beetle Eater dislodges about 61% of adults. However, 13% of the dislodged adults fall to the ground, resulting in a real harvest of 48%. Among the early larval instars (L1 and L2), 42% of the beetles are collected and 3% fall to the ground, while in the later instars (L3 and L4), 50% of the beetles are collected and 23% fall to the ground. The larvae on the ground can climb back up the plant and cause additional damage.

To explain these results, Misener and Boiteau (1993) used a video camera with special software to determine the terminal velocity of the Colorado potato beetle in free fall, at four stages of its life cycle. Terminal velocities for adults and fourth, third and second instars were 9.4, 9.5, 7.3 and 5.9 $m\ s^{-1}$ respectively, which is relatively slow. Vacuum collectors, like the one used by Boiteau et al. (1992), should be able to easily capture beetles dislodged from the plants. The presence of individuals on the ground suggests that other factors are involved which limit the device's effectiveness.

Consequently, it is important to control the physical and mechanical parameters governing the loss of efficiency. The proportion of individuals that remain clinging to the plant after the vacuum collector has passed is determined mainly by physical and behavioral differences between the different phases in the species' life cycle. Misener and Boiteau (1993) measured the force required to remove a Colorado potato beetle from a leaf at different life stages, using an electronic scale attached to a microcomputer. The removal force was as high as 40 mN for adult beetles, 30 mN for third and fourth instar larvae and 10 mN for second instar larvae. The adults held on tightest when they were able to grip the leaf veins with their tarsal claws. For larvae, maximum removal force was measured when they were feeding and gripping onto the leaf with both their mandibles and tarsi. It took slightly less force to dislodge adults and larvae clinging to the edge of leaves. The removal force for the first and second larval instars was relatively small. However, young larvae are often found at the top of plants, inside young leaves that have not yet unfurled completely, where they are protected from the vacuum and vibrations. At all life stages, the ratio of insect weight to removal force is less than 1. This explains why individuals of each stage were found on the ground during the field tests after the machine passed (Boiteau et al. 1992). Adults are much more active than larvae and are distributed more uniformly on the plant. They respond to stimuli such as shade, unusual vibrations and probably also to aspiration with a thanatosis reflex. This may explain why, in field tests, a significant proportion of adults were found on the ground despite the high removal forces produced by aspiration.

The effectiveness of vacuum collection systems is highly variable, and varies also from one potato field to the next. The efficiency of vacuum collection in an individual field depends on the relative abundance of the different insect life phases and the variations in weight within each phase. Vacuum collector efficiency can probably be raised by increasing the suction force and the intensity of vibrations transmitted to the plant. According to Boiteau et al. (1992), the maximum (theoretical) efficiency of a collector is 98% for the earlier larval instars, 97% for adults and 77% for the later larval instars.

Once the system has been optimized to capture a maximum number of Colorado potato beetles, it can be adapted to control other potato pests. However, it is important to avoid secondary pest outbreaks. Studies have shown that populations of several other insect species are affected by the use of vacuum collectors. Boiteau et al. (1992) report capture rates of 62% for a number of predator species belonging to the taxa Arachnidae, Chrysopidae and Coccinellidae. Since native insects seem to play a role in controlling the Colorado potato beetle in fields in New Brunswick, these insects must be taken into account when using the collector. Similarly, if an inundative release of predators is planned for a potato field, it should be coordinated with vacuum collector treatments (Cloutier et al. 1995). The Beetle Eater was found to remove potato aphids [*Macrosiphum euphorbiae* (Thomas)] at a rate of at least 56%, showing that the collector has potential for controlling this group of secondary potato pests.

3.1 Effects on Disease Transmission

The mechanical transmission of disease by farm equipment is a significant concern for growers of seed potatoes, which have very low tolerance levels for viruses and viroids. Field trials showed that the vacuum collector did not spread the PSTVd viroid or the PVX virus, even with a large number of infected plants and multiple passes (Boiteau et al. 1992). Although this finding does not have the same validity as would the results of comparative experiments designed specifically to verify this aspect, it is a good starting point.

4 Conclusion

Pneumatic control of insect pests is a promising technology, but needs to be refined in order to be acceptable to farmers growing crops for extremely competitive markets. There are several avenues of research that should be explored further to make pneumatic control devices more suitable for commercial operations. First, a better understanding is required of the target insects' behaviour in the field, particularly escape behavior and the location of insects on plants during treatments. Second, machine effectiveness must be enhanced to ensure that a high percentage of insects is killed during each pass, regardless of the crop and working conditions. This will require improving the hood and turbine designs. Third, the use of repeated passes of vacuum collectors should be investigated. This approach has already been recommended by Boiteau et al. (1992) for infestations of adult Colorado potato beetles. To illustrate this, we simulated the effect of repeated passes with the Biovac on tarnished plant bug populations (1 bug per plant) that were colonizing strawberry fields with 1000, 10 000 or 20 000 individuals ha^{-1} (Fig. 5). At high

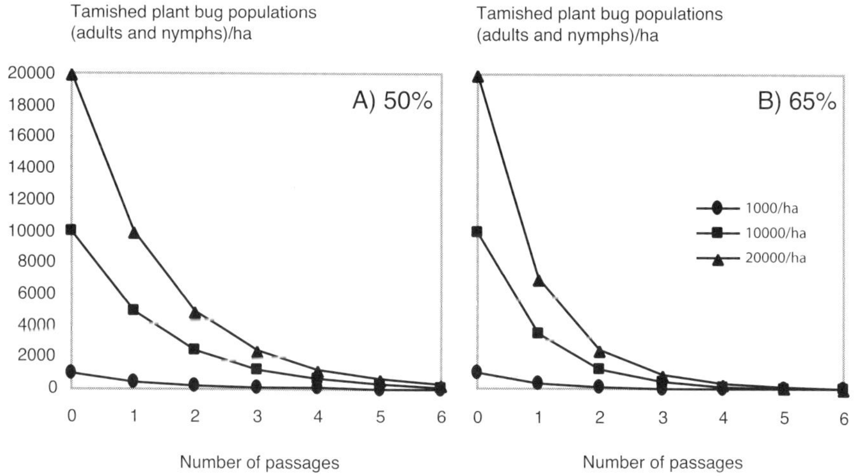

Fig. 5. Simulation of the effect on insect populations of repeated passes, assuming a removal efficiency of **A** 50% and **B** 65% respectively. The other assumptions used are as follows: population densities (individuals ha^{-1})—low (1000), medium (10000) and high (20000)—colonizing a strawberry field with 17778 plants ha^{-1} (125 cm between rows and 45 cm spacing between plants on the row).

plant densities, only 6.6% of the initial population would remain after three or two passes, respectively, when a machine with 50 or 65% removal efficiency is used. However, removal efficiency would be reduced if the subsequent pass was done immediately afterwards, since the Biovac has difficulty in removing bugs that have fallen on the ground. Furthermore, repeated passes could increase soil compaction. In conclusion, at the present time pneumatic control cannot be used as sole control method in agricultural systems. It should instead be envisaged as a complement to other control methods. So far, no studies have been published on the simultaneous or sequential use of pneumatic control with other control methods, such as insecticides or biological control using beneficial insects.

Acknowledgements. We would like to thank Roger Chagnon, Hélène Chiasson and Benoit Rancourt for commenting on an early draft of the manuscript.

References

Boiteau G., Parry R.H., Harris C. R., (1987). Insecticide resistance in New Brunswick populations of the Colorado Potato Beetle (Coleoptera: Chrysomelidae). Can. Entomol. 119:459-463.

Boiteau G., Misener C., Singh R.P., Bernard G., (1992). Evaluation of a vacuum collector for insect pest control in potato. American Potato Journal 69: 157-166.

Boiteau G., Misener G.C., (1996). Response of Colorado Potato Beetles on potato leaves to mechanical vibrations. Can. Agric. Engineering 38: 223-227.

Chagnon M., Gingras J., de Oliveira D., (1989). Effect of Honey Bee (Hymenoptera: Apidae) Visits on the Pollination Rate of Strawberries. J. Econ. Entomol. 82: 1350-1353.

Chagnon R., Vincent C., (1996). A test bench for vacuuming insects from plants. Can. Agric. Engineering. 38:167-172.

Chiasson H., de Oliveira D., Vincent C., (1995). Effects of an insect vacuum device on strawberry pollination. Can. J. Plant. Sci. 75:917-921.

Chiasson H., Vincent C., de Oliveira D., (1997). Effect of an insect vacuum device on strawberry pollinators. Acta Horticulturae 437:373-377.

Cloutier C., Jean C., Bauduin F., (1995). More biological control for a sustainable potato pest management strategy. p.15-52 In Duchesne, R.-M. and G. Boiteau 1995. Lutte aux insectes nuisibles de la pomme de terre/Potato Pest Control. Proc Symp. 31 July-1st August 1995, Québec (QC) Canada, 204 p.

Day W.H., (1991). Can biological control reduce tarnished plant bug damage in strawberry? LISA Small Fruits Newsletter 2(2):9-10.

Dietrick E.J., (1961). An improved backpack motor fan for suction sampling of insect populations. J. Econ. Entomol. 54:394-395.

Duchesne R.-M., Boiteau G. , (1995). Lutte aux insectes nuisibles de la pomme de terre/Potato Pest Control. Proc. Symp. 31 July-1st August 1995, Québec (QC) Canada, 204 p.

Glynn M., (1989). Battling the bugs. California Strawberry (April 16), 18.

Grossman J., (1989). Update: strawberry IPM features biological and mechanical control. IPM Practitioner 11(5):1-4.

Hare J.D., (1990). Ecology and management of Colorado potato beetle. Annu. Rev. Entomol. 35:81-100.

Hillsman K., (1988). Pest vacuums: Innovative Equipment Sweeps Up Bugs. The Grower 21(12): 30-31.

Kelton L.A., 1975. The Lygus bugs (Genus *Lygus* Hahn) of North America (Heteroptera: Miridae). Mem. Entomol. Soc. Canada 95, 101 p.

Khelifi M., Laguë C., Lacasse B., (1992). Test bench for pneumatical control of pest insects. American Society of Agricultural Engineers Paper No. 92-1599. St-Joseph, MI., 8 p.

Khelifi M., Laguë C., Robert J.-L., St-Pierre C., (1994). Numerical modelling of airflow inside and around hoods arranged in different geometries. Canadian Society of Agricultural Engineers Paper no-94-409, 35 p. (mimeo).

Kovach J., (1991). Vacuuming strawberries to control the tarnished plant bug. Reports from the 1990 IPM research, development and implementation projects in fruit. Cornell Univ. IPM Publ. 204: 29-30.

Lachance P., (1990). Les aspirateurs à insectes. La Terre de chez Nous 61(29):49.

Mailloux G., N. Bostanian J., (1988). Economic injury level model for tarnished plant bug, *Lygus lineolaris* (Palisot de Beauvois)(Hemiptera:Miridae) in strawberry fields. Environ. Entomol. 17: 581-586.

McGregor S.E., 1976. Insect Pollination of Cultivated Crop Plants. USDA/ARS Handbook 496. Washington D.C.

Misener G.C., Boiteau G., (1993). Suspension velocity of the Colorado Potato Beetle in free fall. Am. Potato J. 70: 309-316.

Misener G.C., Boiteau G., (1993). Holding capacity of the Colorado potato beetle to potato leaves and plastic surfaces. Can. Agric. Engineering 35: 27-31.

Misener G.C., Boiteau G., (1995). Removal of insect pests from potato using a vacuum collector. Zemedelska Technika, 41:145-149.

Pichette J. (1991). Effet de l'enlèvement des fleurs sur la productivité d'un cultivar de fraisier à jours neutres et ses principaux ravageurs. M. Sc. Thesis in biology, Université du Québec à Montréal, Montréal, Canada, 173 p.

Pickel C., Zalom F.G., Walsh D.B., Welch N.C., (1994). Efficacy of Vacuum Machines for *Lygus hesperus* (Hemiptera: Miridae) Control on Coastal California Strawberries. J. Econ. Entomol. 87:1636-1640.

Pickel C., Zalom F.G., Walsh D.B., Welch N.C., (1995). Vacuums provide limited Lygus control in strawberries. California Agriculture (March-April):19-22.

Rancourt B., Vincent C., De Oliveira D. (2000). Circadian activity of *Lygus lineolaris* (Hemiptera: Miridae) and effectiveness of sampling procedures in strawberray fields. J. Econ. Entomol. 93: 1160-1166.

Snodgrass G.L., (1993). Estimating absolute density of nymphs of *Lygus lineolaris* (Heteroptera: Miridae) in cotton using drop cloth and sweep-net sampling methods. J. Econ. Entomol. 86, 1116-1123.

Southam W.J., (1990). Ag-Vac Agricultural Insect Vacuums, Watsonville, CA, 3p.

Stockwin W., (1988). Sweeping away pests with BugVac. American Vegetable Grower 36 (11): 34-38.

Street R.S., (1990). Is Vacuum Pest Control for Real? Agrichemical Age 34 (2): 22-23, 26.

Taskdall G., Sorum O., 1971. Capsids (Heteroptera, Miridae) in strawberries and their influence on fruit malformation. J. Hortic. Sci. 46: 43-50.

Tingey W.M., Pillemer E.A., 1977. Lygus bugs: crop resistance and physiological nature of feeding injury. ESA Bulletin. 23: 277-287.

Van Driesche R.G., Hauschild K., (1987). Potential for Increased Use of Biological Control Agents in Small Fruit Crops in Massachusetts, pp. 22-32 in R. G. Van Driesche and E. Carey (eds.) Opportunities for Increased Use of Biological Control in Massachusetts. University of Massachusetts Research Bulletin 718, Amherst, MA, 140 p.

Vincent C., Lachance P., (1993). Evaluation of a tractor-propelled vacuum device for the management of tarnished plant bug populations in strawberry plantations. Environ. Entomol. 22: 1103-1107.

Vincent C., Chagnon R., (2000). Vacuuming tarnished plant bug on strawberry: a bench study of operational parameters versus insect behavior. Entomol. Exp. Applic. 97:347-354.

Vincent C., de Oliveira D.D., Bélanger A., (1990). The Management of Insect Pollinators and Pests in Quebec Strawberry Plantations. pp. 177-192 in N.J. Bostanian, T. Wilson and T.J. Dennehy (eds). Monitoring and Integrated Management of Arthropod Pests of Small Fruit Crops. Intercept Ltd., Andover, Hampshire UK.

Zalom F.G., Pickel C., Waldh D.B., Welch N.C., (1993). Sampling for *Lygus hesperus* (Hemiptera: Miridae) in Strawberries. J. Econ. Entomol. 86: 1191-1195.

Pneumatic Control of Colorado Potato Beetle

Benoît LACASSE, Claude LAGUË, Paul-Martin ROY, Mohamed KHELIFI,
Steeve BOURASSA and Conrad CLOUTIER

1 Introduction

The objective of this chapter is to provide basic information on the pneumatic control of the Colorado potato beetle (CPB), *Leptinotarsa decemlineata* (Say), and to analyse the important physical processes that are involved in this particular type of insect control technique.

The overall efficiency of a pneumatic control system against insect pests depends upon the following parameters: (1) maximum insect dislodging and destruction in order to minimise the number of successive passes required in the field and the associated risks for the crop and the soil; (2) maximum field capacity and field efficiency translating into operating widths as large as possible and travel speed as high as possible; (3) minimum power requirements; (4) simple design and operation; (5) significant impact on secondary pest insects populations and minimal impact on the populations of beneficial insects, and (6) minimum crop damage.

2 Dislodging

When potato plants are shaken, adult CPB will often readily drop off the plants and simulate death on the soil for a period of time; this naturally occurring phenomenon is known as thanatosis (Paulian 1988). CPB exposure to an airstream may also induce thanatosis. As a result, a pneumatic control system must not induce vibrations to the crop plants prematurely, otherwise adult CPB may drop on the soil unharmed before being exposed to the airstream.

De Vries (1987) and Misener and Boiteau (1991) have shown that the holding force that adult CPB can develop is larger when the CPB are on the underside of a potato plant leaf than when they are on top of the leaves (0.035 N vs. 0.012 N). The presence of veins on the underside of the leaves and their rougher surface can partly explain this phenomenon (De Vries 1987). The morphology of the CPB legs may also play a role. When CPB walk on the top surface of the leaves, they mostly use the end of the tibia and the first tarsomere of each of their legs; the other tarsomeres being held above the leaf surface (Pelletier and Smilowitz 1987). This reduces the number of tarsal hooks that the insect uses to maintain itself on the leaf. When an adult CPB is crawling along the edges or on the underside of a leaf, it must, however, use more tarsal hooks to maintain its grip and avoid falling. Dislodging adult CPB pneumatically may thus be more difficult when the wind and temperature conditions induce the insects to move to the underside of the potato plants leaves.

For CPB larvae, the force required to pull them off the plants does not depend upon their location on the leaves (Misener and Boiteau 1991). These authors also determined that third (L3) and fourth (L4) instar larvae maintain their grip on the leaves mainly through their mandibles [0.01 N(L3) to 0.03 N(L4)] rather than by their tarsi [0.005 N(L3) to 0.01 N(L4)]. The legs of CPB larvae are shorter and less adapted to rapid motion than those of adults. As a result, CPB larvae move more slowly and may not have the time to reach a leaf vein in order to improve their grip on the potato plants when they are shaken.

The additive effects of different stimuli on improved dislodging efficacy can also be considered. Boiteau and Misener (1996) studied the effects of vibration on CPB dislodging from potato leaves. Male and female adults and larvae responded similarly to the various combinations of frequencies and amplitudes used in the study. Insect dislodging was maximum for frequencies of 20 Hz or more and for amplitudes larger than 0.6 mm. It was reduced when the insects were located on the edge of the leaves rather than on the surface. Most of the insect dislodging occurred within the first 2 s of exposure to the vibration.

3 Capture of the Dislodged Insects

After the CPB have been pneumatically dislodged, they must be directed into some sort of collecting or destruction device before they fall to the ground. Capture efficacy depends largely on the machine configuration and calibration, the spatial distribution of beetles on the plants, plant resistance to airflow and the aerodynamic properties of the insects. The collecting device must capture the dislodged insects and store or destroy them, or direct them to a central collection system.

4 Collection Efficacy

The collection efficacy (ηcoll) of a pneumatic control system depends upon two phenomena: (1) the dislodging efficacy (ηdisl), corresponding to the ratio between the numbers of dislodged insects and the total initially present ont the crop plants, and (2) the capture efficacy (ηcapt), wich represents the proportion of dislodged insects that are either collected or destroyed by the system:

$$\eta\text{coll} = \eta\text{disl} * \eta\text{capt} \qquad (1)$$

According to Harcourt (1971), large potato beetle larvae (L3 and L4) dislodged by rain can usually climb back onto the plants. Collection efficacy provides an accurate indication of the effectiveness of pneumatic control against a given insect population. For the particular case of CPB, Eq. (1) can be modified to take into account the fact that a significant proportion of the small larvae (L1 and L2) that have been dislodged by the pneumatic system but not collected or destroyed will, however, not be able to climb back on the potato plants and will thus die. This is not the case for L3 and L4 larvae or for adult CPB that may readily climb back onto the plants once they have been dislodged (Harcourt 1971).

5 Climatic Factors and Spatial Distribution

The effectiveness of pneumatic control of the CPB may be affected by several climatic variables, particularly temperature, which influences the insects' spatial distribution. May (1982) has shown that the CPB move in different parts of the plant in order to regulate their body temperature. For example, when the ambient temperature is around 25 °C, most of the adults will be on the upper surface of leaves, whereas at higher temperatures, they will move to the shade on the top third of the plant. Even though this type of behaviour is not as common with the larvae, they tend to spend more time on the upper surface of leaves and on the outside portion of the foliage at low temperatures, where they are more exposed to the sun. According to de Wilde (1950) (cited in Hurst 1975), low temperatures weaken the larvae, particularly the L1 and L2, making them more likely to be dislodged from the plant by rain.

6 Airstream Width and Speed and Travel speed

De Vries (1987) carried out field tests with two prototypes in which the airflow was directed horizontally across the rows of potato plants, with a collecting device located on the other side of the plants. At airspeeds between 15 and 30 m s^{-1}, an average dislodging rate of 77% (n = 47 adults) was obtained, while capture efficacy remained around 30%. De Vries (1987) concluded that better collection efficacy could be obtained by improving the collecting device.

De Vries (1987) also identified three phenomena that could explain the low collection or dislodging rates: (1) some insects may be protected from the airflow by leaves or stems; (2) plant parts may deviate the trajectory of some dislodged insects from collector direction; and (3) some insects could fall to the ground before reaching the collector.

According to both De Vries (1987) and Khelifi (1996), blowing air onto the insects has several advantages over suction. Positive pressure airstreams can be more effectively directed and concentrated than negative pressure airstreams. For a given power, it penetrates plants with dense foliage at greater speeds to dislodge and carry the beetles out of the foliage more efficiently.

Box 1. Collection Efficacy

Collection efficacy (ηcoll) is an index used to evaluate the effectiveness of pneumatic systems. It represents the percentage of insects collected during one pass.

A proportion of the insects are dislodged from the plant by the airflow and shaking of the plant. Of these, a portion will be carried to the collection system (by the airflow and/or gravity). Collection efficacy can be divided into two distinct components: dislodging efficacy (ηdis) and capture efficacy (ηcapt).

Collection efficacy can be expressed as:

$$\eta\text{coll} = \eta\text{dis} * \eta\text{capt}$$

$$\frac{\text{Insects collected}}{\text{Insect present initially}} = \frac{\text{Insects dislodged}}{\text{Insect present initially}} * \frac{\text{Insects collected}}{\text{Insect dislodged}}$$

The two components can therefore be analysed separately to obtain a better understanding of the potential weaknesses in the system. If the insects remain on the plant, dislodging must be improved and if the insects are dislodged but not captured, the capture system must be modified.

Dislodging efficacy can be used as an index of the potential that a pneumatic system holds for controlling an insect affecting a given crop. It is the first element that must be examined before building a field prototype. Capture efficacy reflects the system's ability to capture the dislodged insects, independently of dislodging itself.

Most insects undergo a number of life stages, each comprising anatomical and behavioural changes. Therefore, a separate evaluation of the vulnerability of each stage (egg, larva, adult) would be useful for enhancing the effectiveness of treatment.

Khelifi et al. (1995a) compared different configurations using positive or negative airstreams or a combination of both. They found that for a 20-s exposure, configurations using a positive airstream directed horizontally were the most effective at dislodging adult beetles. On the other hand, ascending airstreams tended to push the leaves and stems vertically towards the main axis of the plant, forming a screen that protected the CPB from the airflow. Based on these findings, Lacasse et al. (1998a) subjected potato plants to a 7.6-cm-wide horizontal airflow having four different airspeeds (20, 25, 30 and 35 m s^{-1}) and moved them at three different travel speeds (4, 6 and 8 km h^{-1}), to determine the effectiveness of these combinations in the dislodging of adult CPB. In this laboratory study a grid of horizontal bars on which the plant could lie was placed on the side opposite to the air blower. Behind the grid was a vacuum inlet whose airflow rate was calibrated with respect to blower flow rate. Dislodging efficacy was found to increase with airspeed and travel speed, suggesting that the device would perform well in the field. Complete dislodging of the beetles from single-stemmed plants stem was achieved at 35 m s^{-1} at travel velocities of 6 and 8 km h^{-1}. In this test, only half of the beetles dislodged were collected by suction.

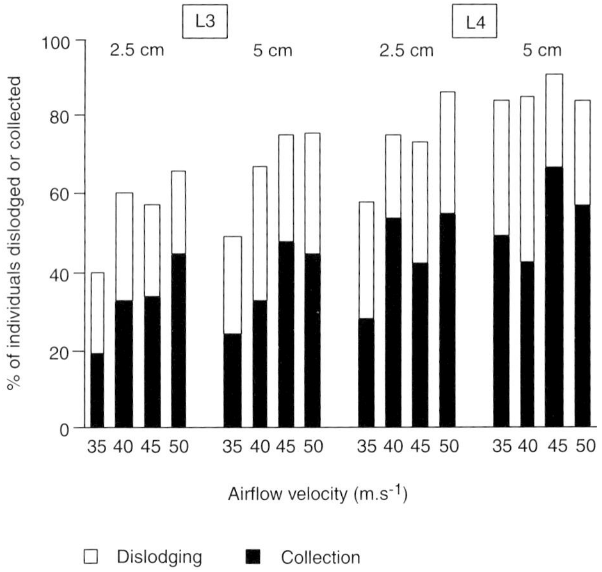

Fig. 1. Dislodging and collection of L3 and L4 CPB larvae at different airflow speed and widths (travel speed = 6 km h^{-1}).

Following these experiments, a prototype single-row collector with a horizontal air blower, grid and collection apparatus was designed and tested in the field on L4 beetles at three airspeeds (20, 30 and 40 m s^{-1}) (measured relative to the plant) and at forward velocities ranging from 2 to 6.8 km h^{-1}. Dislodging and collection efficacy increased with airspeed. For the maximum airspeed value used in the study (40 m s^{-1}), insect dislodging and collection also increased with travel speed, suggesting that the shorter exposure of the plants to airflow was partially compensated by a more vigorous shaking of the plants.

Using a different prototype of pneumatic control system at a travel speed of 6 km h^{-1}, Lacasse et al. (1998a) also studied the effects of various combinations of airspeed (35, 40, 45 and 50 m s^{-1}) and airstream width (2.5 and 5 cm) on the dislodging and collection of CPB in the field. The location from which the dislodged CPB dropped off the plants was also recorded for the different larval instars.

In these tests, increasing the airspeed and the airstream width did not significantly increase the dislodging (= 61%) or collection (= 54%) rates for the adult beetles. This rather low dislodging efficacy was partly caused by the increased density of the potato plant canopy in the field, which partly shielded the CPB from the airflow. The reduction in the width of the airstream (maximum test width of 5 cm versus 7.6 cm in the laboratory) may have also had an effect. Since 88% of the dislodged adults were collected, the problem seemed to be the dislodging efficacy.

Dislodging efficacies were 33 and 45% and collection efficacies were 18 and 24% for L1 and L2 larvae, respectively. For L3 larvae (Fig. 1), higher airspeeds improved dislodging and collection and the airstream width had a significant effect on

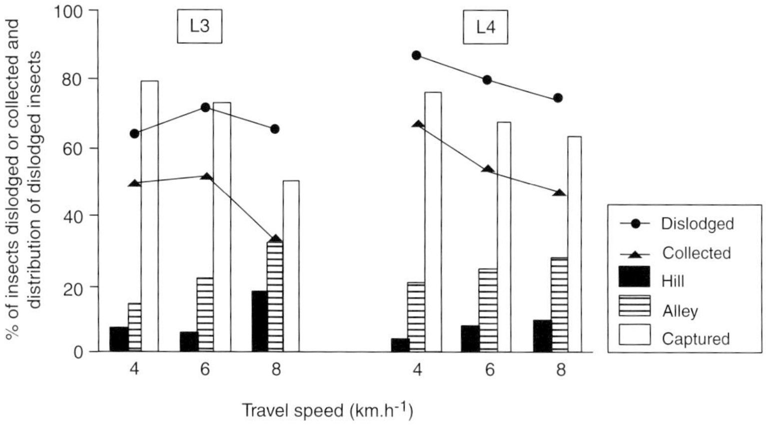

Fig. 2. Dislodging, collection and spatial distribution of L3 and L4 CPB larvae at different travel speeds (airspeed = 45 m s^{-1}, airstream width = 5 cm).

dislodging: 58% of the larvae were captured, 18% fell onto the hill and 24% into the alley. For L4 larvae (Fig. 1), maximizing the airstream width improved dislodging efficacy. Out of all the larvae dislodged, 61% were collected, 15% fell onto the sides of the row hills and 24% into the alley.

Subsequently, the most effective combination (airstream width = 5 cm; airspeed = 50 m s-1) was tested at three different travel velocities (4, 6 and 8 km h^{-1}). No effect could be observed on small larvae due to their low abundance. On average, for L1 larvae, 34% were dislodged and 17% collected, compared with 52 and 33%, respectively, for L2 larvae. For L3 and L4 larvae, increased travel speed reduced dislodging and capture efficacy and increased the proportion of dislodged larvae dropping into the alley (Fig. 2). The dislodging of L4 larvae also decreased with travel speed. On average, about one fourth of the large larvae fell into the alley.

All the tests were carried out on Kennebec potato plants, which have fairly dense foliage. Pneumatic control of CPB would probably be more efficient for potato cultivars whose crop canopies are less dense than that of Kennebec.

7 Effects on Secondary Pests and Natural Enemies

Secondary pests are too often ignored or assumed to be managed by control measures for CPB. It can be assumed that pneumatic control eliminates all the insects present on foliage indiscriminately. Predators, which are by nature more mobile, should be less susceptible than secondary pests.

Alternatives to insecticides, such as mechanical or biological control, may need to be employed in conjunction with pneumatic controls. Studies are currently under way on the use of natural enemies to control CPB (Cloutier and Bauduin 1995; Giroux et al. 1995; Cloutier et al. 1996). Therefore, it is important to evaluate the susceptibility of predators to pneumatic control and determine if there are periods when the net result of using pneumatic control against CPB (eliminating pests but sparing natural enemies) is more positive.

To gather information on these questions, samples of about 500 ml of the insects collected by a prototype four-row pneumatic control machine were analysed. Its configuration consisted of a 2.5-cm wide horizontal airstream with an airspeed of 50 m s^{-1}. The collection apparatus was composed of a horizontal grid and a vertical screen that stopped the dislodged insects and directed them into a container located underneath. The whole unit was allowed to directly slide on the soil surface. Samples were taken on four different dates between July 26 and August 20, 1995, the period when the crop matures and the new generation of adult potato beetles emerges. Samples were examined to identify potential pests and their natural enemies (Boiteau 1983; De Oliveira 1992; Hough-Goldstein et al. 1993; Ferro 1994; Cloutier et al. 1996) as well as other insects not specifically related to potato crops.

The CPB and other pests were abundant in the samples. The July 26 sample contained a total of 5268 arthropods, 84% of which were phytophagous insects and 58.4% potato pests. The taxa identified in the samples are presented in Table 1.

Table 1. Main insect taxa identified in the collector of a pneumatic control system between July 26 and August 20, 1995.

Species or higher taxa	Common name
A) Primary and secondary pests of potatoes	
Leptinotarsa decemlineata (Say)	Colorado potato beetle
Epitrix cucumeris (Harris)	Potato flea beetle
Epitrix tuberis Gentner	Tuber flea beetle
Empoasca fabae (Harris)	Potato leafhopper
Macrosteles quadrilineatus (Forbes)	Aster leafhopper
Lygus lineolaris (Palisot de Beauvois)	Tarnished plant bug
Aphis nasturtii Kalt.	Buckthorn aphid (winged form)
Macrosiphum euphorbiae (Thomas)	Potato aphid
Myzus persicae (Sulzer)	Green peach aphid
B) Potential pests or insects associated with potato weeds	
Acyrthosiphon pisum (Harris)	Pea aphid (winged form)
Trigonotylus coelestialium Kirk.	Plant bug
Systena frontalis (Fabricius)	Red-headed flea beetle
Diptera	Unidentified true flies
Cicadellidae	Leafhoppers
Miridae	Immature mirids
Psyllidae	Psyllids
Poduridae	Springtails
Sminthuridae	Springtails
Thripidae	Thrips

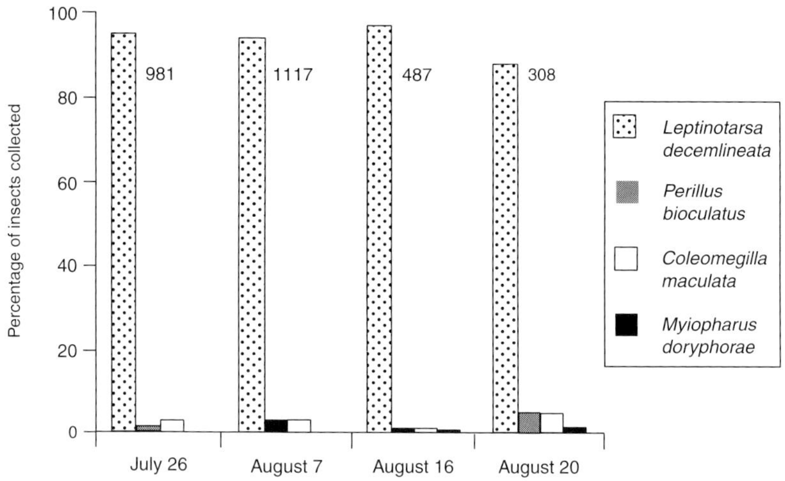

Fig. 3. Proportion of Colorado potato beetles and their specific natural enemies in the pneumatic collection system, by date, Ste-Croix de Lotbinière, Quebec, 1995.

A study of the pest species showed that 22% were CPB. Smaller pests (mainly aphids, thrips and springtails) were also well represented, with the potato aphid accounting for 33% of all pests and the tarnished plant bug 12%.

Natural enemies specific to the CPB included the two-spotted stink bug, *Perillus bioculatus* (F.), which occurred in large numbers (it was being used for experimental biological control on plots roughly 300 m away). A specific parasitoid of the CPB, the tachinid fly *Myiopharus doryphorae*, (Riley) was also present. Aphidophagous insects were the best represented, particularly *Chrysopidae* (*Neuroptera*), but especially ladybugs (30% of captures). Ladybug species included *Coccinella septempunctata* L., *C. quattuordecimpunctata* L., *Coleomegilla maculata* DeGeer and *Harmonia axyridis* (Pallas). A large number of mummified aphids [mainly *Macrosiphum euphorbiae* (Thos.)] were also present, making up 31% of natural enemies. The mummified aphids contained primary parasitoids (Hymenoptera: Aphididae) and hyperparasitoids, whose respective proportions were not determined. Adults of the hyperparasitoid *Asaphes lucens* (Nees) (Pteromalidae) were abundant; they alone accounted for 8% of CPB natural enemies in the July 26 sample. Generalist predators such as spiders and individuals of several families of parasitic hymenopterans were also found.

Temporal variations in the relative proportions of CPB and their natural enemies (*P. bioculatus*, *C. maculata* and *M. doryphorae*) were also studied. In the four samples taken from July 26 to August 20, there were 5 to 20 times more CPB than natural enemies (Fig. 3). The 12-spotted ladybug and the 2-spotted stinkbug were present throughout the period, making up 1–5% of the total of CPB and their natural enemies.

Paying attention to the large numbers of aphids on the sampled foliage, the ratios between the quantities of aphids and natural enemies collected were further investigated (Fig. 4). Although the ratio of natural enemies to pests was below 20% on

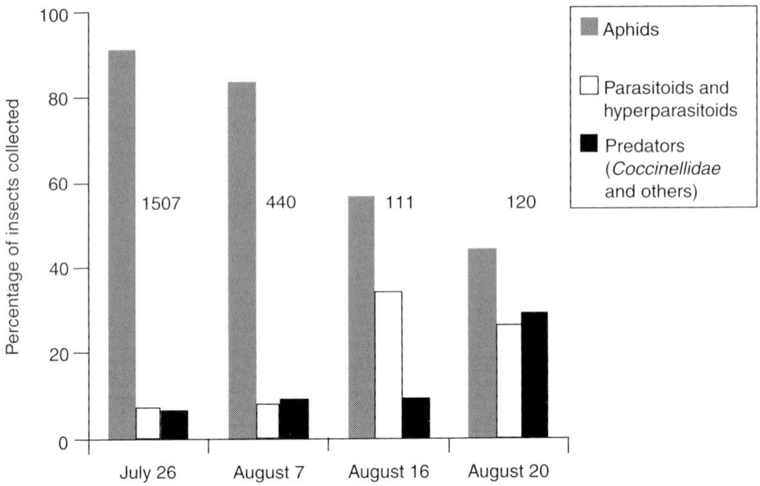

Fig. 4. Proportion of aphids and their natural enemies in the pneumatic collection system, by date, Ste-Croix de Lotbinière, Quebec, 1995.

the first 2 days, it increased in the last two samples, when there was an almost equal percentage of natural enemies and aphids.

As for secondary pests (Fig. 5), the following phenomena were observed: (1) aphid populations decreased gradually from around 45 to 10% during the period; (2) thrips were present in significant numbers only in the July 26 sample; (3) the tarnished plant bug increased from 15% on July 26 to 80% on August 20; and (4) the aster leafhopper increased over the first three samples (from 3 to 40%) but then disappeared almost entirely. The abundance of other secondary pests (of lesser importance) did not vary appreciably.

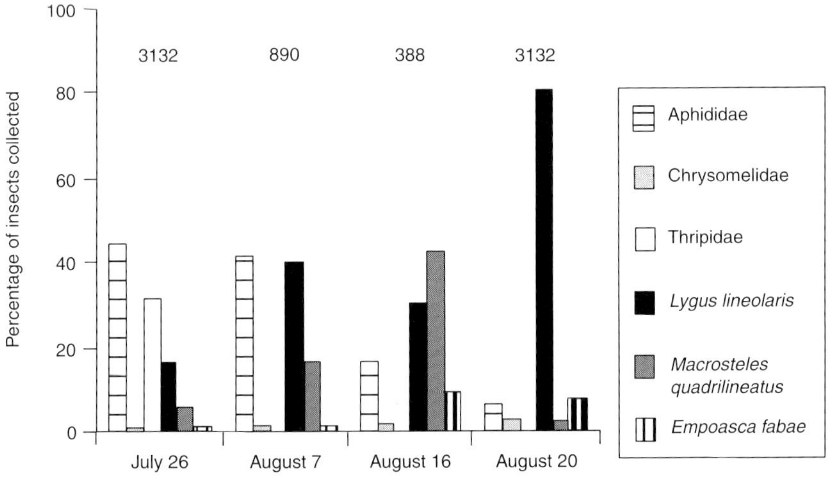

Fig. 5. Proportion of main secondary pests of the potato crop in the pneumatic collection system, by date, Ste-Croix de Lotbinière, Quebec 1995.

The pneumatic control system therefore collects a wide variety of arthropods associated with the potato crop in mid-season. As expected, the samples contained large numbers of CPB, which accounted for a large proportion of the captures.
However, most of the natural enemies of potato pests seemed as vulnerable to pneumatic control as the pests themselves. The presence of natural enemies in the samples throughout the 4-week period indicates that there is no window of opportunity for treatment favourable to natural enemies during this period, when they made up between 5 and 10% of the total of pests and natural enemies. The tachinid fly *M. doryphorae* appeared not to be affected by treatments before mid-August, but this is probably an exceptional case.
One group of natural enemies was not found in the samples: generalist predators with nocturnal habits (e.g. carabid beetles and harvestmen). The absence of such predators in the samples, although they are known to be present (unpubl. data), can be explained by the fact that they were not directly exposed to treatments (Cloutier et al. 1996) which were carried out during the day.

8 Impact of Pneumatic Control on Crops

The nature and importance of the potential damages inflicted on potato crops by pneumatic control systems need to be determined. According to De Vries (1987), who tested a prototype in the field, potato plants can resist to air flows having a velocity of 56 m s^{-1} or less (measured at the outlet of a 0.95-cm-wide nozzle) for short periods of time without permanent damage. An experiment was conducted at Laval University to further confirm these results using the potato cultivar Kennebec. To eliminate interference (i.e. defoliation) by insects, plots were treated chemically throughout the season as required. Pneumatic treatments consisted of: (1) no treatment (control); (2) one pass per week; and (3) two passes per week. During each pass, the plants were exposed to a 2.5-cm-wide airstream at an outlet airspeed of 50 m s^{-1}. At a travel speed of around 3.8 km h^{-1}, the maximum exposure time was 0.024 s. Treatment frequency was found to have no effect on plant growth or crop yield.

9 Use of Pneumatic Control in Combination with other Control Methods

Along with vibrations that may prompt CPB to drop off plants, other methods of physical control can be used in conjunction with pneumatic control, either concurrently to enhance the performance of pneumatic control, or separately at other times during the season.

9.1 Propane flamers

Propane flamers have a treatment window that complements that of pneumatic systems. Using flamers on plants 10 cm tall or less can eliminate roughly 90% of adult CPB in spring (Moyer 1992) and reduce hatching of egg masses by 30% (Moyer 1992), without reducing crop yields. Combining the pneumatic and thermal control methods for the CPB can therefore be envisaged. Early into the season, emerging adult CPB could be controlled by applying thermal treatments directly onto the crop rows when the potato plants are less than 10 cm tall. After the plants have grown taller, it could be possible to pneumatically dislodge the CPB from the crop plants and to direct the dislodged insects on the ground between the rows where they would be destroyed by flaming.

10 Conclusion

The results of this study on the dislodging of insects and the distribution of the dislodged insects are useful in understanding the crucial elements of a pneumatic system. Although complete dislodging of beetles from plants was achieved in bench tests, in the field, it is more difficult to dislodge adults (61%) than to collect those that have already been dislodged (88%). Larvae dislodging must also be improved, particularly if it is confirmed that L1 and L2 larvae cannot crawl back onto the plants.

Increasing the travel speed of the collector results mainly in a decrease in capture efficacy for large larvae. However, it still may be possible to treat crops at a speed of 6 km h^{-1}, with a collection efficacy of 50% (L3) and 65% (L4). The passive collection system has considerable potential. Losses from the air pipes must be minimized to obtain maximum airspeeds and airstream widths, since both of these factors improve collection efficacy for large larvae. High airspeeds can be used for short periods (1/40 s) without affecting yields. Horizontal airstreams and collectors with direct contact with plants may transmit pathogens and this must be verified.

Although pneumatic control systems do not seem particularly compatible with the use of natural enemies, the system does collect large numbers of secondary pests. A portion of the beneficial insects could perhaps be recovered using a passive form of separation based on differences in behaviour (movements or orientation) between pests and predators when disturbed.

Pneumatic control naturally entails labour and equipment acquisition costs, but may have fewer hidden costs than insecticides. Improvements in dislodging the various instars and the integration of other technologies could make pneumatic control a cost-effective alternative to traditional chemical control. Although it may not be adequate on its own, this alternative to insecticides should be given serious consideration as an option for prolonging the use of the few remaining insecticides that are effective against the Colorado potato beetle.

References

Boiteau G., (1983). The arthropods of potato fields: composition and abundance. Agriculture Canada, Research Branch, Contribution No. 1983-16E. 57 p.

Boiteau G., Misener G.C., (1996). Response of Colorado potato beetles on potato leaves to mechanical vibrations. Can. Agric. Eng. 38:223-227.

Boiteau G., Misener G.C., Singh R.P., Bernard G., (1992). Evaluation of a vacuum collector for insect pest control in potato. Am. Potato J. 69:157-166.

Cloutier C., Bauduin F., (1995). Biological control of the Colorado potato beetle, *Leptinotarsa decemlineata* (Coleoptera: Chrysomelidae) in Québec by augmentative releases of the two-spotted stinkbug *Perillus bioculatus* (Hemiptera: Pentatomidae). Can. Entomol. 127: 195-212.

Cloutier C., Bauduin F., Jean C., (1996). More biological control for a sustainable potato pest management strategy for Canada. pp 15-52 In: Actes du symposium "Lutte aux insectes nuisibles de la pomme de terre" (R.-M. Duchesne and G. Boiteau, Eds.) Agriculture and Agri-Food Canada, Agriculture, Pêches et Alimentation Québec and Union des producteurs Agricoles du Québec.

De Oliveira D.D., (1992). La lutte biologique contre le doryphore de la pomme de terre. pp. 205-219. In C. Vincent and D. Coderre (eds.), La lutte biologique. Gaétan Morin Ed., Boucherville (Qc) and Lavoisier Tech. Doc. (Paris)

De Vries R.H., (1987). An investigation into a non-chemical method for controlling the Colorado potato beetle. M. Sc. thesis, Cornell University. Ithaca, NY. 68 p.

Ferro D.N., (1994). Biological control of the Colorado potato beetle. pp. 357-375 In Zehnder, G. W., M. L. Powelson, R. K. Jansson & K. V. Raman (Eds), Advances in Potato Pest Biology and Management. APS Press, Saint-Paul, Minn.

Giroux S., Duchesne R.-M., Coderre D., (1995). Predation of *Leptinotarsa decemlineata* (Coleoptera: Chrysomelidae) by *Coleomegilla maculata* (Coleoptera: Coccinellidae) : Comparative effectiveness of predator developmental stages and effect of temperature. Environ. Entomol. 24: 748-754.

Harcourt D.G., (1971). Population dynamics of *Leptinotarsa decemlineata* (Say) in eastern Ontario: III. Major population processes. Can. Entomol. 103:1049-1061.

Hough-Goldstein J.A., Heimpel G.E., Bechman H.E., Mason C.E, (1993). Arthropod natural enemies of the Colorado potato beetle. Crop Prot. 12: 324-334.

Hurst G.W., (1975). Meteorology and the Colorado potato beetle. Technical Note No. 137. Secretariat of the World Meteorological Organisation, Geneva, Switzerland. 51 p.

Khelifi M., Laguë C., Lacasse B., (1995a). Resistance of adult Colorado Potato Beetles to removal under different airflow velocities and configurations. Can. Agri. Eng. 37:85-90.

Khelifi M., Laguë C., Lacasse B., (1995b). Potato plant damage caused by pneumatic removal of Colorado potato beetles. Can. Agri. Eng. 37:81-83.

Khelifi M., (1996). Optimisation des paramètres physiques de conception d'un système pneumatique de contrôle du doryphore de la pomme de terre (*Leptinotarsa decemlineata* (Say)). Ph.D. thesis. Laval University, Quebec, 310 p.

Lacasse B., Laguë C., Khelifi M., Roy P.M., (1998a). Effects of airflow velocity and travel speed on the removal of Colorado potato beetle from potato plants. Can. Agric. Eng. 40: 265-272.

Lacasse B., Laguë C., Khelifi M., Roy P.M., (1998b). Field evaluation of pneumatic control of Colorado potato beetle. Can. Agric. Eng. 40: 273-280.

May M.L., (1982). Body temperature and thermoregulation of the Colorado potato beetle, *Leptinotarsa decemlineata*. Entomol. Exp. Applic. 31: 413-420.

Misener G.C., Boiteau G., (1991). Force required to remove Colorado potato beetle from a potato leaf. CSAE-paper No. 91-404. Can. Soc. Agri. Eng., Saskatoon, Saskatchewan.

Moyer D.D., (1992). Fabrication and operation of a propane flamer for Colorado Potato Beetle Control. Cornell Cooperative Extension. Riverhead, NY, USA.

Paulian R., (1988). Biologie des Coléoptères. Lechevalier, Paris.

Pelletier G.C., Smilowitz S., (1987). Specialized tarsal hairs on adult male Colorado potato beetles, *Leptinotarsa decemlineata* (Say), hampered its locomotion on smooth surfaces. Can. Entomol. 119:1139-1142.

Vacuuming Insect Pests: the Israeli Experience

Phyllis G. WEINTRAUB and A. Rami HOROWITZ

1 Historical Overview

Historically, the use of vacuum devices to control insect populations was more a novelty than a necessity; until the 1960s the number of insecticides available for insect control was burgeoning. In fact, the first documented use of a tractor-propelled vacuum device was for mass collection of *Lygus hesperus* Knight for mark-release studies (Stern et al. 1965). Another application was for arthropod sampling methods (Ellington et al. 1984). This machine sampled large acreage cotton in an effort to accurately determine insect population densities, and it achieved mean catch rates of 14-64%.
To say that the state of agriculture today is highly dependent on chemicals to control pest species, and that many of these species exhibit resistance to pesticides, is axiomatic. New and innovative means of controlling pest populations are being seriously explored, albeit slowly. Much of the impetus for movement away from chemical pest control results from newly established government policies. For example, the Chief Scientist of the Ministry of Agriculture established a steering committee and charged them with developing a comprehensive policy for reduction of pesticide use in agriculture in Israel. This was motivated by rising pesticide costs, resistance problems, phytotoxicity, and residues in food, all of which affected marketability of agricultural products in Europe and the United States. Recommendations were made for all agricultural crops, orchards, fields and greenhouses. In addition to offering suggestions for physical, cultural and biological control, it was stressed that the development of new technologies must be encouraged and supported. To complement this policy, the head of agricultural mechanics in the Extension Services conceived the idea of vacuum removal of insects – a new technology not specifically mentioned in the published recommendations – as a physical control method.

2 Challenges to Pneumatic Control of Insects

The most difficult task is to develop a machine versatile enough to be capable of removing a variety of insects/arthropods. Strong flying insects such as grasshoppers, larger

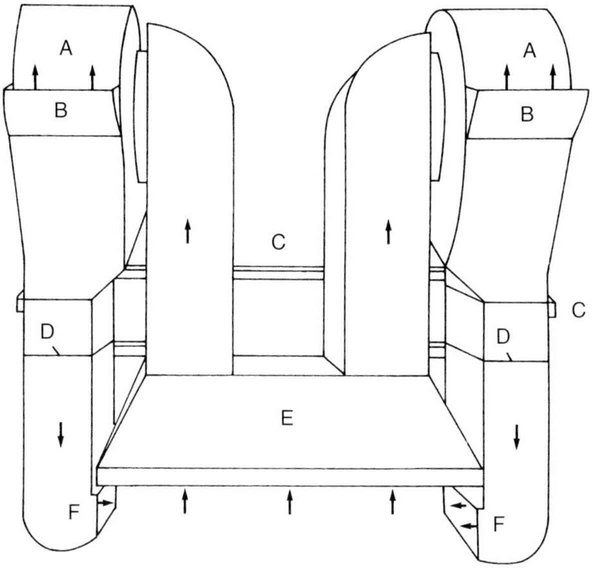

Fig. 1. Scheme of vacuum unit. Arrows indicate direction of air flow. **A** housings for impellers; **B** cover of exit portal for insects and debris; **C** frame to raise and lower suction inlet with respect to outlet air jets; **D** junction point to raise and lower unit (**D** is male, **F** is female); **E** suction inlet; **F** outlet air jets.

flies, beetles, and bees have the ability to simply fly out of the path of the vacuum device. Heavy bodied larvae (such as the Colorado potato beetle, *Leptinotarsa decemlineata* Say) release their hold on the plant and fall to the ground as a defensive measure; and some smaller insects (such as whiteflies) cling very tightly to the plant. Therefore, under normal circumstances only weaker flying insects, true bugs, and those that are easily dislodged from the plant are collected. Furthermore, one cannot simply increase vacuum pressure (see Khelifi et al. 1995) or lower the vacuum unit into the plants to remove insects clinging to plants, since this would cause significant plant damage.

The design of the Israeli machine (Fig. 1) is unique in that it simultaneously employs two actions: blowing air onto plants to dislodge insects and vacuuming from above. Recently, Walklate (1994) showed, using a computer simulation (Computation Fluid Dynamic, CFD) package called FLUENT (Fluent Europe Ltd.), that an outlet jet of air blowing at 30 m s^{-1} affects a much greater area than a suction inlet of the same size and airflow.

In the Israeli unit, air speeds are minimized but effectiveness is maximized by having two blowers directed perpendicularly to, and on either side of a standard bed (1.9 m width, i.e., four rows of celery or two of potato) to dislodge insects from the plants. Air speeds of 40 m s^{-1} are used (commercial garden leaf-blowers blow air at about 60 m s^{-1}). A vacuum opening, located above and encompassing the entire area between the blowers, immediately removes the dislodged insects.

To avoid plant damage and enhance efficiency of insect removal, the vacuum unit

can be raised or lowered by either of two means: by changing the wheel size or by physically raising or lowering the unit within the frame. The unit can be completely raised above the plants by hydraulics once the end of a row is reached or as needed. Insects are exhausted through impeller blades and thus even the smallest are seriously damaged or completely destroyed. The tractor, which propels the unit, moves forward at a speed of about 4 km h^{-1}.
Initial trials showed that strong flying insects flew out of the way as the unit approached; rolling waves of insects were easily observed. To counter this problem, a canopy (the width of the bed, extending 1 m in front of the unit and draping to the ground) was added. Efficiency was increased as insects were trapped on the plants in front of the unit and removed as the vacuum passed over.

3 Evaluating the Efficacy of Insect Control by Pneumatic Removal

In the 1980s, commercial field-scale vacuum machines started appearing on the market, although claims of their effectiveness against *Lygus* spp. were not substantiated. Only recently have serious field trials been performed in the field (Pickel et al. 1995) and laboratory (Chagnon and Vincent 1996) to determine the effectiveness of these machines. Evaluation of these machines involves assessment of adult and immature insect populations, determination of the size/area of trial plots, and analysis of harvest yields. Furthermore, it can be complicated by a lack of grower cooperation when trying to monitor commercial fields. It must also be determined if there is excessive mechanical damage to the plants, or if there is an increase in disease incidence.

3.1 Insect Populations

Over the course of time, we have improved and refined the techniques used for sampling insects in an attempt to assess the true effectiveness of the vacuum unit. All vacuum treatments occurred during daylight hours, so evaluation of nocturnal insects is not included.
The first field trials evaluated the efficiency of leafminer removal in celery. Field evaluations of immature *Liriomyza huidobrensis* (Blanchard) or *L. trifolii* (Burgess) included counting the number of live larvae, dead larvae and empty tunnels within each plot. Although results were obtained, this method was felt to be only adequate, inaccurate, and species could not be determined. This method of evaluation was employed since celery leaves with leafminer tunnels must be removed before marketing.
It has been well established that yellow is the most effective color for trapping a number of adult insects (Affeldt et al. 1983, Yudin et al. 1987). Therefore, yellow

sticky traps coated with Rimifoot were used in an attempt to monitor the efficacy of the vacuum machine on the day of vacuuming. Yellow sticky traps were placed in the field for 1 h before and for 1 h after the field vacuuming was completed. This proved to be an insufficiently sensitive method of monitoring efficiency of the field vacuum unit on the day of treatment.

We then evaluated hand-held vacuum samples taken immediately before and after the field vacuuming. An Echo leaf blower was modified to be a vacuum unit by switching the air intake and exhaust ports; air velocity was 60 m s^{-1}. After sampling, fine-mesh nylon collection bags were secured and kept on ice packs in a cooler, then stored in a freezer until the insects were evaluated. To reduce variations due to diel activity, all hand-vacuum samples were taken at the same time period throughout a growing season. The non-vacuumed control field was also sampled at the end of the field vacuuming. This sampling method has proven most effective for general purposes, producing adequate numbers of a wide variety of insects. Our particular collection bags are, however, inadequate for thrips and mites, which can crawl out.

Whiteflies (*Bemisia tabaci* Grennadius) present a unique situation to evaluate the efficiency of removal of adult insects on populations of immatures. Leaf samples must be taken. Although whitefly adults are easily removed, their sessile immatures are tightly bound to plant leaves. Only by evaluating the number of immatures can one deter-

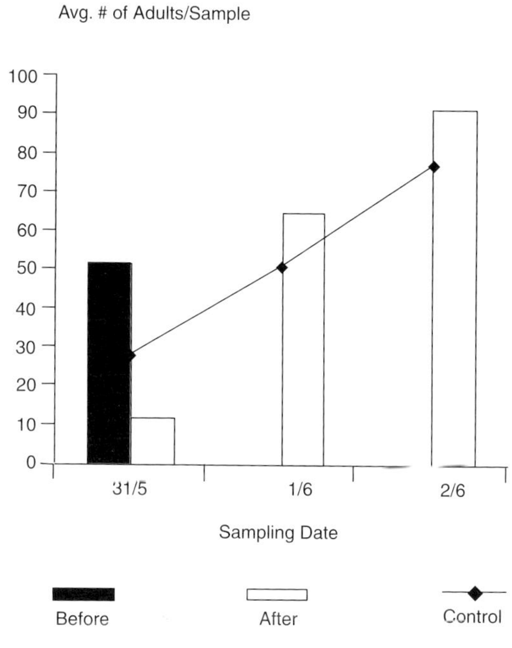

Fig. 2. Reinvasion of *Liriomyza huidobrensis* into vacuumed small-sized potato plots in the spring of 1995. The field was vacuumed on May 31 and fly populations were monitored for 2 subsequent days. Although the fly population in general was increasing, fly populations in small vacuumed plots increased at a faster rate and equalled non-vacuumed control plots after 1 day.

mine if removal of adults is adequate to effect a change in the total population.

3.2 Plot Size

Initially, field trials were performed on relatively small replicated plots – 10-12 x 15 m. Yellow sticky traps, placed in vacuumed, insecticide-treated, or control plots for 1 day once a week, showed no difference in populations of insects between any treatments. To determine why this was occurring, hand-vacuum samples were taken from the check and vacuum-treated plots for 2 consecutive days following the regular field vacuuming. Figure 2 clearly shows what occurred to *L. huidobrensis* populations during the 2 days following field vacuuming. The field vacuum treatment removed a significant number of insects, but the population in the vacuumed plots had recovered to the levels found in non-vacuumed control plots within 1 day. There was no difference between population levels in the vacuum and non-vacuumed control plots. The large edge effect found in small plots allowed easy reinvasion, even though 10 m of non-cultivated, weed-free land surrounded the plots.

When plot size was increased to 20 x 40 m, significant differences in adult capture rates on yellow sticky traps were observed. It was thus concluded that the edge effect plays a significant role, and that plots must be relatively large in order to prevent quick reinvasion of insects, which can compromise results.

3.3 Yield

Because field vacuuming is usually done once a week, which is more often than a tractor would normally enter a field, we were concerned about the possible adverse effects of soil compaction (which could affect quality and yield of root crops), and the possibility of plant damage. However, upon evaluation of potato for two seasons and melon for one season, we found no differences in quality and yield between vacuumed and non-vacuum plots (Table 1). We worked in conjunction with a plant physiologist and found no increase in plant diseases or excessive damage to the plants as a result of tractor/vacuum damage. Furthermore, in the

Table 1. Average yields (kg m^{-2}) in spring 1994 and 1995 potato trials, and percent A and B quality fruit in summer 1996 melon trials.

Plot treatment	Spring 1994	Spring 1995	Summer 1996
Check	7.58	5.37	49%
Vacuum	8.23	5.04	72%
Insecticide		5.37	74%

melon crop, there was distinctly more wilt in the control field, probably due to the excessive numbers of whiteflies.

Table 2. 1992 and 1993 seasonal average number of *Liriomyza trifolii* and/or *L. huidobrensis* per sampling date for each treatment group in celery.

	1992			1993		
Treatment	Live	Dead	Tunnels	Live	Dead	Tunnels
Control	53.6	129.4	125.8	2.8a*	6.0a	48.6a
Vacuum	25.2	66.3	61.3	2.5a	6.3a	39.0a
Commercial	7.1	21.6	31.7	1.2b	3.6b	24.2b

*Numbers followed by different letters within a column indicate significant differences ($P < 0.05$).

3.4 Grower Cooperation

Understandably, new technology can cause apprehension among growers. A multitude of grower questions immediately arise. Will the new control method be sufficient to control pest species ? How long will it take before results are seen ? What if it doesn't work ? etc. In our first attempts to utilize the vacuum unit on commercial farms, the managers acquiesced to only one untreated control plot (thus creating a problem for statistical analysis of data, see Table 2). Furthermore, because growers feared that low-level pest populations would increase in size, they applied insecticides to areas intended to be non-insecticide-treated (control and vacuum-treated). As we worked with various growers, we eventually achieved a level of confidence such that a portion of a commercial field would be dedicated to us and only treated against plant pathogens. This also alleviated the problems of the 'edge effect' seen in small plots.

4 Successful Applications of Pneumatic Insect Removal

4.1 *Bemisia tabaci*

Populations of whitefly in Israel start to increase in number in May and peak in August-September. In spring, potato crops are nearly finished when whitefly first appears. In autumn potato crops, vacuuming on a weekly basis hastens whitefly decline. However, Fig. 3 shows the effects of field vacuuming melons in a commercial sized field during the summer, when the whitefly populations are large. Field vacuuming significantly reduced whitefly populations (χ^2, $P < 0.001$) and maintained them at levels significantly below those of the check and insecticide treated fields. Very similar results were obtained when the trial was repeated in melon 2 years later.

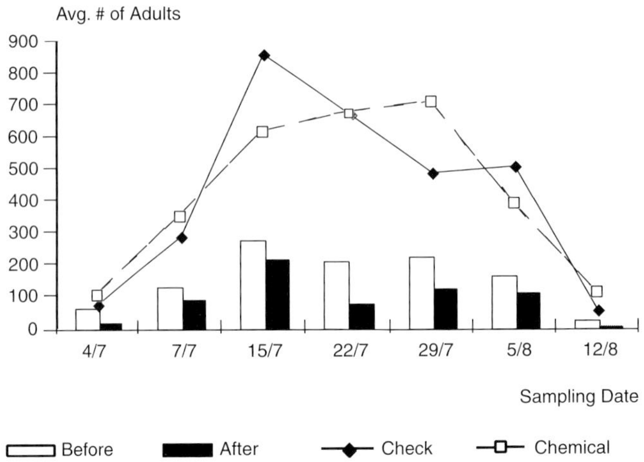

Fig. 3. Effect of field vacuuming on whitefly (*B. tabaci*) populations in melons during the summer of 1996. Immediately before and after the field vacuuming, 1-m-row samples were taken with a hand vacuum. To reduce variations due to diel activity, all samples were taken at the same time period throughout the trial. The lines showing the results from the check and insecticide fields are from the "before" samplings. Field vacuuming occurred once a week on dates indicated.

4.2 *Empoasca* spp. Leafhoppers

The leafhopper populations (*Empoasca* spp.) in Israel can be found in all seasons, although at lower levels during the autumn, winter and early spring months. Although their numbers increased substantially during the spring potato season, they were not high enough for statistical analysis. In attempts to evaluate these insects, we grew celery in the summer, in a small-plot situation. All vacuum treatments significantly reduced the leafhopper population (Weintraub et al. 1996); however, although the population in the vacuum plots was less than the check plots on all sampling dates, the difference was not significant.

4.3 *Liriomyza huidobrensis* and *L. trifolii*

Liriomyza huidobrensis has gradually replaced *L. trifolii* in Israel; however, initial field vacuuming trials in celery were against a mixed population of these leafminers. The results of the first 2 years' (shown in Table 2) vacuum trials shows that field vacuuming of adults reduced the number of adults and larvae. The commercial field (i.e. insecticide-treated) had the lowest number of leafminers because

L. trifolii was predominant in the field and insecticides were effective.
In trials against *L. huidobrensis* in spring potatoes, there were significant reductions in the fly population after field vacuuming. However, because the leafminer is a strong flier and plots were relatively small and close together, population levels did not remain low from one sampling period to the next.

4.4 Parasitoids

Diglyphus isaea Walker is the primary parasitoid of agromyzid leafminers in Israel. It is an ectoparasitoid, developing inside tunnels mined by the leafminer larvae. Its populations basically follow the rise and fall of those of the leafminer, although *D. isaea* also develops on other pests. Although field vacuuming reduced the number of parasitoid adults, the effect did not last from week to week since new adults continued to emerge from the leaves.
Eretmocerus mundus Mercet and *Encarsia lutea* (Masi) are the two primary parasitoids of the whitefly, *B. tabaci*, in Israel. They are endoparasitoids that develop in the nymphal stages of the whitefly. The number of the adult parasitoids was reduced immediately after vacuuming melon; however, there was no significant difference between the number of parasitoids in vacuum-, insecticide- and non-treated plots. As with the *D. isaea*, the immature parasitoids are protected (in the sessile whitefly nymph) and continue to emerge from their hosts.

5 Conclusions

While it is acknowledged that this technology is new and improvements in machinery and methodology need to be made, the potential exists for reduction of the amount of pesticides needed to control pest populations, at least in certain crops or for specific types of insects. It is apparent that trials must be performed on large plots or field-scale to avoid quick reinvasion because of a large edge effect, and that insects which are more sessile in nature are subject to more effective control by vacuuming. We achieved significant population reductions with all insects evaluated; population reductions of leafhopper and whitefly lasted from week to week.
One concern with field vacuuming is that there might be extensive physical damage to the plants as a direct result of the tractor and blowing/vacuuming actions on the plants, or to the yield (as in potato crops) as a result of soil compression. It was apparent from visual observations and comparison of yield results that the plants were not significantly damaged. Furthermore, there was no observable increase in plant diseases (such as *Phytophthora* spp.) which would have invaded injured plants. Another concern is that populations of naturally occurring predators and parasitoids might also be reduced. Our data on *D. isaea* did show population reductions, but because this is a parasitoid found inside the tunnels of leafminers,

its overall population was not affected from week to week. Further, our work with parasitoids of *Bemisia* spp. (*Encarsia* spp. and *Eretmocerus* spp.) showed that their numbers were not adversely affected on a week-to-week basis.

While we do not envision that this form of mechanical control will ever be the sole means of insect control in a field situation, we can foresee its benefits when used in integrated pest management programs. Predator/parasitoid complexes usually cannot overcome high pest populations. However, their efficacy is greatly improved if field vacuuming prior to introducing natural enemies reduces pest populations. Field vacuuming is likewise fully compatible with chemical control measures, reducing pest populations either instead of a regular pesticide treatment or immediately before application.

Acknowledgements. We wish to acknowledge contributions from the following individuals: Y. Arazi, Research and Development, Sha'ar Hannegev Enterprises; Dr. S. Gan-Mor, E. Tzvieli, R. Regev, Dr. A. Gamliel, at the Institute of Agricultural Engineering, Volcani Institute, Israel; G. Tzafrir and S. Shmueli of Extension Services; A. Shapira of Pach-Taas, Ltd. and the Chief Scientists of Ministries of Agriculture and of Arts and Sciences, Jerusalem, Israel. This chapter is contribution No. 2242-E, 1998 series, from the Agricultural Research Organization, The Volcani Center, Bet Dagan, Israel.

References

Affeldt H.A., Thimijan R.W., Smith F.F., Webb R.E., (1983). Response of the greenhouse whitefly (Homoptera: Aleyrodidae) and the vegetable leafminer (Diperta: Agromyzidae) to photospectra. J. Econ. Entomol. 76:1405-1509.

Chagnon R., Vincent C., (1996). A test bench for vacuuming insects from plants. Canad. Agric. Eng. 38:167-172.

Ellington J.J., Kiser K., Cardenas M., Duttle J., Lopes Y., (1984). The Insectavac: a high-clearance, high-volume arthropod vacuuming platform for agricultural ecosystems. Environ. Entomol. 13:259-265.

Pickel C., Zalom F.G., Walsh D.B., Welch N.C., (1994). Efficacy of vacuum machines for *Lygus hesperus* (Hemiptera: Miridae) control in coastal California strawberries. J. Econ. Entomol. 87:1636-1640.

Stern V.M., Dietrick E.J., Mueller A., (1965). Improvements on self-propelled equipment for collecting, separating, and tagging mass numbers of insects in the field. J. Econ. Entomol. 58:949-953.

Walklate P.J., (1994). Aerodynamic methods for controlling insects. Vine Weevil Workshop Conf. Proc., 6 June 1994, Rochester, Kent, U.K. p. 1-6.

Weintraub P.G., Arazi Y., Horowitz A.R., (1996). Management of insect pests in celery and potato crops by pneumatic removal. Crop Protection 15:763-769.

Yudin L.S., Mitchell W.C., Cho J.J., (1987). Color preference of thrips (Thysanoptera: Thripidae) with reference to aphids (Homoptera: Aphididae) and leafminers in Hawaiian lettuce farms. J. Econ. Entomol. 80:51-55.

Current Status and Prospects for the Use of Physical Control in Crop Protection

Bernard PANNETON, Charles VINCENT, Francis FLEURAT-LESSARD

Often a new paradigm emerges, at least in embryo, before a crisis has developed far or been explicitly recognized. T. S. Kuhn, 1970

Physical control in crop protection goes back a very long time. At the dawn of agriculture, it is easy to imagine our distant ancestors pulling weeds from their small plots. With the rapid advances that have occurred in the physical, chemical and biological sciences since the late 19th century, agriculture has been transformed from a strictly empirical activity, largely based on tradition and aimed primarily at staying off famine, to a quantitative form of agriculture focussed on producing a certain amount of food. During this transition, which has been sustained at an increasing rate over the last 50 years, physical control methods have been set aside because of the tremendous success of chemical control. It is only natural that some people should view the use of physical control methods as a step backward to those distant ancestral practices. The many different examples in this book illustrating the effectiveness of physical control provide a clear picture of the technological changes that have occurred over the past 50 years and underscore the new opportunities that now exist for the application of physical control techniques. Thanks to various refinements and greater precision in the implementation of such methods, physical control now has all the necessary attributes to be part of integrated pest management strategies.

The different methods of physical control that can be used against crop pests have some common characteristics. One of the characteristics that differentiates physical tactics from the other control methods (Table 1) is the absence of persistence. In almost every case, the effect of a treatment is limited to the period of application. When treatment stops, the stressor disappears immediately or dissipates quickly and the only damage that subsists is that which has already been caused. By contrast, chemical and biological control agents continue to have an effect after the treatment has been applied. From the standpoint of exercising control over the treatment and its secondary effects, the absence of a residual action is an advantage. However, this characteristic can also be regarded as a drawback, since it means that the treatment has to be repeated every few days to control crop pests that emerge subsequently and are active for a few days or a few weeks. In such a case, persistent chemicals constitute a much more convenient approach for users, although they are often undesirable from an environmental standpoint.

Table 1. Comparison of control methods for crop protection.

Characteristic	Method		
	Chemical	Biological	Physical
Advent	20th century	20th century	With agriculture
Registration	Required	A few cases	Never
Supporting sciences	Analytical chemistry, chemical synthesis, biology	Biology, biotechnology, ecology	Engineering (mechanical, electrical, electronic)
Scientific references	Very abundant	Abundant	Few
Residual action (residues and persistency)	Yes (variable)	Yes, if reproduction occurs	Negligible
Possibility of combining with another method	Yes (sometimes difficult with biological methods)	Yes	Yes
Active or passive method	Active	Active	Active and passive
Application to large acreage	High	Low	Low to moderate
Application to crops with a high profit margin per hectare	High	Moderate to high	Moderate to high
Safety for crop	Moderate to high (phytotoxicity)	High	High (passive) Low (active)
Labour requirements	Low	High	Medium to high
Work rate (hectares treated per hour)	High	Variable	Low (active) High (passive)
Site of action	Photosynthetic system, nervous system (few genes involved)	Systems allowing adaptation to biotic stresses	Systems allowing adaptation to abiotic stresses
Environmental or toxicological requirements, safety	High and costly	Moderate (e.g. virus)	Low (exception: electromagnetic radiation)
Geographic impact	Drift, runoff, evaporation, food chain	Colonization of non-target habitats by parasites or predators	Restricted to area treated (exception: electromagnetic radiation)
Energy requirements	High for production	Low	Low (passive) High (active)
Machinery required	Ground or aerial sprayer	Little or none	Many types of equipment, few machines are suited for more than one purpose
Current market	US$32 billion (FF 192 billion)	About 1.5% of the chemical pesticides market	Negligible

In addition to being restricted to the time of application, the impact of a physical control method is limited spatially. Mechanical, pneumatic, electrical and thermal energies are dissipated locally over a distance of up to a few metres from the site of application. Electromagnetic radiation, which propagates over considerable distances and is subject to numerous restrictions (reserved frequency bands, maximum power, absence of interference), is an exception. Some pesticides have the unfortunate ability to disperse over considerable distances and become a non-point source of contamination, thereby presenting a level of risk that is difficult to quantify. Similarly, many biological control agents can disperse or become dispersed outside the treatment area. Although the risk of contamination is low or non-existent in this case, the dispersal itself represents a loss of efficiency and effects on the ecosystems surrounding the treated area must be taken into consideration.

The modes of action of physical control techniques differ markedly from those of chemical or biological control methods. Pesticides (chemical pesticides and biopesticides) function by inducing biochemical reactions, while biological control methods other than the use of biopesticides are based on alterations in the biotic environment of the target pest. If this environmental manipulation short-circuits the target organism's mechanisms of adaptation to biotic stresses, the control tactic will be successful. Physical control affects the mechanisms of adaptation to abiotic stresses. The various modes of action offered by the broad array of physical control measures allow diversification and open up promising opportunities for managing pesticide resistance in noxious organisms.

Abiotic stresses are non-specific. Plants, animals, microorganisms, in short all organisms living in a homogeneous climatic zone, are exposed to the same abiotic stresses, both qualitatively and quantitatively. This probably explains the lack of selectivity exhibited by many methods of physical control. Two different plants that thrive in the same environment have to adjust to the same changes in abiotic conditions. Weeds and cultivated plants therefore have comparable capacities for adaptation. Therefore, devising selective treatments is a problem. On the one hand, we need to gain greater insight into mechanisms of adaptation to abiotic stresses in order to identify avenues for improving treatment selectivity; but we also need to work on enhancing the selectivity of physical control methods by developing more sophisticated machines equipped with appropriate sensors and controls to ensure that the treatment suppresses the target pest without harming the crop. Both of these aspects are illustrated with the example of flame weeding in corn (Chap. 3). With the flame weeding approach, the phenological stage of the crop offering the greatest resistance to thermal stress provides selectivity, and proper positioning of the burners in relation to the crop plants enhances this selectivity.

Only a few passive physical techniques are available for crop protection (see Chap. 9, 15 and 16). All of the chemical and biological measures are active techniques. Passive methods should be employed whenever possible, because they extend treatment over a lengthy period. For example, mulching done to keep weeds in check is effective as long as the mulch is kept in place.

Most physical methods of control can be used in a crop protection program incorporating both chemical and biological controls. A potential problem occurs when physical barriers are still in place during chemical or biological treatments. Chemical and biological methods are sometimes incompatible, particularly in production systems eschewing chemical pesticides. In the latter case, only biological and physical methods can be applied. Although not economically significant at present, organic crops represent a growing market segment. This is a niche that will definitely provide leverage for the development of physical control measures.

The regulatory framework for physical control differs markedly from that for agrochemical products. First, many physical techniques are subject to rules concerning their use (i.e. the registration process), which are designed to protect users and the general public. Sometimes, such as with the use of propane gas, specialized training is required. The use of electromagnetic radiation is constrained by telecommunications regulations, some of which stem from international agreements. In the case of microwave energy, for instance, only a handful of frequencies have been set aside for industrial, scientific and medical applications (see Chap. 11, Table 2). With regard to the regulatory framework for physical control technologies, it is completely defined a priori. In short, the equipment employed must meet the applicable standards (mostly related to user safety). With chemically or biologically based methods, the difficulty of anticipating secondary effects precludes the establishment of comprehensive specifications which would be known a priori. This explains the need for increasingly costly test protocols designed to evaluate pesticide safety from the standpoint of human health as well as ecotoxicology.

A number of factors tend to complicate the implementation of physical control methods, and physical tactics cannot be readily compared with crop protection systems based solely on the use of an agricultural sprayer to apply pesticides in liquid form. For agricultural operations, this system entails low variable costs and fixed costs. In contrast, the equipment used for physical control is often very specific: cultivators for weeds, vacuuming device for Colorado potato beetles, and so on. Very few physical control tools offer the operational versatility that would allow them to perform several types of crop protection operations. Integration efforts, such as those described in this book, in which researchers have sought to design burners for use in controlling Colorado potato beetles, killing weeds, performing top-killing and dealing with mildew (Chaps. 2,3 and 4), are needed to enable physical control tools to penetrate the crop protection market.

Compared with traditional chemical control, present methods of physical control are more labour intensive and have a lower operational yield (hectares treated per hour). That is one of the main reason why physical control techniques have had very little success in penetrating the field crop market. Given these circumstances, crops with a high profit margin per hectare represent an obvious market for physical control methods. This is all the more true when one considers that crop protection treatments often account for a large percentage of the inputs in that market segment. From the viewpoint of implementation, physical methods compare favourably with biological methods (other than biopesticides), which often entail labour-intensive field observations and are difficult to apply in a field crop setting.

In the post-harvest sector, some staple products of the processing industries are protected from pests either through the use of persistent insecticides (which leave residues that remain active for several weeks or even months), or fumigant insecticides, such as methyl bromide. There is a definitive trend at present to decreasing pesticide residues at all stages in post-harvest processing of agricultural products, particularly in cereals which are a staple food for all of humanity. Furthermore, methyl bromide, which for a long time has played an effective and leading role as a pest control tool in food processing plants and in the eradication of quarantine pests, was recently identified as an ozone-depleting substance. As a result, all industrialized countries are currently phasing out most uses of methyl bromide and this substance is to be banned completely by the year 2005. This situation has prompted a quest for new alternatives to the use of chemical control. Some advanced physical control technologies have gained renewed popularity as efforts get under way to implement integrated pest management in the post-harvest sector. Mechanical barriers to keep pests out, combined with physical suppression techniques, are the cornerstone of the systemic approach adopted by the most dynamic countries with a view to replacing methyl bromide (e.g. the Food Safety Enhancement Program (FSEP) of the Canadian Food Inspection Agency). In parallel, the program to reduce pesticide residues also draws to a great extent on physical procedures for suppressing post-harvest pests (thermal treatment using microwaves or radio-frequencies, inert atmospheres, etc.). Over the past 5 years, intensive research has been devoted to elucidating the physiology of the physical stress induced in insects found in food products with a low moisture content. These studies, which are aimed at devising practical solutions, have enriched the otherwise scarce scientific literature on this topic.

The size of the corpus of scientific and technical literature in any field provides an indication of the amount of research that has been done. In the area of chemical control, tens of thousands of scientific and technical articles (more than 50 000) have been written over the past 50 years or so. There are many specialized journals that disseminate the results of research on pesticides (e.g. Pesticide Science, Weed Science, Journal of Toxicology and Environmental Health, Pesticide Biochemistry and Physiology), and there are many books on the subject (e.g. Coats 1982; Hayes and Laws 1991; Tomlin 1994; Ware 1994). Chemical control is therefore a mature area of research combining the expertise of chemists and biologists.

In the field of biological control, several thousand articles have been published over the past 50 years. Specialized journals exist in this field as well, such as Biological Control, Biocontrol Science and Technology. There are quite a few books on biological control, such as the ones by Debach (1974), van den Bosch (1982), Vincent and Coderre (1992), Hokkanen and Lynch (1995), Van Driesche and Bellows (1995) and others. As well, specialized monographs have been published by New (1991), Godfray (1994) and Jervis and Kidd (1996). On the whole, biological control is viewed favourably by the general public. This field draws on the expertise developed in many branches of science: biology, population dynamics, animal behaviour, ecology, physiology and so on.

Physical control should be an integral part of integrated pest management (Chap. 1, Fig. 1). In the classic textbooks on integrated pest management (Riba and Silvy 1989; Metcalf and Luckmann 1994), physical control is barely mentioned from either a theoretical or practical standpoint. A book by the US National Academy of Sciences (1969) on integrated pest management contains only one chapter on physical control. There are approximately 4000 scientific and technical articles on the use of physical control for crop protection, but no books currently exist on this topic. Obviously, physical control has not benefited from the same research and development efforts as chemical and biological control. This approach did not emerge as a true alternative to conventional pesticide use until the early 1990s, when it began to be studied using means equivalent to those in biological control: models were built to identify the impacts of physical stresses on insect pests and weeds and to link them to physiological mechanisms in an effort to understand the lethal process from a mechanistic modelling viewpoint. Physical control involves a variety of scientific and technical challenges, and many teams of researchers have been set up in an effort to solve them. As pressure for sustainable agriculture increases, new research groups will form and new companies will be created with the goal of marketing the resulting technologies. Since physical control offers promising opportunities for cutting down on the use of synthetic pesticides, the development of new physical techniques can go a long way toward achieving the pesticide reduction objectives that have been set by many countries and organizations. In this context, stakeholder agencies should actively support the development and implementation of physical control methods within integrated pest management programs for crop protection.

In the current context of protection of crops and post-harvest food products, the contribution of physical control to integrated protection is considered insufficient. However, we hope to have shown that some highly effective methods are currently available and that they are compatible with IPM programs. Furthermore, in certain specific situations, the physical control measures can be the primary component of such a strategy. It should be kept in mind that intensive cropping systems are bound to lessen in extent gradually, given the need to adapt to price reduction policies, consumer demands in relation to food safety and security and the declining number of new pesticide registrations in many crops. Research associated with the expansion of physical control applications and with process engineering will help to bring about this evolution, which now seems inevitable.

References

Debach P., (1974). Biological Control by Natural Enemies. Cambridge University Press, Cambridge, U.K., 323 p.

Coats J.R. (ed.), (1982). Insecticide Mode of Action. Academic Press. New York, 470 p.

Godfray H.C.J., (1994). Parasitoids, Behavioral and Evolutionary Ecology, Princeton University Monographs in Behavior and Ecology, Princeton, N.J., 473 p.

Hayes W.J., Laws E.R. (eds.), (1991). Handbook of Pesticide Toxicology. Academic Press, San Diego, (Vol 1: p. 1-496), (Vol 2: p. 497-1123)(Vol 3: p. 1125-1576).

Hokkannen H.M.T., Lynch J.M. (eds.), (1995). Biological Control, Benefits and Risks. Cambridge University Press, Cambridge, U.K., 304 p.

Jervis M., Kidd N. (Eds.), (1996). Insect Natural Enemies. Practical Approaches to their Study and Evaluation. Chapman and Hall, New York, 491 p.

Kuhn T.S., (1970). The Structure of Scientific Revolutions. 2nd ed. University of Chicago Press, Chicago, 210 p.

Metcalf R.L., Luckmann W.H., (1994). Introduction to Insect Pest Management, 3rd ed. Wiley Interscience, New York. 650 p.

National Academy of Sciences, (1969). Principles of Plant and Animal Pest Control, Vol. 3: Insect Pest management and Control, Washington, D. C., 508 p.

New T.R., (1991). Insect as Predators. The New South Wales University Press, Kensington, NSW, Australia, 178 p.

Tomlin C., (1994). The pesticide manual: a world compendium. 10th ed., Bath Press, Bath, U.K.. 1341 p.

Riba G., Silvy C., (1989). Combattre les ravageurs des cultures: enjeux et perspectives. INRA Editions, Paris, 230 p.

Van den Bosch R., Messenger P.S., Gutierrez A.P., (1982). An Introduction to Biological Control. Plenum Press, New York, 247 p.

Van Driesche R.G., Bellows T.S., (1996). Biological Control, Chapman & Hall, New York, 539 p.

Vincent C., Coderre D. (eds.), (1992). La lutte biologique, Gaëtan Morin Editeur (Montréal) et Lavoisier Tech Doc (Paris), 671 p.

Ware G.W., (1994). The Pesticide Book, Fresno, Ca., Thomson Publ., 384 p..

Glossary

Acclimation (cold): In a living organism, result of a physiological adaptation in response to changes in surrounding temperature. During a cold acclimation phase, insects synthesise and accumulate various compounds known as cryoprotectants, in their tissues. Inside the insect body, the primary role of these sugars and polyols is to lower the temperature where ice crystals start to form spontaneously.

Adventive (adj.): Refers to plants growing on cultivated land which were not intentionally established through seeding or planting. (syn. weeds).

Apterous (adj.): Wingless.

Biopesticide: Pesticide whose active ingredient is composed of, or extracted from or derived from a living organism. For example, the most widely sold bioinsecticide in the world, *Bacillus thuringiensis* var. *kurstaki*, is a mixture of bacteria, protein crystals and cellular debris. This mixture results from an industrial fermentation process.

Climacteric: The climacteric stage is an important stage in the ripening process of several fruits. It is characterised by a sudden and significant increase in the respiration rate at the onset of ripening.

Cold stupor: Cessation of motor activity and slowing of vital functions in an animal caused by exposure to cold temperatures. Stored-grain insects, for example, stop moving when the temperature is about 4 °C or lower, but they can survive for several weeks or even months at temperatures around this threshold.

Commensal (adj.): Denotes a species living in a beneficial association with another species, which is not harmed by the relationship.

Cooling front: During grain cooling through ventilation, there is a sharp boundary between the chilled grain and the unchilled portion of the grain bulk. The term "cooling front" aptly illustrates the progression in space of the cooling operation of the grain. As the cooling front moves, all the grain behind the cooling front has been chilled. Silothermometry is used to track the position of the cooling front in the bin during chilling operation. (syn. cold transition zone).

Degree-day accumulation: It is often useful to be able to predict periods of insect emergence and activity. Since the development of these poikilothermic organisms is a function of temperature and occurs within a specific range of temperatures (between a lower temperature limit and an upper limit), their development can be modelled. For example, in temperate zones, entomologists use simple physiological models based on the use of a reference temperature below which no development can occur. Degree-days are computed by integrating over time, the difference between suitable mean temperatures (typically daily means) and the reference temperature. For instance, if an insect begins developing at a base temperature of 5 °C and mean temperatures of 3, 10, 16, 22 and 18 °C have been recorded, the temperature differences of 0, 5, 11, 17 and 13 °C add up to 46 degree-days for this time period. The degree-days are correlated to the physiological development of insect species. Degree-days are also used to model plant growth by selecting appropriate reference temperatures.

Dielectric material: Perfect dielectric materials do not conduct electricity but respond to electromagnetic fields by storing energy within their internal structure. In alternating fields, the internal structure of a dielectric is continously changing at the same frequency as the electromagnetic signal. During this processes, internal friction or losses occur and heat is generated. The amount of heat generated is a function of the properties of the dielectric and of the frequency of the electromagnetic field. The dielectric properties of a material are the permittivity (measure of the material's capacity to polarise in an electrical field) and the dielectric loss factor (determines the material's rate of heating in an electrical field). (See Chap. 7).

Dislodging: Action of taking an insect off a plant. Once insects have been dislodged using a pneumatic control system, the air stream must direct them to a collecting device before they fall to the ground.

Drag coefficient: Dimensionless parameter. It is the ratio of the drag force exerted on a body in relative motion with respect to a surrounding fluid, to the inertia of the body. With the exception of very small spherical bodies, the drag coefficient is calculated using fairly complex empirical equations.

Entoleter: Apparatus designed to mechanically injure insects by impact. It is named after the Trade Mark given by the company that manufactures the machine. See "impact disinfestor" in this glossary.

Epidemiology: Epidemiology in plants (as in human and veterinary medicine) is the study of disease outbreaks and the factors that affect their onset and progression. A detailed understanding of the mechanisms underlying the development and spread of plant diseases is essential for developing more effective and more environmentally friendly control methods.

Flame weeding: Physical control method where weeds are exposed to high temperatures either through direct contact with heated gases or through thermal radiative transfer. The thermal shock induced causes the expansion and bursting of cells along with the coagulation of proteins. Direct combustion of plant material is not involved.

Fluidized bed: Bed of small particles which results when a fluid, usually a gas, flows upward at a velocity high enough to buoy and suspend granules or heavy particles. Substances that are fluidized in this manner can be transported over considerable distances (pneumatic transport) or undergo thermal transfer or rapid treatment in a fixed bed (e.g., puffing of cereals or seed treatment).

HACCP process: Quality control concept that originated in the United States (HACCP stands for Hazard Analysis and Critical Control Point). In the agri-food sector, the purpose of a HACCP system is to ensure consistent standards of safety and quality in manufactured products. A holistic and rational approach is employed which focuses on identifying the risks of poor quality products, evaluating them and establishing preventive measures for managing those risks, which may be microbiological, physical or chemical in nature.

Heat stupor: Physiological state of an insect exposed to a temperature higher than the heat stupor point, which generally ranges from 42 to 48 °C depending on the species. This is the temperature at which the insect's motor activity completely stops.

Hypercarbia: Atmosphere whose composition has been modified through the addition of carbon dioxide.

Ice-nucleating bacteria: Found widely in nature, these bacteria promote the crystallisation in organisms with a high moisture content by raising the temperature corresponding to the supercooling point (see Chapter 6) by a few degrees. They are responsible for triggering freezing in plants and are used primarily in the manufacture of artificial snow. The most commonly employed species is *Pseudomonas syringae*. Once ice-nucleating bacteria have been ingested and end up in the gut of an insect, they cause a significant reduction in tolerance to freezing.

Impact disinfestor: Machine that uses centrifugal force to hurl flour or wheat against plates that revolve on a central shaft at high speed and against the metal housing of the machine. The impact of the infested grain or flour hitting the revolving plates is sufficient to kill all insect life in the product.

Magnetron: Electronic device using a resonant cavity to generate electromagnetic signals in the microwave range.

Modified atmosphere: Storage atmosphere whose composition differs from that of normal air owing to the removal or addition of gases (nitrogen, oxygen, carbon dioxide or argon), either in pure form or as a mixture.

Mulching: Operation where foreign material (e.g. plastic film) is layed on the ground around the stems of cultivated plants.

Mycochrome: In a fungus, photoreceptor excited at specific wavelengths of light in the 300–500 nm range. Excited mycochromes can either induce or inhibit sporulation.

Normalised airflow rate (grain aeration): Volume of air introduced at atmospheric pressure per unit time per unit volume of the grain mass. Typical units are $m^3/h/m^3$. In grain cooling, it is estimated that about 10 nights of ventilation at a rate of 7 $m^3/h/m^3$ are required to bring the temperature in near equilibrium with ventilating air temperature. This translates to using 1000 m^3 of air for cooling 1 m^3 of grain (Lasseran 1994 - see Chap. 6)

Persistence: Property of a crop protection agent to remain intact or active. Quantitatively, it is measured as the duration of the persistence. Persistence of chemical pesticides varies from a few hours to several days following application. In biological control, persistence refers to the length of time control agents (parasites or predators) will survive following an inundative release. In physical control, the persistence of a treatment is often nil. Persistence is a function of the chemical or organism used and of the environmental parameters (e.g. degradation by ultraviolet radiation, heat, rain).

Photomorphogenesis: Development of plant structures under the influence of light.

Photonic flux: Represents the number of photons reaching a plant per unit area and per unit time.

Photoreversion: Process whereby an active phytochrome is converted to inactive phytochrome following exposure to far-red light.

Photoselectivity: Process whereby plastic mulches absorb, reflect or transmit different wavelengths depending on the colour, content and type of material used.

Phytochrome: Light sensitive pigment in plants. Some respond specifically to the photoperiod that control morphogenesis in plants.

Pneumatic conveyor: Equipment for conveying bulk material using a high airflow rate in pipes. Industrial machines for unloading ships proceed by suction. Models designed for farm use have conveying rates ranging from 4 to 7 $t.h^{-1}$, whereas the models designed for unloading ships operate at a rate of 10 to 60 $t.h^{-1}$. When a pneumatic conveyor is used to empty a grain storage bin, mortality among stored-grain insects ranges from 50 to 100%.

Primary versus secondary pest: On a given crop, several insect species are present, some of which cause more damage than others. If left untreated, primary pests will cause serious damage to plants year after year. By contrast, secondary pests cause negligible damage.

Retention force: Force exerted by an insect grabbing on a surface. This force is used to counteract the effect of gravity and of other forces such as the inertia force induced when leaves rustle under the action of wind. In pneumatic control methods, suction and drag forces resulting from the airflow must overcome the retention force.

Ruderal (adj.): Relates to plants that grow in modified but uncultivated habitats (e.g., along highways and roads, on roofs, abandoned sites).

Sessile (adj.): Refers to any directly attached object that has no stem or pedicle. Some insects are sessile throughout all or part of their life cycle, and generally remain attached to a structural part of a plant. In many scale insects, for example, the immature forms and females are sessile (immobile), whereas the males are mobile.

Sheeting: Semi-forcing technique in which a semi-permeable plastic sheeting is layed flat over cultivated plants.

Solarization: Method involving the use of transparent plastic covers layed flat on the soil for several weeks prior to cultivation. Heat builds up under the covers, increasing the soil temperature to a level that eliminates most crop pests, including weed seeds and pathogenic microflora.

Specific air flow (aeration of grain): The volume of air moved through one m^3 of grain per hour during aeration of grain. Approximately 7 $m^3.h^{-1}$ of ambient air per m^3 of grain is needed to cool grain to the ambient air temperature over about 10 nights of aeration (or about 1,000 m^3 of air to cool one m^3 of grain, Lasseran 1994).

Stale seedbed: This technique involves tilling the soil as for seedbed preparation, and then allowing weed seeds to germinate. Once the weed seedlings have emerged, the soil is tilled again to prepare the seedbed or another stale seedbed, thereby destroying any weeds that have germinated. For the technique to be successful, the time period between tillage operations must promote maximum seedling. Delayed seeding minimises competition with the crop plants that become established and reduces the need for weed control during the season.

Stupor (cold stupor or heat stupor): cessation of motor activity associated with temperatures at both ends of the thermobiological scale. Among stored-grain insects, cold stupor occurs below 4 °C and heat stupor at temperature higher than 45 °C. Insects die after a few minutes' exposure at 45 °C, but at 4 °C they can survive for a few days to a few months.

Thanatosis: Behavioural strategy whereby an insect falls from a plant and feigns death. This may enable the insect to avoid predator attack or escape from some perceived danger.

Waveguide: Just as an electrical wire conducts electricity from one point to another, a waveguide is used to direct electromagnetic waves to a specific location. A waveguide serves as a connection between the electromagnetic wave source and the point being served. The characteristics of waveguides (shape, size and material) depend on the frequency of the electromagnetic waves to be conducted. They are generally cylindrical or rectangular in cross-section and made of materials that permit reflection without excessive loss of electromagnetic energy. The size of the waveguide is directly related to the length of the electromagnetic waves. For example, visible light is channeled through optic fibres with a very small diameter (a few dozen microns), whereas microwaves require metallic guides ranging up to a few dozen centimetres in diameter. Losses through absorption or reflection limit the connection distance (hundreds of kilometres for optic fibres and a few metres for microwave guides). The power required to transmit electromagnetic radiation is also a factor limiting the size of waveguides.

Viral disease: disease caused by a virus.

Virulence or aggressiveness of a pathogen: The virulence of a particular pathogen strain denotes its ability to cause disease in a given plant variety or cultivar. A given strain may be virulent for some cultivars and non-virulent (unable to cause disease) in other cultivars of the same cultivated plant species. A distinction is generally made between the terms "virulence" and "aggressiveness." Aggressiveness refers to the difference in severity of symptoms that a virulent strain induces in a host plant. Whereas an isolate is either virulent or non-virulent for a given cultivar (qualitative assessment), a virulent isolate may be more or less aggressive (quantitative).

Subject Index

Q

R

U

V

W

XYZ

List of Contributors

Joseph Arul
Laval University,
Department of Food Science and Nutrition and Horticultural Research Centre, Pavillon Paul-Comtois, Sainte-Foy, Quebec, Canada G1K 7P4

Mebarek Baka
University of Constantine,
Natural Sciences Institute, Route Ain-El Bey, Constantine 25000, Algeria

Serge Bégin
Quebec Department of Agriculture, Fisheries and Food Research Branch, 200 Chemin Ste-Foy, Quebec, Qc, Canada G1R 4X6

Diane L. Benoît
Horticulture Research and Development Centre,
Agriculture and Agri-Food Canada,
Saint-Jean-sur-Richelieu,
430 Gouin Blvd., Quebec,
Canada J3B 3E6

G. Boiteau
Potato Research Centre,
Agriculture and Agri-Food Canada,
Fredericton, New Brunswick,
Canada E3B 4Z7

Steeve Bourassa
Laval University,
Biology Department,
Faculty of Sciences and Engineering,
Sainte-Foy, Quebec, Canada G1K 7P4

Joe Calandriello
Poly Expert Inc., 850 avenue Munck,
Laval, Quebec, Canada H7S 1B1

Marie-Thérèse Charles
Laval University,
Department of Food Science and Nutrition and Horticultural Research Centre, Pavillon Paul-Comtois,
Sainte-Foy, Quebec, Canada G1K 7P4

Conrad Cloutier
Laval University,
Biology Department,
Faculty of Sciences and Engineering,
Sainte-Foy, Quebec, Canada G1K 7P4

Daniel Cloutier
Institut de Malherbologie
CP 222, Ste Anne de Bellevue
Quebec, Canada H9X 3R9

Bruce G. Colpitts
University of New Brunswick,
Department of Electrical and Computer Engineering,
Fredericton, New Brunswick, Canada

Jocelyn Douhéret
Laval University,
Department of Phytology,
Faculty of Agricultural and Food Sciences, Pavillon Paul Comtois,
Sainte-Foy, Quebec,
Canada G1K 7P4

Sylvain L. Dubé
R&D Phytologie International Inc.,
A-2643 Jean-Brillant,
Sainte-Foy, Canada G1W 1E9

Raymond-Marie Duchesne
Quebec Ministry of Agriculture,
Fisheries and Food,
Direction de l'environnement et du développement durable,
200, Chemin Sainte-Foy, Quebec, Qc,
Canada G1R 4X6

Paul G. Fields
Cereal Research Centre,
Agriculture and Agri-Food Canada,
195 Dafoe Rd., Winnipeg,
Manitoba, Canada R3T 2M9

Francis Fleurat-Lessard
INRA, Laboratoire sur les Insectes des Denrées stockées,
Domaine de la Grande Ferrade, BP 81,
33883 Villenave d'Ormon Cedex, France

Jacques Gill
Laval University,
Faculty of Agricultural and Food Sciences, Department of Soil Science and Agri-Food Engineering,
Sainte-Foy, Quebec, Canada G1K 7P4

Rami Horowitz
Department of Entomology, ARO,
Gilat Experiment Station, MP Negev,
85280, Israel

M. Khelifi
Laval University,
Faculty of Agricultural and Food Sciences,
Department of Rural Engineering,
Sainte-Foy, Quebec,
Canada G1K 7P4

Zlatko Korunic
Hedley Technologies Inc.
14 Greenwich Dr., Guelph
Ontario, Canada N1H 8B8

Benoit Lacasse
Laval University,
Faculty of Agricultural and Food Sciences,
Department of Soil Science and Agri-Food Engineering,
Sainte-Foy, Quebec,
Canada G1K 7P4

Claude Laguë
Laval University,
Faculty of Agricultural and Food Sciences,
Department of Soil Science and Agri-Food Engineering,
Sainte-Foy, Quebec,
Canada G1K 7P4

Christophe LaHovary
University of Guelph,
Crop Science Department,
Guelph, Ontario,
Canada N1G 2W1

Martin Lanouette
Laval University,
Department of Phytology,
Faculty of Agricultural and Food Sciences, Sainte-Foy, Quebec,
Canada G1K 7P4

Jean-Marc Le Torc'h
INRA, Laboratoire des Insectes des Denrées Stockées,
Domaine de la Grande-Ferrade, BP 81
33883 Villenave-d'Ornon Cedex,
France

Maryse Leblanc
Institut de Recherche et de Développement en Agroenvironnement
Quebec Department of Agriculture
Fisheries and Food,
3300 Sicotte
Saint-Hyacinthe,
Québec, Canada J2S 7B8

Gilles Leroux
Laval University,
Department of Plant Science,
Faculty of Agricultural and Food Sciences, Sainte-Foy, Quebec,
Canada G1K 7P4

Jacques Lewandowski
Opto Labs
24, rue Dufresne
Saint-Luc,
Quebec, Canada J2W 1K9

Rohanie Maharaj
University of the West Indies
St. Augustine's Campus,
Trinidad, West Indies

Julien Mercier
DNA Plant Technology
6701 San Pablo Avenue
Oakland, CA 94608
USA

Marie Mermier
INRA, Station de Bioclimatologie
Site Agroparc
84914 Avignon Cedex 9, France

Nicolas Morison
INRA, Station de Zoologie
Site Agroparc
84914 Avignon Cedex 9, France

Philippe C. Nicot
INRA, Station de Pathologie Végétale
Domaine St-Maurice, BP 94
84143 Montfavet Cedex, France

Bernard Panneton
Agriculture and Agri-Food Canada,
Horticultural Research and
Development Centre,
430 Gouin Blvd.,
St-Jean sur Richelieu,
Quebec, Canada J3B 3E6

Yvan Pelletier
Potato Research Centre,
Agriculture and Agri-Food Canada,
Fredericton, New Brunswick,
Canada E3B 4Z7

Guy Péloquin
Envir Aqua Inc.
1316, rue de la Couronne, Val-Bélair,
Quebec, Canada G3K 2J5

Paul-Martin Roy
Laval University,
Faculty of Agricultural and Food Sciences,
Department of Soil Science and Agri-Food Engineering, Sainte-Foy, Quebec,
Canada G1K 7P4

Robert S. Vernon
Pacific Agri-Food Research Centre,
Agriculture and Agri-Food Canada,
Agassiz, British Columbia,
Canada V0M 1A0

Clément Vigneault
Horticulture Research and
Development Centre,
Agriculture and Agri-Food Canada,
Saint-Jean-sur-Richelieu, Quebec,
Canada J3B 3E6

Charles Vincent
Agriculture and Agri-Food Canada,
Horticultural Research and
Development Centre,
430 Gouin Blvd., St-Jean sur
Richelieu, Quebec, Canada J3B 3E6

Phyllis G. Weintraub
Department of Entomology, ARO,
Gilat Experiment Station,
M.P. Negev, 85280, Israel

Druck: Strauss Offsetdruck, Mörlenbach
Verarbeitung: Schäffer, Grünstadt